Textbook of COSMETICS

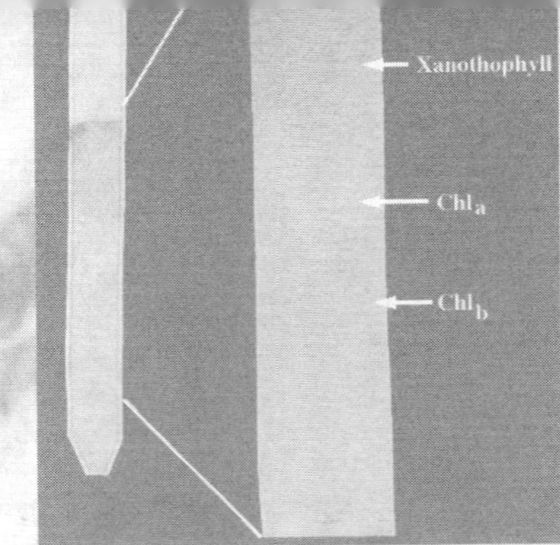

A thing of beauty is a joy for ever:
Its loveliness increase; it will never
Pass into nothingness; but still will keep
Bower quiet for us, and a sleep
Full of sweet dreams, and health and quiet breathing.
—John Keats (British poet)

"...products nobody needs—but wanting them is human nature. Today, the desire to look better, smell better, and thus feel better causes consumers worldwide—mostly women—to spend an estimated \$65 billion annually on personal enhancement—cosmetics"

—from *Encyclopedia Britannica*

Textbook of COSMETICS

Rajesh Kumar Nema M Pharm, PhD
Director
SD College of Pharmacy and Vocational Studies
Muzaffarnagar, UP

Kamal Singh Rathore M Pharm
Senior Lecturer
BN Girls College of Pharmacy
Udaipur, Rajasthan

Bal Krishna Dube M Pharm, PhD
Director
TIT College of Pharmacy
Bhopal, MP

CBS Publishers & Distributors Pvt Ltd

New Delhi • Bengaluru • Chennai • Kochi • Kolkata • Lucknow • Mumbai
Hyderabad • Jharkhand • Nagpur • Patna • Pune • Uttarakhand

Disclaimer

Science and technology are constantly changing fields. New research and experience broaden the scope of information and knowledge. The authors have tried their best in giving information available to them while preparing the material for this book. Although, all efforts have been made to ensure optimum accuracy of the material, yet it is quite possible some errors might have been left uncorrected. The publisher, printer and the authors will not be held responsible for any inadvertent errors or inaccuracies.

Textbook of
Cosmetics

First Edition: 2009

Reprint: 2015, **2024**

ISBN: 978-81-239-1761-0

Published by **Satish Kumar Jain** and produced by **Varun Jain** for

CBS Publishers & Distributors Pvt Ltd
4819/XI Prahlad Street, 24 Ansari Road, Daryaganj, New Delhi 110 002, India.
Ph: 011-23289259, 23266861 Website: www.cbspd.com
e-mail: delhi@cbspd.com

Corporate Office: 204 FIE, Industrial Area, Patparganj, Delhi 110 092
Ph: 011-4934 4934 Fax: 011-4934 4935 e-mail: publishing@cbspd.com;publicity@cbspd.com

Branches

- **Bengaluru:** Seema House 2975, 17th Cross, K.R. Road, Banasankari 2nd Stage, Bengaluru 560 070, Karnataka, India
  Ph: +91-80-26771678/79 Fax: +91-80-26771680 e-mail: bangalore@cbspd.com
- **Chennai:** 7, Subbaraya Street, Shenoy Nagar, Chennai 600 030, Tamil Nadu, India
  Ph: +91-44-26680620, 26681266 Fax: +91-44-42032115 e-mail: chennai@cbspd.com
- **Kochi:** 42/1325, 1326, Power House Road, Opp KSEB, Ernakulam 682 018, Kochi, Kerala, India
  Ph: +91-484-4059061-67 Fax: +91-484-4059065 e-mail: kochi@cbspd.com
- **Kolkata:** 147, Hind Ceramics Compound, 1st Floor, Nilgunj Road, Belghoria, Kolkata 700 056, West Bengal, India
  Ph: +91-33-25633055/56 e-mail: kolkata@cbspd.com
- **Lucknow:** Basement, Khushnuma Complex, 7-Meerabai Marg (Behind Jawahar Bhawan), Lucknow 226 001, UP, India
  Ph: +0552-4000032 e-mail:tiwari.lucknowi@cbspd.com
- **Mumbai:** PWD Shed. Gala no. 25/26, Ramchandra Bhatt Marg, Next to JJ Hospital Gate no. 2, Opp. Union Bank of India, Noorbaug, Mumbai 400 009, Maharashtra, India
  Ph: 022-66661880/89 e-mail: mumbai@cbspd.com

Representatives

• **Hyderabad**	0-9885175004	• **Jharkhand**	0-9811541605	• **Nagpur**	0-8692091830
• **Patna**	0-9334159340	• **Pune**	0-9664372571	• **Uttarakhand**	0-9716462459

Printed at: Sanjay Printers, Sahibabad, UP, India

Preface

Cosmetics have been used in human societies from ancient times. There is a concept of sixteen modes of beautification *(Solah Shringar)* from head to toe mentioned in literature and followed in the traditional Indian families. The desire to look beautiful is a primary human weakness and is as old as the human race. Cosmetics are the substances especially prepared to enhance beauty and increase the attractiveness of the person. In recent years awareness about the use of cosmetic products has increased by leaps and bounds. A plethora of cosmetic products are available in the market due to increased publicity of the beauty products generated by several beauty contests and hordes of TV channels promoting cosmetic products. Cosmetic products are now available for all age groups and for both the sexes.

The book starts with a brief overview and updating of the main information used in the formulation and usage of cosmetic products. Influence of other ingredients like preservatives, colorants, flavoring agents, etc. are also discussed in detail. A large part of the book is devoted to skin cosmetics, herbal cosmetics, hair cosmetics, etc.

Cosmetic science and technology is a very interesting field with a lot of challenges that changes rapidly. It is our hope that this book can be the starting point for students of cosmetic science and pharmaceutical dosage forms for those interested in expanding the knowledge within this area. Students and practitioners in the field of beauty culture will also find it useful.

It gives us immense pleasure in writing this book on cosmetic science and technology. One of the key purposes of writing this book is that there was a great dearth of good books available on this subject as only a few publications are available. This book is prepared by taking a view of the syllabi of various universities in India and abroad.

A special emphasis has been given to herbal cosmetics because herbals are a part of our life and their uses are increasing day-by-day not only in India but all over the world.

This book has been prepared to provide a comprehensive presentation of the various aspects of cosmetic science that are of particular significance in the formulation and evolution. This book gives a concise information on the cosmetics used for skin, eye, lips, hair, nail, etc. and general approaches also taught. It is hoped that it will prove useful to all cosmetic scientists, students and academicians.

Rajesh Kumar Nema
Kamal Singh Rathore
Bal Krishna Dube

to

Venus

Goddess of Beauty

"Beauty without virtue is like a rose without scent."
— Alexander Pope

Acknowledgements

We feel privileged to express our deep sense of gratitude to everyone who helped us at each and every step from inception to the publication of this book. We gratefully acknowledge the moral support and constant encouragement provided by the staff of BN College of Pharmacy and BN Girls College of Pharmacy, and our family members in writing this book. This was really a driving force behind this book.

In preparing the subject matter, we have freely consulted a number of textbooks and various internet sites. We humbly acknowledge quoting some textual matter from these titles and sites to make certain points clear for the ultimate benefit of our students/readers.

We are highly thankful to our Secretary, Vidhya Pracharini Sabha, Dr. M.S. Rathore, Managing Director, Shri Tej Singh Ji, and Director, B.N. College of Pharmacy, Dr. A.C. Rana, for their help and suggestions.

We are profusely thankful to Dr. G. D. Gupta, Dr. S.S. Sisodia, Dr. Y.S. Tanwar, Dr. C.S. Chouhan, Dr. M.S. Ashawat, Mr. D.S. Rathore, Mr. P.S. Naruka, Mr. R.P.S. Rathore, Mrs. Meenakshi, Mr. Sudhir B., Mr. Kumar Gaurav, Mr. Dhiraj, Mr. Bhupendra, Mr. Amul, Mr. Deepak, Mr. Narendra, Mr. Manoj, Mrs. Sunita and Mr. Vishal, who were always available to us for their skills.

We thank Mr. Hement for finishing the typing work neatly in record time and for bearing with our last-minute corrections and changes at various stages.

We are deeply indebted to the publishers for their cooperation in bringing out this book as an international edition.

At last we all should not forget to mention that without Almighty's blessing no work can ever be done.

Comments and suggestions from all our readers are welcome for further improvement of this book.

Rajesh Kumar Nema
Kamal Singh Rathore
Bal Krishna Dube

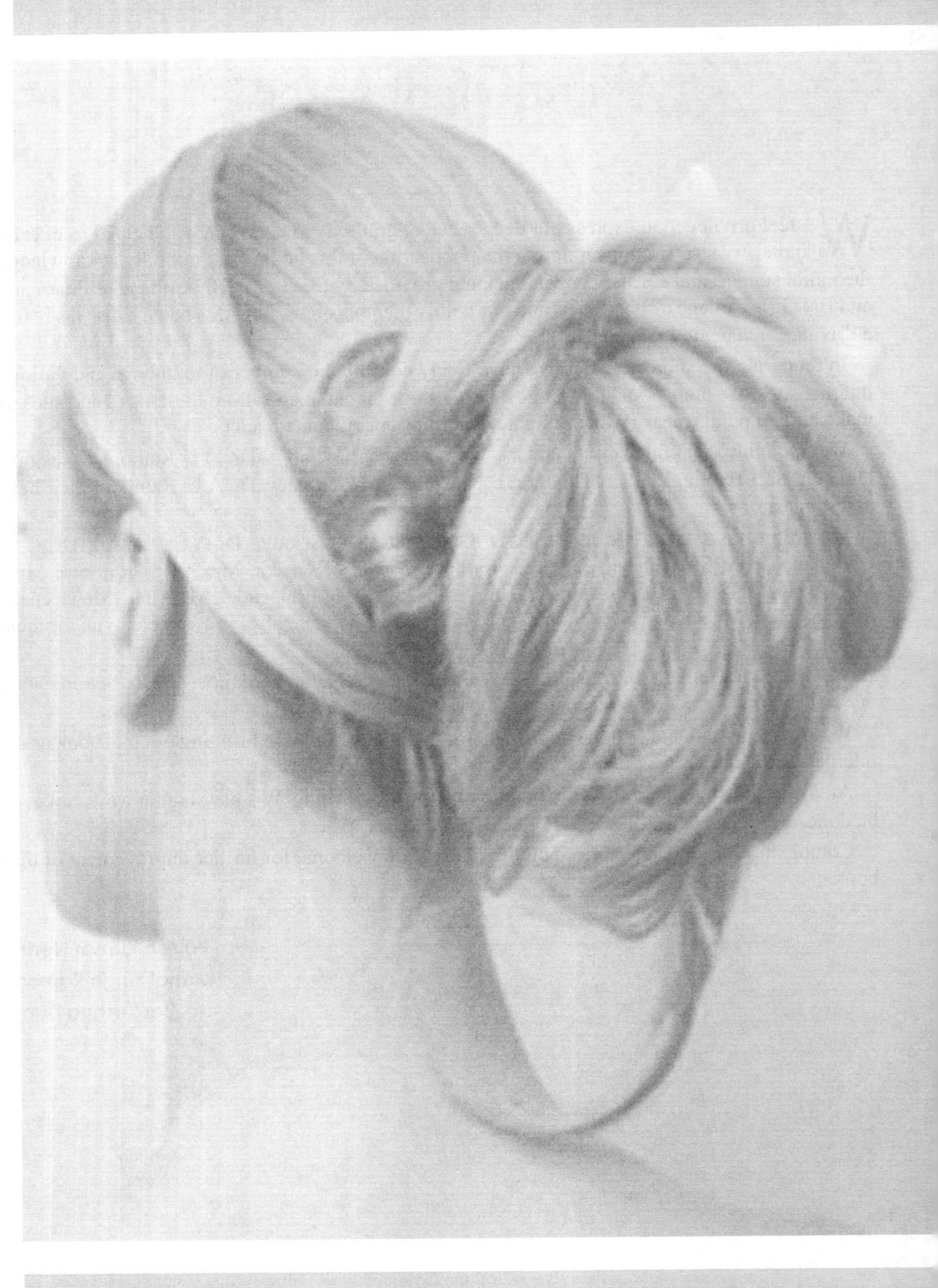

Contents

Contents

Abbreviations

μm: micrometer
Å or AU: Angstrom units
ACD: Allergic contact dermatitis
Ad: made upto
AHA: Alpha hydroxy acids
ASCC: Australian Association of Cosmetic Chemists
BET: Bacterial endotoxin test
BHA: Butylated hydroxy anisole
BHT: Butylated hydroxy toluene
BIS: Bureau of indian standards
BSE: Bovine spongiform encephalopathy
CAS: Chemical abstract service
CCTFA: Canada Cosmetics, Toiletries and Fragrance Association
CFR: Code of Federal Regulations
CFU: Colony forming units
cGMP: Current good manufacturing practices
CHIC: Cosmetics harmonization and international cooperation
CI: Colour index
CIR: Cosmetic identification review/cosmetic ingredient review
CLA: Comprehensive licensing system
CMC: Carboxy methyl cellulose
COEX Tubes: Co-extrusion tubes
COLIPA: Comite' de Liaison des Associations Europe'enes de l' Industrie de la Perfumeric des Products Cosmetique et de Toilette (European Cosmetics, Toiletries and Perfumery Association)
CPSC: Consumer product safety commission
CSC: Coconut shell charcoal
CTFA: Cosmetics, toiletries and fragrance association
D&C: Drug and Cosmetics
DEA: Diethanolamine
DHA: Dihydroxyacetone
DHT: 5α-dehydrotestosterone
DIN: Drug identification number
DME: Dimethylether
DOPA: dihydroxyphenylalanine
DSC: Differential scanning calorimetery
ECETOC: European Centre for Ecotoxicology and Toxicology of Chemicals
ECVAM: European Centre for Validation of Alternative Methods
EDTA: Ethylene diamine tetra acetic acid
EINECS: European Inventory of Existing Commercial Chemical Substances
EGF: Epidermal growth factor
ELINCS: European List of Notified Chemical Substances
EVOH: Ethylene vinyl alcohol
FD&C: Food, drug and cosmetics
FDA: Food and Drug Administration
FDAMA: Food and Drug Modernization Act
FP&L / FPLA: Fair Packaging and Labeling Act
FTA: Federal Trade Commission
GNRAS: Generally not recognized as safe
GOT: Glutamate oleate transaminase
GPT: Glutamate pyruvate transaminase
GRAS: Generally recognized as safe
HLB: Hydrophillic lipophillic balance
HPFB: Health products and food branch
HPLC: High performance liquid chromatography
HPTLC: High performance thin layer chromatography
ICH: International Conference on Harmonization
INCI: International Nomenclature of Cosmetics Ingredients
INN: International Non-proprietary Name (recommended by WHO)
ISI: Indian Standard Institute
ISO: International Organization for Standardization
IUPAC: International Union of Pure and Applied Chemistry
JND: Just noticeable difference
JSCI: Japanese Standards of Cosmetic Ingredients
JTU: The Jackson Turbidity Units
LDH: Lactate dehydrogenase
LDPE: Low density polyethylene
LO: Lysyl oxidase
MED: Minimum erythema dose
MLT: Microbial limits
MNC: Multinational company
MRP: Maximum retail price
MRUC: Media research users council
NDA: New drug application
NIST: National Institute of Standards and Technology
NMF: Natural moisturizing factor
NOAEL: No observed adverse drug effect level
NPM: Non prescription medicines
NTP: National Toxicology Program
NTU: Nephelometric turbidity units

Abbreviations

o/w: oil in water
OT: Odour threshold
OTC: over the counter
PABA: Para amino benzoic acid
PCD: Product category designation
PDP: Principal display panel
PEG: Polyethylene glycol
Ph.Eur.: European Pharmacopoeia
POE: Polyoxyethylene
PPD: p-phenylenediamine
ppm: parts per million
PVA: polyvinyl alcohol
PVP: Polyvinyl pyrrolidone
q.s.: Quantity sufficient
QC: Quality control
QSAR: Quantitative structure activity relationship
r.p.m.: Revolutions per minute
RO: Reverse osmosis
ROS: Reactive oxygen species
SCCNEP: Scientific Committee for Cosmetics and Non-food Products
SEM: Scanning electron microscope
SLS: Sodium lauryl sulphate
SPF: Sun protection factor
TCU: True colour units
TDS: Total dissolved solids
TEA: Triethanolamine
TEM: Transmission electron microscopy
TEWL: Transepidermal water loss
TFC: Total fat content
TGA: Thermogravimetric analysis
TGF: Transforming growth factor
TiO_2: Titanium dioxide
TLC: Thin layer chromatography
TMA: Thermo mechanical analysis
TOC: Total organic carbon
TVC: Total viable count
UHST: Ultra high short term treatment
UNSCEAR: United Nations Scientific Committee on the Effects of Atomic Radiation
US FDA: United States Food and Drugs Administration
UVA: Ultra Violet A radiations (wavelength [λ] between 290 and 320 nm)
UVB: Ultraviolet B radiations (wavelength between 320 and 400 nm)
VOC: Volatile organic content
w/o: Water in oil
ZnO: Zinc oxide
ZPT, ZPTO: Zinc pyrithion or zinc pyridinium thiol N-oxide

Part 1

Introduction to Cosmetics

1. Introduction and Classification
2. Ingredients of Cosmetics
3. Present Scenario of Cosmetics

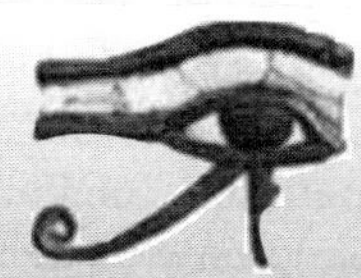

Introduction and Classification

Beauty is a greater recommendation than any letter of introduction.

— Aristotle

COSMETICS: HISTORICAL BACKGROUND

The use of cosmetics as substances to enhance or protect the beauty of the human body dates back to Vedic and Puranik period. Earlier human race of tribal era used animal parts, vegetable leaves, flowers, colour stones, shells, etc. to adorn their bodies. The ancient ayurvedic literature is full of herbal cosmetic formulations. Nowadays cosmetic are considered to be one of the essential commodities of life. A subset of cosmetics is called make-up, which refers primarily to coloured products intended to alter the user's appearance. Cosmetics include skin-care creams, lotions, powders, perfumes, lipsticks, fingernail polishes, eye and facial makeup, permanent waves, hair colours, deodorants, baby products, bath oils, bubble baths, and many other types of products. Their use is widespread, especially among women in Western countries. These are over the counter (OTC) and some time non prescription medicine (NPM) products.

Traditionally the people used home-made preparations as cosmetics to gratify their senses. Three of our five senses smell, sight and touch find expression in cosmetic usage. The cosmetics were used extensively by the more privileged people. From beginning, there is the tendency to have flawless skin and enhance sense of well-being through beautiful outlook. Cosmetics are mainly used for the two purposes, i.e. enhancing personal appeal of human beings and care of body parts.

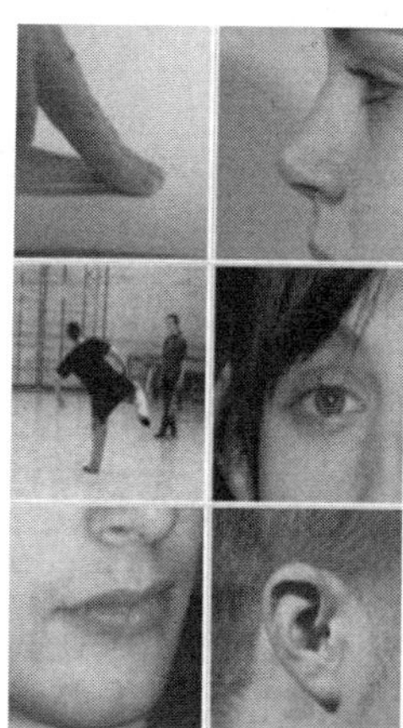

The use of kajal for beautifying the eyes, the use of kohl or kajal has a long history in Hindu culture. The use of traditional preparations of kohl on children and adults has been considered to have health benefits, and protects from evil eyes, though in the United States it has been linked to lead poisoning and is prohibited. Avleha and reetha as a substitute for shampoo, use of flowers for preparation of perfumes and substitute for aroma therapy are just few examples to substantiate the above documented statements. In the ancient Greek and Egyptian cultures also stress has been laid on preserving the beauty. Coloured earth, malachite green, copper ore, lamp black, myrrh, etc. were used to beatify body in various ways. At the time of Queen Cleopatra cosmetics was at its peak.

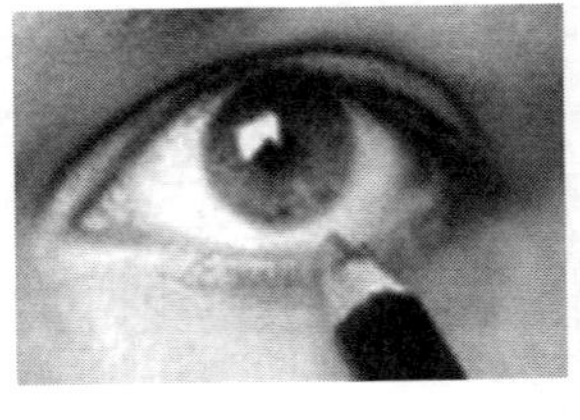

In Greek mythology Venus is considered as the goddess of beauty.

CHRONOLOGY OF DEVELOPMENT OF COSMETICS

The history of cosmetics spans at least 6000 years of human history, and almost every society on earth.

Circa 4000 BC The first archaeological evidence of cosmetics usage is found in Ancient Egypt.

Old civilizations In Indus, Romans, Greek, Chinese, Mesopotamia civilizations, cosmetics were associated with religious practices and used cosmetics containing mercury and often lead. The cosmetic uses of kohl and henna have their roots in North Africa. In Egypt some high priests were recognised as medical practitioners and everything related to body and healthcare was associated with medicine, their practice was freely adulterated with astrology magic, mysticism and religion. So it can be said that cosmetics were unearthed from tombs, temples and other religious places in early period.

Circa 3000 BC Chinese people began to stain their fingernails with gum arabic, gelatin, beeswax and egg.

India had a medical code since 1000 BC in the ayurveda and used the native raw material in medicine; also in religious rites and for aesthetic use to alleviate the rigours of the hot climate excavations in the Indus valley have yielded cosmetics pots of clay, stone, ivory, faience and alabaster from the third to second millennium BC. They were believed to be kohl pots. Helena and lamp black were used for eye make-up. Most popular being sandal wood perfumed body oil to give a long lasting odor. Fragment oils were used in the hair for both men and women dyed grey hair, eyebrows were painted black and brought fairly close together. There was no mention of soap being used to wash the body but clay was mentioned that was called sopocosmetics. Barley flour and butter were used as a cure of pimple and pumice stone used for whitening the teeth.

Circa 460–400 BC Hippocrates (father of medicine and pharmacy) dissociated medicine from magic religion and superstitions, he also advocated importance of diet, exercise, sunlight, baths and massage for good health and beauty. Theophrastus was probably the earliest Greek who wrote on the subject of perfumery. He also described the raw

material (various parts of the plants, leaves, twigs, root, wood, fruit, gum or mixture of several parts) from which perfumes were prepared.

Gupta dynasty (India) Cosmetic hygiene advanced, ladies used home made creams, oils, pastes, hair dyes, perfumes abundantly. Cosmetics made up of camphor, myrrh, saffron, aloes, sandal, vermillion (kumkum), turmeric various spices, herbs, resins, fats, etc. were given to the world by India.

Circa 50 BC Arts of cosmetics and cosmetology applying and compounding developed so much in reign of Queen Cleopatra of Alexandria.

Circa 40 BC Materia Medica prepared by Pedanios Dioscorides, Galen prepared first cooling ointment (*Ceratum refrigerans*) and cold cream.

Circa 4th or 5th centuries Henna had been used in India as a hair dye, or in the art of mehandi, in which complex designs were painted on the hands and feet, especially before a Hindu wedding and on festivals.

Circa 529 AD Plato established Academy of Athens for exchange of ideas from all over the world especially to cosmetics and perfumery.

Circa 936–1013 AD Abu'al-Qassim al-Zahrawi or Abulcassis, a cosmetologist and physician, wrote medical encyclopedia Al-Tasreef, in 30 volumes. Chapter 19 was devoted to cosmetics.

Circa 14th century Henri de Mondeville, a Norman separated cosmetics from medicine.

Paracelsus introduced Iatrochemistry and introduced a large number of chemical substances in drugs and cosmetics.

Circa 15th century Andre Le Fournier, Dean of Faculty of Medicine at University of Pans, France published a comprehensive text on beautification.

Jean Liebant, French physician and agronomist wrote several books and notes on general health, toxicology and cosmetics.

1492 AD Vasco da Gama gave new route of India to the world, the advantage for trafficking in coveted spices, medicines, cosmetics and other luxuries from India to Europe.

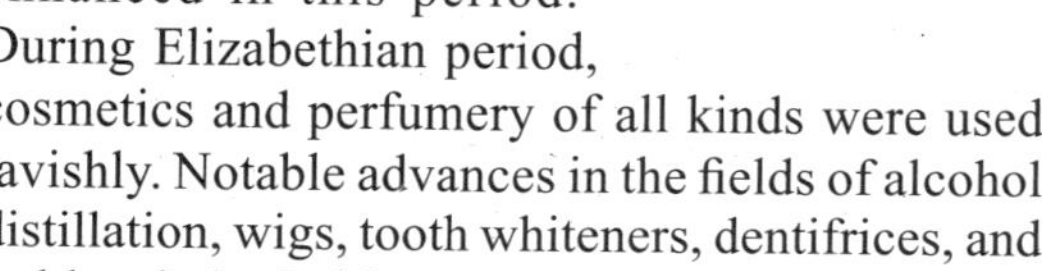

Circa 16th century AD renaissance period in Europe medicine and cosmetics enhanced in this period. During Elizabethian period, cosmetics and perfumery of all kinds were used lavishly. Notable advances in the fields of alcohol distillation, wigs, tooth whiteners, dentifrices, and golden hair fashion.

Queen Elizabeth I of England was one well-known user of white lead, with which she created a look known as the Mask of Youth.

Franngalo Paladini produced a cream and toilet vinegar widely used by the ladies of Tuscan court. Thomus Tusser, in point of good husbandry, gave a list of 21 herbs suitable for strewing. Iron oxide and cinnabar (mercury sulphide) were used for rouge and lead carbonate for face powder. Egg white used to give fashionable glaze to the face. Most species to whiten teeth included ground bricks, cattle bone, red and white coral, eggshells, alum, pumice, etc. victims of spots and pimple treated with sulphur and turpentine for an hour then rubbed them with fresh butter.

1641 AD first soap made in England, law of gallantry had recommended that entire body be bathed occasionally, the hands be washed at least everyday, the face almost as often and head from time to time.

17th and 18th centuries Cosmetics still made at home raw materials purchased from pharmacies and druggists. Development of cosmetics in America contributed greatly through newspapers and legitimate pharmacies.

1760 AD Men used rouge, blackened their eyebrows and started to wear wigs.

19th century Period of extraordinary activities and accomplishment in the field of science and technology.

1800s, Queen Victoria publicly declared makeup improper. It was viewed as vulgar and acceptable only for use by actors.

1818 AD Hydrogen peroxide was discovered to be used as bleach hair.

1856 AD Discovery of borax in California, important ingredient for cold cream.

1861 AD Thomas Graham worked out on colloids and emulsion formation.

1888 AD First cosmetic deodorant was invented by an unknown inventor from Philadelphia, and was trademarked under the name Mumm®.

The 20th century make-up became fashionable in the United States of America and Europe owing to the influence of ballet and theatre stars and the most influential new development of all was that of the movie industry Hollywood. Among those who saw the opportunity for mass-market cosmetics were Max Factor, Sr., Elizabeth Arden, and Helena Rubinstein.

1905 AD Charles Nessler, a German invented mechanical means of putting permanent curls in hair with borax.

1907 AD Modern synthetic hair dye was invented by Eugene Schueller, founder of L'Oréal.

1914–1919 AD World War I created many problems for cosmetic and medical industry. After the First World War, the flapper look came into fashion for the first time, and with it came cosmetics: dark eyes, red lipstick, red nail polish, and the suntan, invented as a fashion statement by Coco Chanel. Previously, suntans had only been sported by agricultural workers, while fashionable women kept their skins as pale as possible.

1920 AD Beauty saloons flourished in America and Europe.

1922 AD King Tutankhamen's tomb was opened in at pyramids, cosmetics were found inside that were still fragrant and perfectly usable. Palettes were also often found in pyramids, these were originally used for grinding and mixing face and eye powders.

1923 AD Poucher's book on cosmetics first of its kind in English published.

1936 AD Eugene Schueller invented sunscreen lotion.

Post World War II 1939–1945 and after: Period of growth of chemical and pharmaceutical industry. Availability of synthetic colours and pigments and detergents revolutionised cosmetics industry.

1940 AD Brushless non-lathering shaving cream was introduced.

1952 AD Roll-on deodorant launched. Aerosol deodorant developed.

The popularity of cosmetics in the 20th century has increased rapidly. Especially in the United States, cosmetics are being used by teens (especially teen girls) at a younger and younger age. Many companies have catered to this expanding market by introducing more flavoured lipsticks and glosses, cosmetics packaged in glittery, sparkly packaging and marketing and advertising using young girls.

With change of time the focus has shifted from perfect image to simple one. The field of cosmetics is now no longer (confined to females) predominantly catering to females only. The women feel inspired when they have a mental feeling that they are beautiful. Cosmetics are finding greatest acceptance in daily use of males also. People are highly style-quotient and they spruce-up on all occasions through cosmetic usage.

The use of perfumes, deospray, sun screen lotions, lip guards, talcum powders, bathing perfumes are just few stances to stress the dominating role of cosmetics in life of males also.

The use of cosmetics, with time, has transcended the barriers of age. These are now being used from small kids in form of baby powder and baby soaps. To younger generation including teenagers in form of perfumes, deosprays, beauty creams, to middle aged and elder in form of age defying lotions, enriching creams, anti-wrinkle creams.

Certain cosmetics have transcended the barriers of age and gender and have assumed the role of protestants as they are now being used in form of lip gels, cold cream, sunscreens.

The cosmetics are now finding greater distribution among all the economic sections of our society. The acceptance of cosmetics as appearance enhancers by the people from lower strata to upper privileged has created a congenial environment for the growth of cosmetic industry. The use of bathing soap, toothpaste as essential commodities stresses the importance of cosmetics in day-to-day life.

Cosmetics are the external preparations, which are applied to skin, hair, nails for various purposes like protecting, covering, colouring, beautifying, cleansing and nourishing. Cosmetics may be classified according to the body part over which they are applied.

2000 and after Nanotechnology is being applied in cosmetics.

Cosmetics enhance a person's appearance. It is especially useful for people who like to have fun. Cosmetics are considered essential for todays life style. The cosmetics help in addressing the sun-burnt and damaged skin. The cosmetic industry claims to use healthy and tested products but many critics argue that the cosmetic industry's claims are not reliable. It should be remembered that cosmetics improves the appearance of a person but it does not succeed in making him look entirely different. But, cosmetics have the capability to alter a person's appearance.

The overwhelming population in the world believes that cosmetics have a positive effect on a person's appearance. The U.S. consumers spend more than \$3.4 billion on the cosmetic products.

People are advised to use those cosmetics that would suit their eyes, hair and skin. Cosmetics help a person to attain various kinds of looks. It helps a person to attain the natural look, classic look or an upscale look. The 'concealer' helps the person to cover blemishes. The foundation of the make-up should match the colour of the skin.

The following are the positive effects of cosmetics on a person's appearance

- It helps in increasing the moisture level of the skin.
- It helps in reducing the fine lines.
- It hides the blemishes and the dark circles.
- It tones the skin and enables improvement of the skin.
- The cosmetics give beautiful and a clean appearance.
- The cosmetics reduce wrinkles and puffiness.

The negative effects of the cosmetics on a person's appearance:

- Cosmetics cannot put a stop to the ageing process as the wrinkles return after a certain period of time.
- It was established that the cosmetics were contaminated with molds and other micro-organisms.
- The claims made by the cosmetic companies are not necessarily true. Various companies claim that their product is 'hypoallergenic'. This may not be true, as the product is not tested or regulated by anybody.
- The cosmetics do not contain sufficient extracts from the plants so they are not healthy to use.

- There is no proof that the vitamins are beneficial to the skin. It is quite a possibility that the vitamins do not have any impact on the skin.

CLASSIFICATION OF COSMETICS

A detailed classification is given below:

1. **Cosmetics for Skin**

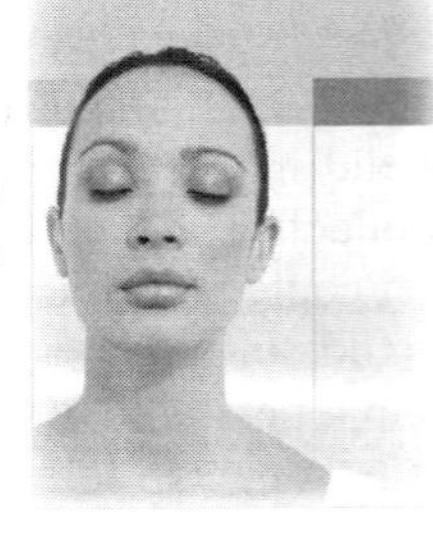

   Cosmetics used for skin can further be divided into two groups

   a. *Those which are applied for beautifying and protecting, e.g.*
      - *Skin creams, lotions, milks and emulsion:* Used to moisturize the face and body, e.g. vanishing cream, cold cream, cleansing cream, foundation cream, massage cream, all purpose cream, anti-wrinkle cream, bleach cream, moisturizing lotions, hand lotions, face wash, anti-ageing products, lip balm, lip salve.
      - *Face powders, face packs and masks*
        - *Powders:* Used to set the foundation, giving a matte finishes, e.g. face powders, and face packs, body powders or talcum powder.
        - *Compacts:* Round cake of face powders, cake.
      - *Astringents and skin tonics:* Antiperspirants, astringent lotions, astringent emulsion, stick astringents, astringents in gel form.
      - *Sunscreen, suntan and anti-sunburn preparations:* Anti-sunburn cream, prickly heat powders, sunscreen lotion (to protect the skin from damaging UV radiation).
      - *Coloured make-up preparations:*

        - *Lips make-up* Lipsticks, lip gloss, lip pencil, lip-duo.
        - *Cheeks make-up* Rouges, blush or blusher (used to colour the cheeks and emphasize the cheekbones. This comes in powder, cream and gel forms).

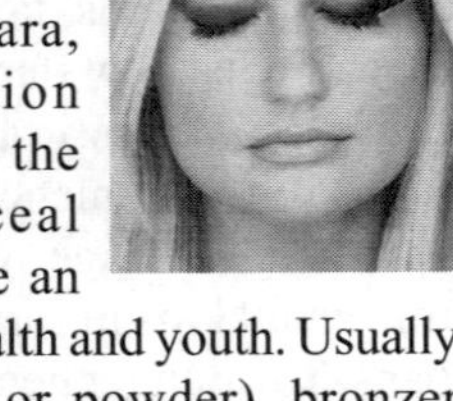

        - *Eye make-up preparations:* Eye shadows, mascara, kohl, foundation (used to colour the face and conceal flaws to produce an impression of health and youth. Usually a liquid, cream or powder), bronzer (used to create a more tanned or sun-kissed looks).
      - *Concealer:* A type of thick opaque make up used to cover pimples, various spots and inconsistencies in the skin.
      - Skin lighteners or bleaches
      - Anti perspirants and deodorants
      - Bath preparations.

   b. *Cosmetics which are applied on skin for nourishment of skin*
      - *Creams* Vitamin creams, moisturizer, hormones cream, fruit juice cream.
      - *Lotion* Treatment products to repair or hide skin imperfections (acne, dark circles under eyes, wrinkles, etc.).

2. **Cosmetics for Hair**

   Cosmetic used for hair can be classified into following groups

   a. *Preparation for cleansing, e.g. shampoos* Clear liquid shampoo, liquid cream or lotion shampoo, solid cream and gel shampoo, oil shampoo, powder shampoo, aerosol shampoo, dry shampoo, baby shampoo, antidandruff and medicated shampoo.

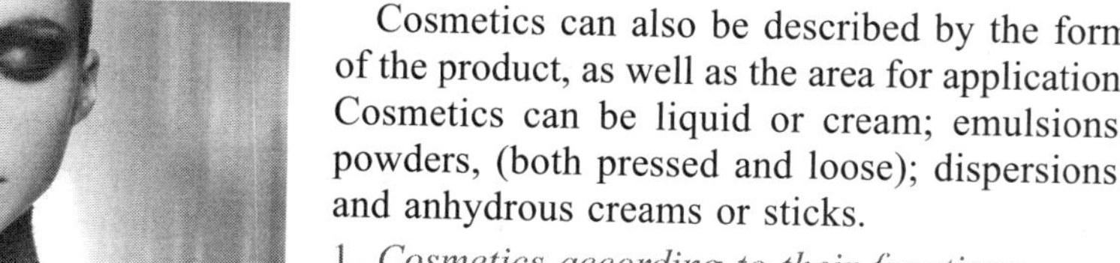

b. Hair grooming preparation, e.g. waving lotions, hair setting lotions, sprays.
c. Hair straighteners.
d. Hair tonics and conditioners.
e. Antidandruff preparations.
f. Hair removers.
g. Hair colouring and bleaching preparations.
h. Preparation for colouring of eyelids and eyebrows—Mascara, eye liners, eye shadows.

3. The **decoration of nails** is in term of shine is done by applying abrasive materials like stannic oxide and powdered silica. In general cosmetics applied on nails can be classified as:
   a. Nail lacquers and polishes
   b. Enamel removers
   c. Nail bleaches
   d. Nail white
   e. Cuticle removers
   f. Nail cream
   g. Nail strengtheners.

4. **Cosmetics for Dental Care**

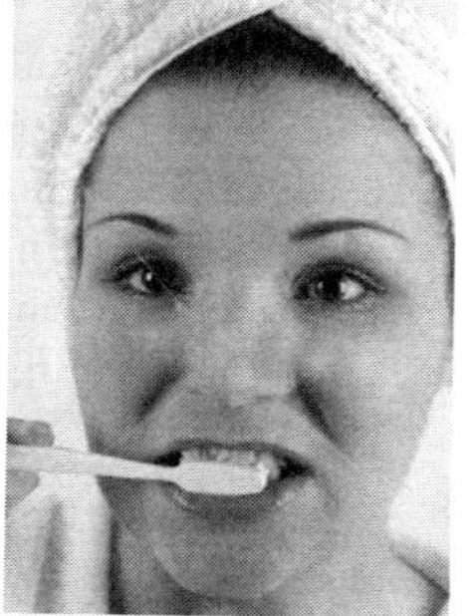

These preparations are normally used from the point of view of hygiene for dental and oral cavity. White and shiny teeth and good breath are considered as a part of cosmetic personality. All kind of preparations for cleansing of teeth are manufactured and marketed by cosmetics companies. These preparations can be classified in two groups.

1. Preparations used for cleansing of teeth, e.g. Dentifrices
   a. Toothpowders
   b. Toothpastes.
2. Proportion which are used as hygienic product for oral cavity
   a. Mouth washes
   b. Gargles

Cosmetics can also be described by the form of the product, as well as the area for application. Cosmetics can be liquid or cream; emulsions; powders, (both pressed and loose); dispersions; and anhydrous creams or sticks.

1. *Cosmetics according to their functions*
   a. *Decorative function* Make-up, blushers, rouge, nail polish, eyelashes, lipsticks, etc.
   b. *Corrective function* Dry cream and heavy face powder.
   c. *Protective function* Cold cream, lip balm, suntan preparations.
   d. *Curative function* Anti-perspirants, hair preparation, deodorant, mouthwash.
   e. *Dressing and washing purpose* Various toiletries (perfumes, after shave lotion, body wash, deodorant, etc.)
2. *Cosmetics according to their site of action:*
   a. *Skin* Powder, cream, lotion, deodorant cleansing preparation.
   b. *Hair* Shampoos, tonics, hair dressings, hair waving, beard softener, shaving preparation, hair remover, brilliantine, depilatories.
   c. *Nails* Nail polish, polish remover and other manicure preparations.
   d. *Teeth and mouth* Dentifrices (tooth powder, tooth paste and gel), mouth washes.
   e. *Borderline and kindred products* Eye products, foot powder and application, insect repellent.
3. *Cosmetics according to physical nature*
   a. *Aerosols* Hair set, perfumes, hair and body sprays.
   b. *Cakes* Rouge, compact shaving cake, toilet soap.
   c. *Emulsion* Cold cream, cleansing cream, vanishing cream.
   d. *Jellies* Hand jelly, wave set jelly, brilliantine jelly.
   e. *Mucilage* Hand lotion, wave set.
   f. *Oil* Hair oil, body oil, brilliantine.
   g. *Paste* Tooth paste, deodorant paste.
   h. *Powder* Face powder tooth powder, body powder.
   i. *Soap* Shampoo soap, shaving soap.
   j. *Solution* After shave lotion, hair sets, shampoos, astringents.

k. *Stick* Lipsticks, deodorant stick.
l. *Suspension* Liquid powder, lotion.

4. *Extreme cosmetics*
   Prescription or surgical cosmetic procedures
   - Cosmetic contact lenses
   - Cosmetic coloured contact lenses.

   In major cosmetic surgery many techniques, such as
   - Microdermabrasion,
   - Chemical or physical peels and
   - Removing the oldest, top layers of skin cells are done. The younger layers of skin left behind appear more plump, youthful, and soft.

   Permanent application of pigments (tattooing) is also used cosmetically.
5. *Other classification of cosmetics are:*
   1. *Baby products*
      a. Baby shampoos
      b. Lotions, oils, powders and creams
      c. Other baby products.
   2. *Bath preparations*
      a. Bath oils, tablets and salts
      b. Bubble baths
      c. Bath capsules
      d. Other bath preparations.
   3. *Eye makeup preparations*
      a. Eyebrow pencil
      b. Eyeliner
      c. Eye shadow
      d. Eye lotion
      e. Eye makeup remover
      f. Mascara
      g. Other eye makeup preparations.
   4. *Fragrance preparations*
      a. Cologne and toilet waters
      b. Perfumes
      c. Powders (dusting and talcum, excluding aftershave talc)
      d. Sachets
      e. Other fragrance preparations
   5. *Hair preparations (non-colouring)*
      a. Hair conditioner
      b. Hair spray (aerosol fixatives)
      c. Hair straighteners
      d. Permanent waves
      e. Rinses (non-coloring)
      f. Shampoos (non-coloring)
      g. Tonics, dressings, and other hair grooming aids
      h. Wave sets
      i. Other hair preparations.
   6. *Hair colouring preparations*
      a. Hair dyes and colours (all types requiring caution statements and patch tests)
      b. Hair tints
      c. Hair rinses (colouring)
      d. Hair shampoos (colouring)
      e. Hair colour sprays (aerosol)
      f. Hair lighteners with colour
      g. Hair bleaches
      h. Other hair colouring preparations.
   7. *Makeup preparations (not for eye)*
      a. Blushers (all types)
      b. Face powders
      c. Foundations
      d. Leg and body paints
      e. Lipstick
      f. Makeup bases
      g. Rouges
      h. Makeup fixatives
      i. Other makeup preparations.
   8. *Manicuring preparations*
      a. Basecoats and undercoats
      b. Cuticle softeners
      c. Nail creams and lotions
      d. Nail extenders
      e. Nail polish and enamel
      f. Nail polish and enamel removers
      g. Other manicuring preparations
   9. *Oral hygiene products*
      a. Dentifrices (aerosol, liquid, pastes and powders)
      b. Mouthwashes and breath fresheners (liquids and sprays)
      c. Other oral hygiene products.
   10. *Personal cleanliness*
       a. Bath soaps and detergents
       b. Deodorants (underarm)
       c. Vaginal douches
       d. Feminine hygiene deodorants
       e. Other personal cleanliness products.
   11. *Shaving preparations*
       a. Aftershave lotion
       b. Beard softeners

c. Men's talc
d. Preshave lotions (all types)
e. Shaving cream (aerosol, brushless and lather)
f. Shaving soap (cakes, sticks, etc.)
g. Other shaving preparations.

12. *Skin care preparations*
a. Cleansing (cold creams, cleansing lotions, liquids and pads)
b. Depilatories
c. Face and neck (excluding shaving preparations)
d. Body and hand (excluding shaving preparations)
e. Foot powders and sprays
f. Moisturizing creams and lotions
g. Night creams and lotions
h. Paste masks (mud packs)
i. Skin fresheners
j. Other skincare preparations.

13. *Suntan preparations*
a. Suntan gels, creams and liquids
b. Indoor tanning preparations
c. Other suntan preparations,

Coloured cosmetics: Facial foundations, facial powders, pressed powder, base, blush, rouge, acne cosmetica, cosmetic-associated acne, sunscreen and acne, self-tanning products, sunless tanning, lipstick, camouflage makeup, scar, scarring, mascara, eye shadow, etc.

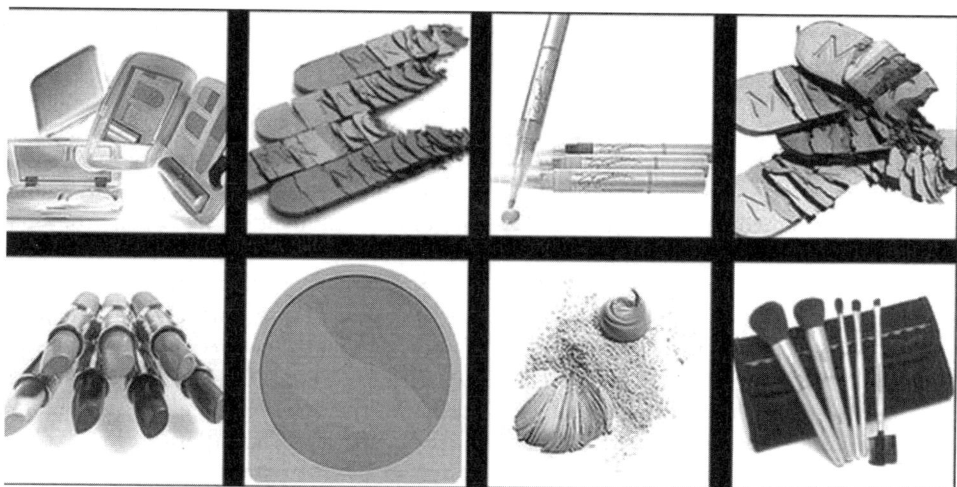

2 Ingredients of Cosmetics

> *There is no cosmetic for beauty like happiness.*
> — Blessington, Marguerite Gardiner, Countess Johann Wolfgang Von

Cosmetics in a variety of forms date back to early civilizations, with the need to improve ones personal appearance being an important factor in attracting a mate. Over the years the ingredients have changed dramatically. The realization of the dangers of many common ingredients also greatly impacted the growing industry.

The ingredients of cosmetics come from a variety of sources but, unlike the ingredients of food, are often not considered by most consumers. Cosmetics often use vibrant colours that are derived from some unexpected sources, ranging from crushed insects to rust. Many new techniques have allowed manufacturers to synthesize such colours and the use of animals (or parts thereof) has been declining for many years.

To protect the public the ingredients used in cosmetics are highly regulated and frequently (though controversially) tested on animals. Still there have been injuries and adverse reactions caused by the use of a variety of ingredients throughout history.

MOST COMMONLY USED COSMETICS INGREDIENTS

- **Purified water:** Water is most frequently used raw material as solvent or as a vehicle for many ingredients in manufacturing of cosmetics. There should not be any type of contaminants in water either inorganic (magnesium, zinc or calcium impurity) or presence of microorganism. Therefore, treatment of raw water from mains (wells, ponds, river, lakes, sea) is necessary for uses in cosmetics, otherwise, many problems like formation of insoluble residues, foul odour, visible colonies of moulds, fungi or bacteria (*E. coli, P. auregenosa, S. aureus, coliforms*) may occur. Purification of minerals from water takes place by means of distillation, ion-exchange systems, reverse osmosis and purification of microorganism by means of chemical treatments (1 to 4 ppm of chlorine is used), heat treatment (heat at 120°C and cooled rapidly), filtration (membrane filters of pore size 0.22μm or less used to retain microbes), UV radiations (below 300 nm have lethal effects on microbes).
- **Preservatives:** Preservatives are used to prevent spoilage of cosmetics by oxidation of oils and fats and microbial growth. Most cosmetic preparation are likely to deteriorate if no preservative is added. Water promotes the growth of microorganisms, therefore all preparations which contain water must include preservatives. Preservatives in cosmetics are the second most common cause of skin problems

also, they prevent bacteria and fungus from growing in the product and protect products from damage caused by air or light. But preservatives can also cause the skin to become irritated and infected. Some examples of preservatives (antimicrobial agents and antioxidants) are:

Table 2.1: Preservatives used in cosmetics and pharmaceuticals

Preservatives	*Examples*
Alcohols	Ethanol
	Isopropanol
	Chlorbutenol
	Phenylethyl alcohol
	Phenoxyethanol
Aldehydes	Formaldehyde
	Gluteraldehyde
	Cinnamic aldehyde
Esters	Methyl-*p*-hydroxy benzoate
	Ethyl-*p*-hydroxy benzoate
	Propyl-*p*-hydroxy benzoate
	Butyl-*p*-hydroxy benzoate
	Benzyl-*p*-hydroxy benzoate
Mercury compounds	Phenyl mercuric acetate (PMA)
	Phenyl mercuric nitrate (PMN)
	Phenyl mercuric gluconate
	Nitromersol
	Thimerosal
Organic acids	Benzoic acid
	p-Chloro benzoic acid
	o-Chloro benzoic acid
	p-Hydroxy benzoic acid
	Formic acid
	Propionic acid
	Sorbic acid
	Salicylic acid
	Vanillic acid
Phenolic compounds	Phenol
	Cresol
	o-Phenyl phenol
	bis-Phenol
	Methyl chloro thymol
	p-chloro-*m*-cresol
	p-chloro-*m*-xylenol
	Dichloro-*m*-xylenol
Surface active agents	Benzalconium chloride (BAC)
	Benzathonium chloride
	Cetyl trimethyl ammonium bromide (CTMAB)
	Cetyl pyrimidium chloride
	Dimethyl di-iododecenyl-ammonium bromide
Miscellaneous	DMDM hydantoin
	Imidazolidinyl urea

Ideal preservative should possess following characteristics:

- It must be compatible with formulation.
- It should provide sustained anti-microbial action.
- It should be soluble enough to effective concentration.
- It should be non toxic, non-irritant or non-allergic.
- It should be colourless and odourless.

Table 2.2: Names and usual strength of antimicrobial agents.

Name	*Concentration %*
Benzoic acid	0.5
Esters of *p*-hydroxy benzoic acid Methyl Ethyl Propyl Butyl Benzyl	0.1
Salicylic acid	0.5
Chlorbutanol	0.5
Formaldehyde	0.5
Phenyl mercuric acetate	0.002
Phenyl mercuric nitrate	0.002
Benzyl alcohol	0.1
Phenoxyethanol	0.5
Hexamethylene tetramine	0.2
Alkyl trimethyl ammonium bromide	1.0
Chlorphenacin	0.2
Benzalkonium chloride	0.25
Gluteraldehyde	0.1
Bromopol	0.1
Dihydroacetic acid	0.5
4-isopropyl-*m*-cresol	0.1
4-chloro-3,5-xylenol	0.1

- **Oils, fats and waxes:** These are extensively used as bases of cosmetic preparation like creams, bath oils, ointment, hair oils, soaps, lipsticks, lotions, brilliantines, salves, shampoos, etc.
  A number of compounds derived from mineral sources are used in cosmetics, e.g. mineral oils, petrolatum, and other paraffins; chemically they are hydrocarbons.
  Oils from vegetable and mineral sources are used in cosmetics.
  - **Vegetable oils:** Almond oil, arachis or groundnut oil, castor oil, coconut oil, olive oil, sesame oil.

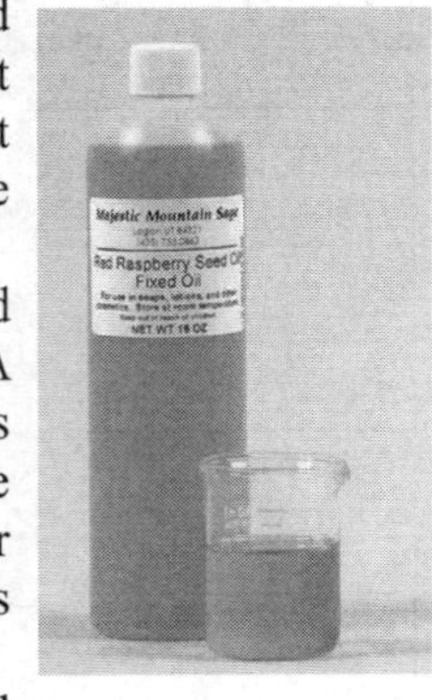

  - **Mineral oils:** Heavy and light liquid paraffin. A variety of fatty materials from animal, vegetable and marine sources or may be synthetic sources are used in cosmetics.
  - **Fatty acids:** Stearic acid **triple pressed** (mixture of 55% palmitic acid and 45% stearic acid), oleic acid (used for pearly sheen in creams and lotions but used with preservative due to rancidity problem).
  - **Fatty acid esters:** They are oily and have low viscosity, when applied to skin, they deposit a thin, oily film which is not tacky or greasy, e.g. butyl stearate, isopropyl stearate, isopropyl palmitate, isopropyl myristate, etc. used in creams and lotions.
  - **Fatty alcohols:** They are hydrophobic in nature and produce an occlusive film which helps in hydration of dry skin. Lauryl and myristyl alcohols are widely used in creams and lotions. Cetyl and steryl alcohols are used as emollient.
  - **Lanolin and its derivatives:** It is good emollient because of its hydrophobic and adhesive character. Lanolin alcohols (mixtures of sterols, triterpene alcohols and aliphatic alcohols), acetylated lanolin, propionyl lanolin derivatives and polyoxyethylene lanolin, etc. used in cosmetics.
  - **Soft paraffin:** It is a purified mixture of hydrocarbons obtained from petroleum yellow and white soft paraffins (petrolatum or petroleum jelly) are used in various cosmetic preparations as emollient.

  Some of the waxes which are widely used in cosmetics: beeswax, spermaceti, paraffin wax (hard and soft), carnauba wax, ceresin and ozokerite wax.

- **Fragrances or perfumes:** Fragrances and preservatives are the main ingredients in cosmetics. Fragrances are the most common cause of skin problems. More than 5,000 different kinds of perfumes are used in products. Products marked **fragrance-free** or **without perfume** means that no fragrances have been added to make the product smell good. There are many products which are sold for their perfumes, e.g. perfumes or sprays, toilet waters, after shave lotions, cologne, aromatic cosmetics, etc.

- **Colourants:** Colours (natural, inorganic and coal tar) are important ingredients in cosmetic preparations. FDA does require safety testing for colour additives used in cosmetics. Cosmetics may only contain approved and certified colours. FD&C, D&C, or external D&C listed on cosmetic labels.
  1. FD&C – colour that can be used only in foods, drugs, and cosmetics.
  2. D&C – colour that can be used only in drugs and cosmetics.
  3. External D&C – colour that can be used only in drugs applied to the surface of the skin and cosmetics.

Cosmetics for the Indian market confirming to Schedule Q of Drug and Cosmetics Act 1940 and IS 4707 (Part I):2001.

Colour additives are subject to a strict system of approval under U.S. law [FD&C Act, Sec. 721; 21 U.S.C. 379e]. Except in the case of coal-tar hair dyes, failure to meet U.S. colour additive requirements causes a cosmetic to be adulterated [FD&C Act, sec. 601(e); 21 U.S. Code 361(e)]. All colour additives must meet the requirements for identity and specifications stated in the Code of Federal Regulations (CFR).

The FD&C Act Section 721(c) [21 U.S. C. 379e(c)] and colour additive regulations [21 CFR Parts 70 and 80] separate approved colour additives into two main categories: those subject to certification (sometimes called **certifiable**) and those exempt from certification. In addition, the regulations refer to other classifications, such as straight colours and lakes.

These colour additives are derived primarily from petroleum and are sometimes known as **coal-tar dyes** or **synthetic-organic** colours.

Note: Coal-tar colours are materials consisting of one or more substances that either are made from coal-tar or can be derived from intermediates of the same identity as coal-tar intermediates. They may also include diluents or substrata. Today, most are made from petroleum.

Except in the case of coal-tar hair dyes, these colours must not be used unless FDA has certified that the batch in question has passed analysis of its composition and purity in FDA's own labs. If the batch is not FDA-certified, do not use it.

An example is "FD&C Yellow No. 5." Certified colours also may be identified in cosmetic ingredient declarations by colour and number alone, without a prefix (such as "Yellow 5").

Colours exempt from certification: These colour additives are obtained primarily from mineral, plant or animal sources. They are not subject to batch certification requirements. However, they still are considered artificial colours, and when used in cosmetics or other FDA-regulated products, they must comply with the identity, specifications, uses, restrictions, and labeling requirements stated in the regulations [21 CFR 73].

Straight colour: "Straight colour" refers to any colour additive listed in 21 CFR 73, 74 and 81 [21 CFR 70.3(j)].

Lake: A lake is a straight colour extended on a substratum by adsorption, co-precipitation, or chemical combination that does not include any combination of ingredients made by a simple mixing process [21 CFR 70.3(l)]. Because lakes are not soluble in water, they often are used when it is important to keep a colour from **bleeding,** as in lipstick. In some cases, special restrictions apply to their use. As with any colour additive, it is important to check the Summary of Colour Additives and the regulations themselves [21 CFR 82, Subparts B and C] to be sure you are using lakes only for their approved uses.

Organic lake colourants: Lake pigments are water-insoluble colouring agents used in foods, drugs and cosmetics. Lakes are special kind of colour additive prepared by precipitating a soluble dye onto an approved insoluble base or substratum. Aluminum lake colours produced by the adsorption of water soluble dye onto a hydrated aluminum [$Al_2(OH)_3$] substrate rendering the colour insoluble in water. Lakes are available with a pure dye content ranging from less than 1% to more than 40% and with moisture level of 6–25%. Lakes are insoluble in most solvents, although some bleeding or leaching may be observed in solvents, in which unlaced dye is soluble. Properties of lakes that enhance their usefulness include their opacity, their ability to be incorporated into products in the dye state, their relative insolubility, and their superior stability towards heat and light. These can be used to colour the end product either directly or by coating onto the surface of the product.

In general lakes are more stable than the corresponding water soluble dyes, producing brighter, more intense colours and are most suitable for products containing lipids or products with low moisture content.

Inorganic colourants: These colourants are one or a combination of various synthetically prepared iron oxides, including the hydrated forms. The naturally occurring oxides are unacceptable as colour additives because of the difficulties frequently encountered in purifying them. The commonly used forms are yellow hydrated oxides, the brown, red, and black oxides. Primary colours and blend are available in a variety of forms including powder, fine grinds and dispersions, e.g. iron oxides, carbon black, titanium dioxide, aluminum hydrate, ultramarines, zinc oxide, chromium oxide.

Colour muds are thick pastes manufactured by dispersing lake colours in capric/caprilic triglycerides for use in lipsticks. These facilitate to avoid any grinding of the colourants used in lipsticks.

Treated pigments in powder form have very low surface tension and excellent hydrophobicity. These powders are easier to disperse and impart waterproofness with good affinity to skin. They also reduce oil absorption in the formulation allowing higher loading and producing more natural finish.

Thixotropic gel and base offer a great advantage to produce the nail varnish for any stage. Thixotropic gel and base, NC chips, NC colour solutions, ready-to-fill pigmented base, etc. available for nail enamel base to make process trouble-free and reduce production time.

Fluorescent colours : Only the following fluorescent colours are approved for use in cosmetics, and there are limits on their intended uses: D&C Orange No. 5, No. 10, and No. 11; and D&C Red No. 21, No. 22, No. 27 and No. 28 [21 CFR 74.2254, 74.2260, 74.2261, 74.2321, 74.2322, 74.2327 and 74.2328].

Glow-in-the-dark colours: Luminescent zinc sulfide (ZnS) is the only approved glow-in-the-dark colour additive [21 CFR 73.2995].

Halloween makeup: These products are considered cosmetics [FD&C Act, Sec. 201(i); 21 U.S.C. 321(i)] and are therefore subject to the same regulations as other cosmetics, including the same restrictions on colour additives.

Liquid crystal colours: These additives, which produce colour motifs in a product through diffraction, are unapproved colour additives. Their use in cosmetics is therefore illegal [FD&C Act, Sec. 601(e); 21 U.S.C. 361(e)].

- **Humectants or moisturizer:** Humectants are hygroscopic materials. The cutaneous permeability barrier is localized in the *stratum corneum* interstices, and it is mediated by the lamellar bilayers enriched in cholesterol, free fatty acids, and ceramides. Moisturizers make the *stratum corneum* softer and more pliant by increasing its hydration. Drying out of cosmetics can occur due to vaporization of water. Aqueous solutions of humectants can reduce the rate of loss of moisture.

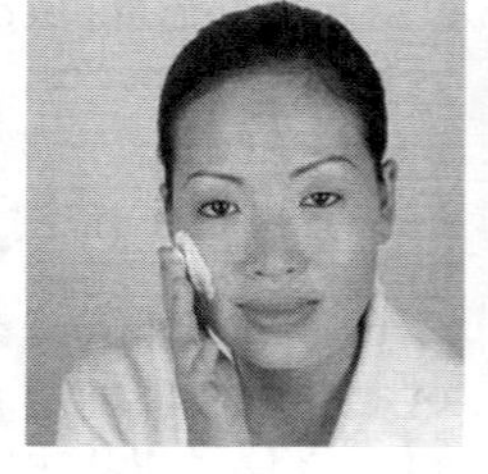

Ideal characteristics of humectants are:

- It must absorb moisture from atmosphere and remain the same under normal conditions of atmospheric humidity.

- It should not solidify or crystallize at normal conditions.
- It should not be too costly.
- It should be compatible with wide range of raw materials.
- It should be colourless, non-toxic, non-irritant and should have good odour and taste.
- It should be non-corrosive to the packaging material.

Various types of humectants used for cosmetic preparation are the following:

1. **Inorganic humectants:** Limited use in cosmetics because of corrosive nature and compatibility problems. Calcium chloride is used in some cosmetic preparations.
2. **Organic humectants:** These are widely used in cosmetics. Typical examples are polyhydric alcohols, their esters and ethers which are most likely to be used in cosmetic preparations, e.g. ethylene glycol, glycerol, sorbitol, propylene glycol, polyethylene glycol (of varying molecular weight grade) and other examples like: glycerine, alpha methyl glycerine, di-ethylene glycol, di-propylene glycol, ethylene glycol, glucose, mannitol, polyoxyethylene glycerine, polyoxyethylene sorbitol, propylene glycol glucoside, sorbitan, sorbide, sodium lactate, triethanolamine, triethanolamine lactate, urea, etc.
3. **Metal-organic humectants:** In general they are also not widely used in cosmetics due to incompatibility problems, corrosive nature and bad taste. Only sodium lactate has some use in skin creams.

- **Surfactants:** Surfactants or **surface active agents** lower interfacial tensions between boundaries of the system. They also stabilize one or more interfaces by formation of absorbed layers. These effects are used in formulation of cosmetics. There are various properties for which surfactants are used in cosmetics like: emulsification, foaming, detergency, wetting and solubilization. All these properties are not found in one single surfactant, these are shared to some degree by all the surfactants and depending on these properties and the end use of cosmetics, selection of surfactants are done.

All surfactants are amphipathic molecules which means they have two parts, one, a hydrophobic part (non-polar group, a hydrocarbon chain, ring or mixture of two) and other a hydrophilic part (polar group such as carboxylic, sulphate or sulphonate group).

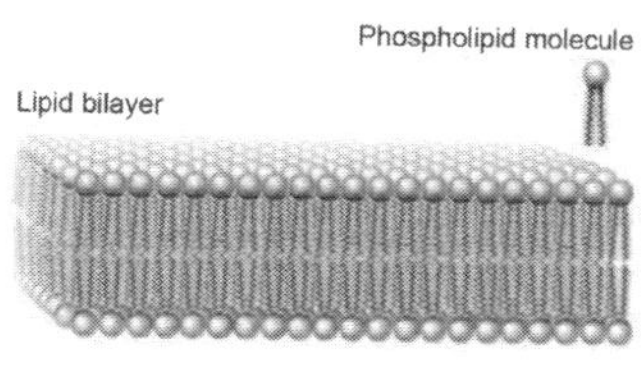

Surfactants are classified on the basis of their physical properties, chemical structure or on the basis of their use and most important is their ionic behaviour in aqueous solution. Surfactants are classified according to ionic behaviour as:

1. **Anionic surfactants:** Soaps (sodium oleate).
2. **Cationic surfactants:** Nitrogenous and non-nitrogenous cationic surfactants (quaternary ammonium salts, phosphonium and sulphonium salts).
3. **Non-ionic surfactants:** Alkanolamides, polyethylene glycol der., polyethyleneimine der.
4. **Ampholytic surfactants:** Alkylamino acids, acylamino acids, alkylimidazolines.

- **Herbal products:** Detailed discussion is given in chapter herbal cosmetics.

COSMECEUTICALS

Cosmeceuticals is the novel approach where wellness meets beauty in cosmetics. The word is a portmanteau of the words **cosmetic** and **pharmaceutical**. Some products can be both cosmetics and drugs. This may happen when a product has two uses. For example, a shampoo is a cosmetic because it is used to clean the hair. But, an antidandruff treatment is a drug

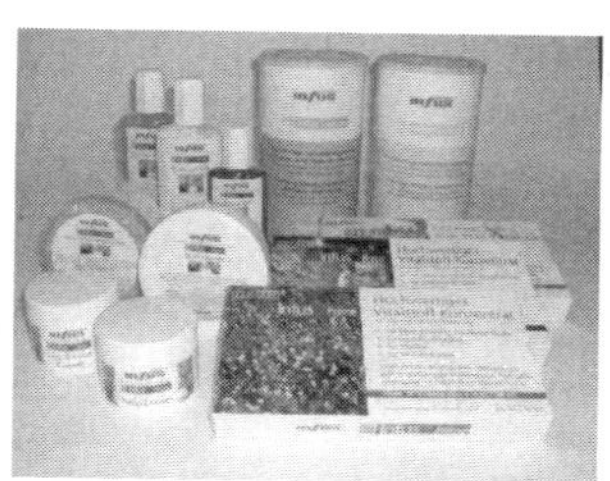

because it is used to treat dandruff. So an anti-dandruff shampoo is both a cosmetic and a drug. Other examples are:

- Acne products
- Antimicrobial products
- Antiperspirants products
- Astrigent products
- Antiperspirants products
- Oral care products
- Skin protectant products
- Sun-screen products
- External analgesics products
- Deodorants that are also antiperspirants.
- Moisturizers and make-up that provide sun protection.
- Toothpastes that contain fluoride.

These products must meet the standards for both cosmetics (colour additives) and drugs.

Like cosmetics cosmeceuticals are also applied topically, but they contain ingredients that influence the biological function of the skin. Cosmeceuticals improve appearance, but they do so by delivering nutrients necessary for healthy skin. Cosmeceuticals are used for different purposes such as skin care, hair care, moisturizer, scar management, hair removal, sun care, anti-ageing and antibacterial, etc.

Cosmeceuticals may contain purported active ingredients such as vitamins, phytochemicals, enzymes, antioxidants, and essential oils. However, these ingredients may not necessarily be effective, and if they are effective, the cosmeceutical may not have the active ingredient(s) in an effective formulation or at effective concentrations.

Some cosmetic makers use the term **cosmeceutical** to refer to products that have drug-like benefits. Cosmeceuticals offer a more affordable and less invasive alternative to plastic surgery. FDA does not recognize this term. A product can be a drug, a cosmetic, or a combination of both. But the term **cosmeceutical** has no meaning under the law.

HYPOALLERGENIC COSMETICS

Hypoallergenic cosmetics are products that makers claim cause fewer allergic reactions than other products. Women with sensitive skin, and even those with **normal** skin, may think these products will be gentler. But there are no federal standards for using the term hypoallergenic. The term can mean whatever a company wants it to mean. Cosmetic makers do not have to prove their claims to the FDA.

Some products that have **natural** ingredients can cause allergic reactions. If you have an allergy to certain plants or animals, you could have an allergic reaction to cosmetics with those things in them. For example, lanolin from sheep wool is found in many lotions. But it is a common cause of allergies too.

BODY ART

Body art is made on, with, or consisting of, the human body. The most common forms of body art are:

- Body painting,
- Branding,
- Full body tattoo,
- Scalpelling,

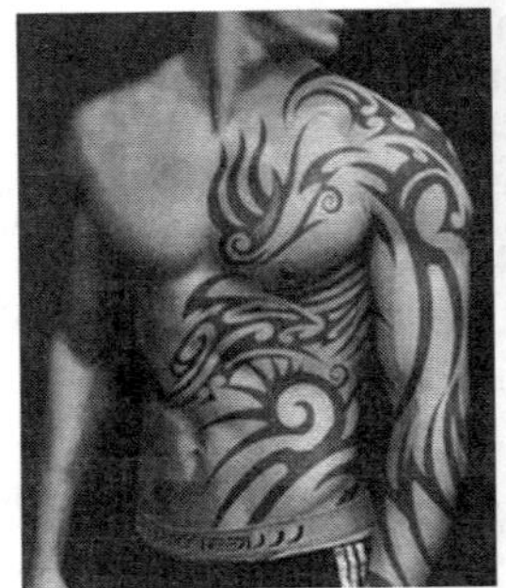

- Scarification,
- Shaping (e.g. tight-lacing of corsets), and
- Tattoos and body piercing.

In India more and more people are also becoming regular user of the cosmetic preparations. In the Indian market cosmetics for legal control and its manufacturing has been governed under Schedule Q of the Drug and Cosmetics Act 1940 and IS 4707 (Part I): 2001.

3 Present Scenario of Cosmetics

> *Beauty rests on necessities.*
> — Ralph Waldo Emerson

Usage of cosmetics has increased tremendously, resulting into an increased production, import distribution and sale of cosmetics. Cosmetics are luxury articles from past few years. The commercialization of the home made remedies have created boom in the cosmetic industry of late in last decade or so. The various cosmetic product ranging from soap, talcum powders to cream, lotions, and face mask are now easily available at every place at a premium price.

The commercialization of products has changed the attitude of the masses and can be considered to be trend setting in their own respect.

It is to be stressed here that commercialization of product is one of prime factors responsible that the industry is having tremendous growth rate.

An important observation of present scenario of cosmetics is that the primary target of the major Multi National Companies (MNC's) is the girls in the age group 16–25 years. They represent their potential buyer segment and the industry is trying to woo them by all trendy means and use of visual media.

The various factors that have contributed to the present scenario of cosmetics can be summarised as follows:

GLOBALISATION

Since the time immemorial materials used for beautification or improvement of appearance are known as cosmetics. But from the last decade or so has seen the opening up of economy to the global players. The cosmetic majors have taken this opportunity to establish themselves and to capture maximum market share of the relatively untapped large growth potential.

Opening up of economy has led to the introduction of world-class quality cosmetic products and greater options are now available to masses for selecting from wide range of products. Moreover, the increase in purchasing power has resulted in the growth of cosmetic industry.

COMMERCIALIZATION OF CONCEPT OF BEAUTY

The practice of adornment or improvement of appearance continued unabated across the centuries. The various **beauty pageants/contests** held at national and international levels have changed

Table 3.1: Age-wise classification of motives

Motive	*Below 20*	*20–35*	*35 above*	*Total*
For better appearance	24	20	14	58 (29%)
For health and hygiene	16	28	24	68 (34%)
For personal reason	22	16	10	48 (20%)
Uncertain	10	8	8	26 (13%)
	72	72	56	200

the concept of beauty in the modern era. The various women who have emerged winner in such contest become the role model for the masses. The cosmetic giants are using them for the promotion of their range of products.

The masses out of their shear liking for their role models are using the various products to enhance their physical appearance.

MAJOR TECHNOLOGICAL REVOLUTION

With major technological revolution in information technology (IT) more and more people are in constant touch with the global market. The people are aware of every new product and the pre-existing products easily available in the market.

The visual media, to an extent, has affected the mentality of all the people and especially that of children of younger generation. The people are now more of consumers with considerably buying potential. The advertisements that are being aired through the cables have thus helped in tapping the market potential.

Angel dusting is the misleading marketing or advertising practice of including a very small amount of an active ingredient in a cosmetic, cosmeceutical, dietary supplement, food product, or nutraceutical, insufficient to cause any quantifiable benefit. The advertising materials may claim that the ingredient is helpful and that the ingredient is contained in the product, both of which are true (with the exception of some extreme cases where the ingredient is so diluted that most units of the product don't even contain a single molecule of the active ingredient). However, no claim is made that the product contains enough of the active ingredient to have an effect—this is just assumed by the purchaser. Thus, while misleading, angel dusting is typically legal.

EDUCATION

The spreading of education among the people also has its share in the success of cosmetic industry. The education has made people more aware about their looks and personality. The people want to be up with time and trendy.

Education has also resulted in better understanding of print and visual media which has thus led to greater understanding of the product. Thus, the product with favourable people show increase in sale volume.

Table 3.2: Education-wise spending

Spreading of Education	*Hr. Sec.*	*Graduate*	*PG*	*(%)*
< 100	4	6	2	12
100–200	10	13	9	32
200–500	26	36	23	85
>500	5	15	1	21
	45	70	36	15

WOMEN LIBERALISATION

With women coming up in every discipline, they have indirectly contributed to growth of cosmetic industry. The working women believe that her appearance should suit the job people. Thus the cosmetics provide a supportive role to them in their development.

Table 3.3: Sex-wise spending on cosmetics

Amount Spent on Cosmetic per month	*Male No.%*	*Female No. %*	*Total No. %*
<100	15 (15%)	5 (5%)	20 (10%)
100–200	24 (24%)	16 (16%)	40 (20%)
200–500	41 (41%)	42 (42%)	83 (83%)
>500	20 (20%)	37 (37%)	57 (57%)

COSMETIC PRODUCTION IN INDIA

The fast growing cosmetic industry is estimated to be worth of about Rs.12 billion. Out of which Rs. 5 billion is in the organised sector and rest being with unorganised sector.

Total cosmetic market is growing at the rate of 20 per cent while market for herbal products is growing at the rate of 70 per cent per annum. The Indian cosmetic market can be segmented on basis of:

A. Product Categories

1. Colours—comprising of lipsticks, nail enamels, foundations, eye makeup, blush on and compacts.
2. Skin care
3. Fragrance
4. Hair care—Hair oils/shampoo.

B. Price Segmentation

1. Premium segment
2. Popular segment
3. Economy segment.

ROLE OF MULTINATIONAL COMPANIES (MNC's)

The entry of MNC's in the field of cosmetics had revolutionized the Indian cosmetic market. In the post globalization period of the market more stress is laid on consumer satisfaction, consumer acceptance and quality at a premium.

To counter the threat posed by multinationals the domestic cosmetic companies have reacted by following the course of action as:

1. Launching new products with speciality after extensive research and constant product improvement keeping in view the consumer's need and expectation.
2. More budget allocation for advertising both in print and visual media. Use of role-models for the product ranges so that there is better consumer acceptance.
3. After product launch, market research and consumer feedback forms and use of questionnaires and facilities for consumer to enquire about their queries.
4. Offering products with quality and cost effectiveness.
5. Multiple targeting based on various categories of market as per the price segmentation.

An important consequence of this market domination by MNC's is that a major portion of proceeding from sale of cosmetics is being grained out of country. This is detrimental to the growth of Indian economy.

The major MNC's with backing of sound financial resources, advanced technology and protective covering of their parent countries are invading the Indian market through joint ventures or their subsidiaries.

The domestic players with relative lack of funds and indifferent attitude of government and its policies thereof are at the verge of being wiped out from the domestic market scenario.

Moreover, the MNC's are dumping their products in Indian markets, thus creating an environment of unhealthy competition. The dumping policy is harming the domestic players as the production cost of their products is higher than the cost at which the MNC's are dumping the products.

From the legal and administrative perspective also the MNC's are welcomed in the market.

Due to better financial standing MNC's are investing heavily on carrying out their marketing policy especially on effective and extensive advertisements and sales promotion technique, better distribution coordination, etc.

The outlook for cosmetic and mien goods industry is dependent on a number of factors:

- Economic conditions;
- Population trends, fashion;
- Scientific progress; and
- Government involvement.

Heretofore, the cosmetics and toilet goods industry has been relatively immune to economic conditions.

Most of the factors that have been responsible for recent growth of the industry will continue to have a favourable effect;

- Increasing population particularly in 15–34 years age group,
- Increasing affluence resulting in greater spending for personal care items,
- More women in the work force, and
- Continuous promotion and advertising alongwith rising exports of cosmetics and toilet goods.

There is increasing pressure to establish tighter control over the industry by government agencies.

Some steps have been taken by trade associations to institute a system of self regulation. The industry has indicated readiness to provide a registration of all ingredient, but not formulae. While there have been no serious charges against the industry, the speed at which, the industry moved towards a serious programs of self regulation should influence future government action with the developments of newer products, the improvements of old products and through imaginative advertising and promotion, the field of cosmetic users and uses, should continue increasing. With increasing population, with more money to spend and more time to spend it, there is every reason to believe that the cosmetic and toilet industry will flourish.

ACTS REGULATING IMPORT, MANUFACTURING AND SALE OF COSMETICS

The import, manufacture, distribution and sale of cosmetics in India are regulated by the Drug and Cosmetics Act, 1940 and Rules 1945. Under the Drug and Cosmetics Act 1940, cosmetic is defined as any article intended to be rubbed poured, sprinkled or sprayed on, or introduced into, or otherwise applied to the human body or any part thereof for cleansing, beautifying, promoting any article intended for use as a component of cosmetic.

Before 1962, in India import, manufacturing and sale of cosmetic was not regulated by any act. However it became necessary to prevent manufacturing and sale of misbranded and spurious product. The word 'cosmetic' was combined with the Drug Act and the name of Act was changed to Drugs and Cosmetics Act, 1940 by Act 21 of 1962.

Soap was included in Act in 1982, as toilet soaps have been covered under the definition of cosmetics.

A cosmetic maker can sell products without FDA approval. FDA does not review or approve cosmetics, or their ingredients, before they are sold to the public. But FDA urges cosmetic makers to do whatever tests are needed to prove their products are safe. Cosmetic makers must put a warning statement on the front labels of products that have not been through safety testing, which reads, **WARNING—The safety of this product has not been determined.**

IMPORT OF COSMETICS

Import of cosmetic is regulated by the rules which are applicable to the drugs import of which requires no license. However import of following cosmetics is prohibited under Chapter III and Section 10 of Drug and Cosmetic Act.

1. ***Misbranded cosmetics***

   A misbranded cosmetic is that which contain:
   - Colours other than those prescribed
   - Which is not labelled in proper prescribed manner
   - Makes any false or misleading claims.
2. ***Spurious cosmetics***

   Spurious cosmetics are those which are:
   - Substitute for other cosmetics.
   - Imported under a name which belongs to another cosmetic.
   - If it bear names of manufacturers which are fictitious or does not exist.
   - If it bear names of manufacturers which are not true manufacturers.

QUALITY CONTROL CONSIDERATIONS

Quality is an inherent property of a product which when compared with a standard; provide a basis for measuring the uniformity of the product. Uniformity of the product is necessary for its acceptability.

Quality control is required to:

- Maintain product's quality at the desired level within limits of acceptability.
- To recognize any possibility of narrowing these limits, maintaining product quality at desired level.
- To reduce cost of production.
- To detect quality changes and to prevent them.

It is the responsibility of the management to define the term **acceptable quality**, but it is the quality control department which has to maintain this quality at the prescribed level. Quality control department should realize that "quality cannot be inspected into a product but must be built in".

Standards

The quality control must be with respect to certain criteria or standards. The quality control department has the responsibility of preparing these standards. Before the implementation the standards along with statement of reasons must be submitted to appropriate key personnel for final approval. The particular standard should be issued only to the department or individuals who are directly concerned with it. In the case of revision, all superseded copies should be recalled. Copies of its active or superseded standards should be kept for future reference, and also as a history of the evolution of formulas and processes.

The Standard Must Cover Following Points

- The formula is statement of ingredients which form the product. Looking to the high confidential nature of the formula, it should be given a special setup to keep it apart from other standards.
- Raw materials specifications should include all the characteristics of materials that are used to formulate the product. Any kind of deviation in doing so may cause failure of the product or undesirable lack of uniformity.
- Operating standards include data and information which are needed by the manufacturer to formulate same product repetitively with minimum variations. A complete and thoroughly understandable description of the actual operations and specific instructions for taking samples, and list of various test to be done on samples must be preserved.
- Finished product standards must include all the characteristics related to the proper performance, durability and safety of the product, e.g. the finished product standards for a liquid must include limits for viscosity, clarity, colour pH and residue on evaporation.

Some products may require to be subjected for biological controls in order to determine the efficacy and safety of finished product, e.g. antiseptic products are subject to bacteriological control and eye washed should be checked for sterility.

- Packaging material standards should be set for everything related with product like bottles, caps, cans, labels, printed inserts, wrapping paper, boxes, etc.
- Testing methods should be designed to provide sufficient accuracy. It is not possible to standardise testing methods so that they will give results of accuracy when used by different persons and in different laboratories. Using too many methods is wasteful and using one which is not fine enough is dangerous.

EVALUATION OF RAW MATERIALS

Evaluation of raw material is the first step in obtaining a product of suitable uniformity. It consists of sampling and testing.

Method used for sampling depends upon number and type of containers, manner of transportation, receiving and storing of raw materials.

The method of testing the raw material is decided on the basis of experience; in general no specific rules can be made. However active ingredients should be given with major attention and impurities of dangerous nature should be detected carefully.

This may require use of even very complicated testing method. The range of testing method may include organoleptic examination to spectrophotometric determination, apart from physical and chemical testing.

After completion of testing a part of sample should be presented in a suitable container for future reference. The quality control standards for raw material evaluation also include the instructions for proper storage of material before use.

Finished Product

The samples of the product must be taken and tested. Methods for test are chosen as per need. It must be assured that the composition of the product conforms to standards. Organoleptic evaluation, i.e. colour, odour, flavour, etc. are very important on finished product. They must be carried out by some trained persons. The samples of each completed batch of product should be preserved.

Packaging Materials

The test methods for evaluation of packaging material depend on nature of material and its use. For a label simple measurement and proof reading may be sufficient but for a container which is running on a high speed production line measurement of all its critical dimensions is required. Sometimes laboratory testing of small samples may not provide correct operational characteristic particularly when material is supplied by a new supplier. In such conditions a large sample of material may be used through entire finishing process as a trial run.

Filling Control

Control of the quality during filling is an important part. The finished package must be clean, undamaged and presentable. Closure must be properly applied. Net contents of the packages must be checked time to time that confirms declaration of contents.

Market Control

The acceptance of the product by the customers or the behaviour of the product as the market is vital question. The answer of the question demand for:

- Taking samples from market at selected intervals and evaluating them for condition of the container and its contents.
- Prompt entertainment and intelligent interpretation of complaints made by customers.
- Careful scrutiny of the product returned back from market. Detection of cause of return and providing a remedy for it.

SAFETY CONCERNS OF COSMETICS

Serious problems from cosmetics are rare. But sometimes problems can happen:

- The most common injury from cosmetics is from scratching the eye with a mascara wand. Eye infections can result if the scratches go untreated. These infections can lead to ulcers on the cornea (clear covering of the eye), loss of lashes, or even blindness. To play it safe, never try to apply mascara while riding in a car, bus, train or plane.
- Sharing make-up can also lead to serious problems. Cosmetic brushes and sponges pick up bacteria from the skin. And if you moisten brushes with saliva, the problem can be worse. Washing your hands before using make-up will help prevent this problem.
- Sleeping while wearing eye make-up can cause problems too. If mascara flakes into your eyes while you sleep, you might wake up with itching, bloodshot eyes, infections, or eye scratches. So be sure to remove all make-up before going to bed.
- Cosmetic products that come in aerosol containers also can be a hazard. For example, it is dangerous to use aerosol hairspray near heat, fire, or while smoking. Until hairspray is fully dry, it can catch on fire and cause serious burns. Fires related to hairsprays have caused injuries and death. Aerosol sprays or powders also can cause lung damage if they are deeply inhaled into the lungs.

- One may not be able to use eye make-up, such as mascara, eyeliner, and eye shadow for as long as other products. This is because of the risk of eye infection. Some experts recommend replacing mascara three months after purchase. If mascara becomes dry, throw it away. Do not add water or, even worse, saliva to moisten it. That will bring bacteria into the product.
- One may also need to watch certain all natural products that contain substances taken from plants. These products may be more at risk for bacteria. Since these products contain no preservatives or have non-traditional ones, risk of infection may be greater.
- If cosmetics are not stored as directed, they may expire before the expiration date. For example, cosmetics stored in high heat may go bad faster than the expiration date. On the other hand, products stored the way they should be can be safely used until they expire.

LATEST TRENDS IN COSMETICS

Today's latest trends in cosmetics reflect the fashion trends-shimmer and shine. There's a cosmetic product for every need. That's basically how cosmetics work—they allow enhancing, covering up, camouflaging, illuminating, emphasizing or minimizing—whatever necessary to create a picture of beauty.

From super glossy lip gloss to candy like eye make up with glitter to make your eyes truly pop. This trend only applies to eyes, lips, and nails, however, since cream and matte finishes is hot for foundations, powders and blush. The colour of eye and lip makeup is quite subtle. Gone are the vibrant colours and eye popping shades of eye shadow. Instead, pale browns, ivories, grays, tans, and other natural colours are all the rage. Lip shades are also a bit more subdued, with a nude lip or a sheer pink gloss being incredibly hot.

Lipstick is out today, with lip gloss taking over the first place in purses all over the country. These lip glosses are made in every colour under the sun and many mimic the appearance of lipstick with a much smoother and glossier look. For individuals looking to care for their lips and enjoy that terrific nude lip look, consider using a simple hydrating stick. The days of pressed powder may be numbered; as today's hot products include cream powders that go on thick but dries sheer and ultra light. This option is great for individuals who would like a bit more coverage than pressed powder offers, but do not want to deal with the mess and hassle of using foundation. Another popular trend along this line is the cream blush, which also goes on like a cream but dries a powder. Also, cream blushes prevent that unnatural colouring that appears on many women's cheeks.

There are "Makeup Specialists/Cosmetician" or beauty advisors on hand to assist you with makeup application, techniques, and colour matching.

With millions of ageing babyboomers, the hottest trend in the cosmetics industry is cosmeceuticals, cosmetics using active ingredients that are found in pharmaceuticals.

These ingredients include cleansing agents, surfactants, emollients, fats, mineral oils, botanical extracts, and enzymes.

Novel Approaches of Cosmetics

Recent advances in nanoparticulate systems show their promise as potential ideal drug delivery systems for poorly soluble, poorly absorbed and labile herbal extracts and phytochemicals. Various newer drug delivery systems and technologies that effectively import natural moisturizing agent from leave-on products, including humectants and lipid materials, also development of specialized nanosystems that are able to encapsulate active ingredients (e.g. sunscreens, fragrances or anti-ageing molecules) to interact with skin strata like nanoencapsulation, nanovesicles, solid lipid nanoparticles, nanoemulsions, microemulsions, nanosuspensions, nanoparticles, nanocrystals, ethosomes, liposomes, quantum dots, cubosomes, niosomes, phytosomes, transferosomes, etc. can be used in cosmetics.

The hydration state of skin is of considerable interest from cosmetic point of view. In cosmetology, the degree of skin hydration particularly the *stratum corneum*, is directly related to its appearance (i.e. smoothness of texture).

NUTRACOSMETICS

Nutracosmetics are emerging class of health and beauty products, which combine the benefits of nutraceutical ingredients with the elegance, skin feel and delivery systems of cosmetics. Herbs and spices have been used in maintaining and enhancing human beauty, since herbs have a lot of properties like sunscreen effects, anti-ageing, moisturizing, antioxidants, anticellulites, and antimicrobial effects. Future studies could be done taking lipomelanins as UV absorbers, microsponge technology and various softgels cosmetic supplements, which could encapsulate oil, liquid extracts, suspension into capsules giving advantage of high bioavailability, accurate content, even shape and nice appearance.

Human genome, i.e. deoxyribonucleic acid has now opened up new opportunities for cosmetology to develop specialized ingredients for skin care.

Human DNA contains all of the information that defines the skin's defences, developmental programming and response mechanisms to external stresses. In addition, the advent of single nucleotide polymorphism analysis (SNPs, read as snips) allow a formulator to begin devising ways to customize cosmetic formulations to a level of sophistication never before seen in this industry. **Dermagenetics** is the use of SNPs to help determine a person's susceptibility to various external threats, such as oxidative stress and allows cosmetic raw material supplier to develop unique formulations which addresses the skin's defensive deficiencies.

Part

2

Cosmetics for Skin

4 Skin and Cosmetics

Beauty is only skin deep.
— William Shakeshpeare

Majority of cosmetic products are applied over skin for different purposes like beautification, protection, etc. Skin care preparations are not new; it is the age old necessity of mankind. Therefore structure and function of skin is important consideration for designing cosmetics.

Skin has been described as the mirror of the body. It reflects the physical, mental and psychological state of the individual. Skin is also a barrier between the body's internal and external environment. It protects the body from physical trauma such as light (UV rays), heat and cold, and microbial injuries. Skin also performed the function of immunological mechanism.

Skin is outer protective layer of body which prevents loss of several internal body fluids and restricts penetration of foreign materials, radiations, etc. from outside. Skin contains several chemicals like keratin, lipids, fatty acids, etc. All these chemical substances have some specific function.

Skin is not merely the largest organ (**integumentory system**) of the body but it also receives the largest number of assaults and insults from the environment as well as from its owner. Environmental allergens, cosmetics, detergents, fomites and microbial contaminants of air such as bacteria and fungal spores, keep impacting on the skin all the time.

THE FUNCTIONS OF THE SKIN

The skin performs several important physiological functions.

a. **Barrier function:** The outer layer of epidermis consists of several overlapping plates of keratin. These are surrounded by a thin film of lipid. The envelop thus produced is both strong and flexible, providing semi-permeable barrier to the outside world. Removal of the horny layer results in a marked increase in water loss, and further stripping of the epidermis leads to significant protein loss. Thus, the skin has

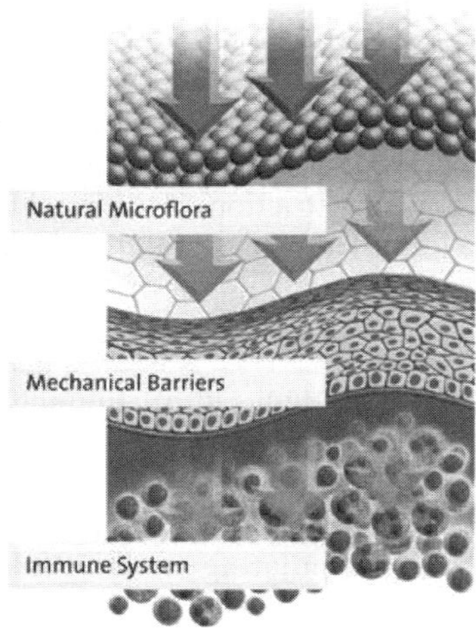

an important role in maintaining the integrity of all that entire lies within or under it. Although the horny layer repels water on first contact, it can also absorb a large amount of water and can survive a significant degree of water loss.

b. **Melanocyte function and pigmentation:** One aspect of the skin's barrier function deserves special mention: The role of melanocyte and melanin in the prevention of damage by ultraviolet radiation. The melanocyte produces melanin granules and transports them into the cytoplasm of surrounding keratinocytes. Melanin absorbs UV radiation and thus protects the nuclei of the basal and spinous cells from DNA damage. In the lower layers of epidermis, the melanin granules are arranged as a shield or umbrella over the nuclei of the basal and spinous cells. In the outer layers, melanin granules are scattered throughout the cells.

c. **Sensation:** We all take the skin's sensory role for granted. It is the organ of touch, hot and cold, sexual arousal or stimuli and we can also feel pressure and pain throughout skin. The sensation, which is unique to the skin, causes considerable distress when it becomes persistent and severe. The skin is very much rich in nerve endings; especially on the fingers, toes, lips and tongue (not, perhaps, strictly 'the skin') and this allows us localize sensations accurately.

d. **Temperature regulations:** Like all warm-blooded animals, humans normally maintain a constant temperature (36.8° C or 98.4° F) for many biological functions. The skin plays a role in this process by alteration to the blood flow through the cutaneous vascular bed (vasodilation=more blood loss, higher heat loss; vasoconstriction=less blood flow, reduced heat loss). The skin also allows the body to cool itself by the evaporation of sweat from the surface.

e. **Immunological surveillance:** The skin is an important site of immunological activity. Depending upon the 'need' for the immunological response, a variety of cells and chemical messengers (cytokines) is involved in recruiting and stimulating both cellular and humoral responses. Epidermal Langerhans cells are constantly 'on the lookout' for antigens in their surroundings in order to trap them and 'present' them to lymphocytes. In certain circumstances, the epidermal keratinocytes themselves can express immunological markers on their surface and produce cytokines. Mast cells in the dermis aid the process by releasing vasoactive chemicals that helps in the recruitment of cells. Tissue macrophages are recruited by vessel dilation and release of chemical attractants. These complex interactions are governed by an individual's genetics.

f. **Biochemical reactions:** The skin is known to be involved in several biochemical processes a vital part of vitamin D metabolism takes place in the epidermis exposure to ultraviolet radiation converts dehydrocholesterol to vitamin D_3. Without this natural process calcium and phosphate absorption is impaired, leading to osteomalacia and rickets.

g. **Social signalling:** Even the briefest study of human history tells us that the skin is important. Socially men and women with pigmented skins have been treated as second class citizens by many societies over the centuries; the presence of a major facial blemish, such as a naevus or purple birthmark (port-wine stain or strawberry mark), often causes even the most open-minded of us to turn our head as we pass by. The presence of 'abnormalities' that are essentially physiologically—such as man balding, excessive hairiness (hirsutism), body odour (bromhidrosis), and the signs of ageing—also can cause misery and anxiety. Many people decorate their skin. Sometimes the decoration is relatively simple (e.g. the wearing of earrings, and the painting of fingers and toe nails); sometimes it is extremely elaborate (e.g. painting for religious and other ceremonies, tattooing).

h. **Excretion:** The skin is minor excretory organ for some substances including:

- Sodium chloride in sweat (excess sweating may lead to abnormally low blood sodium levels).
- Urea.
- Aromatic substances, e.g. garlic and other species.

ANATOMY AND PHYSIOLOGY OF SKIN

The **skin** (word derived from Greek *derma*, and Latin *cutis*) completely covers the body and is continuous with the membranes lining the body orifices. It is body's largest organ and account for 16% of total body weight and approximately 17.5 square feet or 1.5–2 m^2 body surface area and weighs about 2100 gm in adults and it contains glands, hair and nails. Skin varies in thickness average being 1–2 mm but it is 0.5 m on the eyelids and up to 6 mm on palms and soles.

The skin is constantly being renewed and every 28–50 days there is a new skin in healthy well-nourished person. The renewal takes place by shedding of skin's outermost layer. About 40 lb (18 kg) of dead skin is shed by an average adult in his/her lifetime. pH of skin varies from 4–5.6.

Complicating the scene is some patient's tendency of self-medication, especially with regards to their skin problems.

Thus instead of visiting a qualified doctor, most patients:

a. Do a self-diagnosis and apply lotions and ointments advertised in the lay press.
b. Use leftover tubes which had been prescribed earlier for another problem or for another family member.
c. Use a skin preparation recommended by friends or neighbours.

The skin normally is very smooth but due to ageing, exposure to sunlight, cold, dust, abrasions, microbial infection, etc., its smoothness is lost and it becomes rough and thicker. Old age also produces wrinkles on the skin surfaces.

Skin is constantly exposed to several types of chemicals as various formulations are applied on it. Few chemicals may produce harmful effects, which include irritation and allergic sensitization. Not only the consumers but workers too may suffer

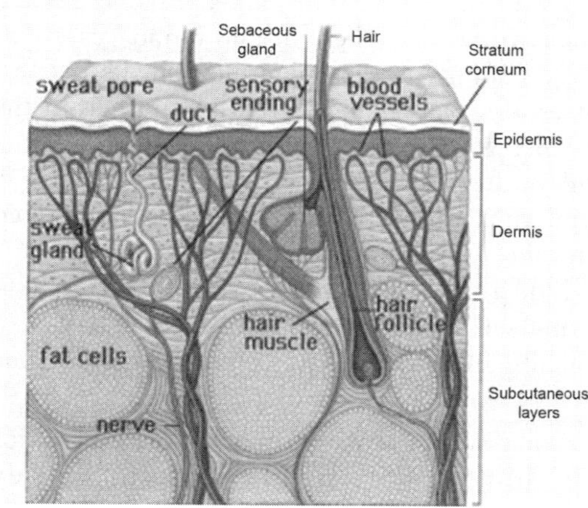

from these harmful effects as they handle large quality of ingredients in cosmetic factories.

Some substances used in cosmetics products may be photosensitive in nature, i.e. the substance may become harmful when activated by light. Such substances are called phytotoxic. Phytotoxic substance may cause allergic stimulation called photo allergic substance.

All the ingredients used in cosmetic products should be carefully evaluated for their phytotoxic or photo allergic nature.

There are two main layers of the skin:

- Epidermis.
- Dermis or corneum.

Subcutaneous tissue containing fat is present below the dermis although it is not part of the skin.

Epidermis

The epidermis is the most superficial layer of the skin and is composed of *stratified epithelium* which varies in thickness in different parts of the

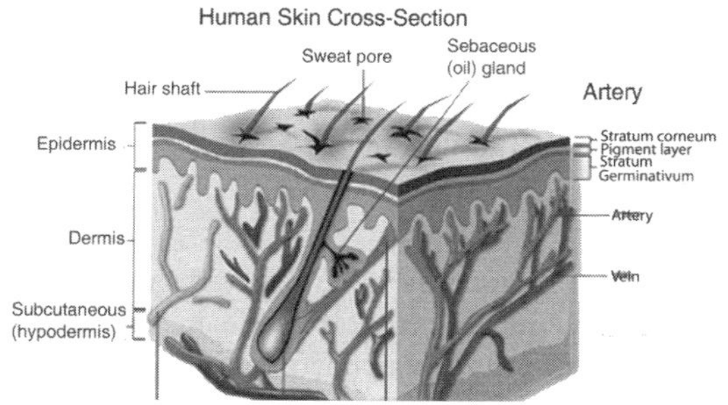

body. It is thickest on the palms of the hands and soles of the feet. There are no blood vessels or nerve endings in the epidermis, but its deepest layers are bathed in interstitial fluid from the dermis, which provides oxygen and nutrients, and is drained away as lymph. Epidermis is approximately 0.4–1.5 mm in thickness compared with the 1.5–4 mm full thickness skin.

Epidermis is the cellular external layer of skin composed mostly of keratinocytes but it also contains other cells. The cells in the epidermis include:

a. **Keratinocytes:** Keratinocytes are the major building blocks of the epidermis. They differentiate from actively dividing basal cells to become anucleate plates of keratin (horny cells) which are then shed.

b. **Langerhans cells:** These are antigen presenting cells and represent one arm of the body's immune system.

c. **Melanocytes:** These are large cells that are interspersed among keratinocytes along the basement membrane (one melanocyte per 10 basal cells). They produce melanin.

d. **Merkel cells:** They represent specialized nerve endings within the epidermis. There are several layers of cells in the epidermis which extend from the deepest *germinative layer* to the surface *stratum corneum* (horny layer). The cells on the surface are flat, thin, non-nucleated, dead cells in which the cytoplasm has been replaced by keratin. These cells are constantly being rubbed off and replaced by cells which originate in the germinative layer and have undergone changes as they progressed towards the surface. Complete replacement of the epidermis takes about 40 days.

e. Hair, secretion from the sebaceous glands and ducts of sweat glands pass through the epidermis to reach the surface. Majority of cells in the epidermis are keratinocytes.

Layers of Epidermis

Epidermis is generally the main target of most skin problems. It is itself made up of several layers. The outermost portion of this layer is casting off regularly. New cells from lower layer continuously replace, it protects the skin from bacterial and fungal attacks as well as from mechanical injury.

The layers of epidermis are described in next column:

Table 4.1: The layers of epidermis

In palms and soles	*In other areas*
Stratum corneum	Stratum corneum
Stratum lucidum	Stratum granulosum
Stratum granulosum	Stratum spinosum
Stratum spinosum	Stratum basale (stratum germinativum)
Stratum basale (stratum germinativum)	

1. **Stratum corneum:** It is the outermost layer of cornified (or horny) cells which provides mechanical protection to the skin and a barrier to water loss. It is composed mainly of dead cells that lack nuclei. As these dead cells slough off, they are continuously replaced by new cells from the *stratum germinativum*. In the human forearm, for example, about 1300 cells/cm^2/hr are shed and commonly accumulate as house dust. Cells of the *stratum corneum* contain keratin, a protein that helps keep the skin hydrated by preventing water evaporation. In addition, these cells can also absorb water, further aiding in hydration and explaining why humans and other animals experience wrinkling of the skin on the fingers and toes (colloquially called **pruning**) when immersed in water for prolonged periods. The thickness of the *stratum corneum* varies according to the amount of protection and/or grip required by a region of the body. For example, the hands are typically used to grasp objects, requiring the palms to be covered with a thick *stratum corneum*. Similarly, the sole of the foot is prone to injury, and so it is protected with a thick *stratum corneum* layer. In general, the *stratum corneum* contains 15–20 layers of dead cells. In reptiles, the *stratum corneum* is permanent, and is only replaced during times of rapid growth, in a process called ecdysis (desquamation) or

moulting. The *stratum corneum* in reptiles contains beta-keratin which provides a more rigid skin layer.

2. **Stratum lucidum:** It is the clear layer of the skin found below stratum corneum on palms and soles.
3. **Stratum granulosum:** It is composed of granular cells containing keratohyalin granules. It reduces water loss from the skin. It is more prominent in palms and soles.
4. **Stratum spinosum:** It contains spinous cells which have spine like appearance. They contain keratin filaments. It is also called transitional layer.
5. **Stratum germinativum:** It contains basically situated mitotically active keratocytes that attach to the basement membrane zone and give rise to cells of the more superficial epidermal layers. Innermost layer of polyhedral cells divide to form new cells. All the glands and keratin structures are derived from this layer.

The surface of the epidermis is ridged by projections of the cells in the dermis called the papillae. The pattern of ridges is different in every individual and the impression made by them is the fingerprint. The downward projections of the germinative layer between the papillae are believed to aid nutrition of epidermal cells and stabilise the two layers, preventing damage due to shearing forces. Blisters develop when acute trauma causes separation of the dermis and epidermis and serous fluid collects between the two layers.

The colour of the skin is affected by three main factors:

- *Melanin:* A dark pigment secreted by melanocytes I, the deep germinative layer, is absorbed by surrounding epithelial cells, the amount varies between different parts of the body, between members of the same races. The number of melanocytes is fairly constant, so the difference in colour depends on the amount of melanin secreted. It protects the skin from the harmful effects of sunlight. Exposure to sunlight promotes synthesis of increased amount of melanin.
- The level of oxygenation of haemoglobin and the amount of blood circulating in the dermis give the skin its pink colour.
- Bile pigments in blood and carotenes in subcutaneous fat give the skin a yellowish colour.

Dermis

This is a layer varying thickness between epidermis and the subcutaneous tissue. It is the inner layer of skin. The dermis is tough and elastic. It is composed of collagen fibres interlaced with elastic fibres. Rupture of elastic fibres occurs when the skin is over-stretched, result in permanent striae or stretch marks, that may be found in obesity and

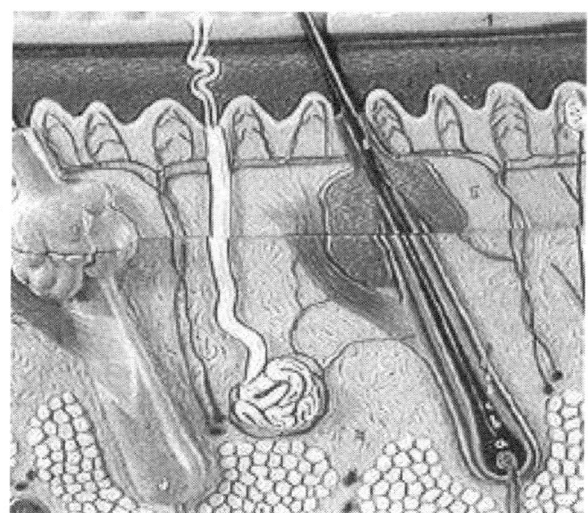

pregnancy collagen fibres bind water and give skin its tensile strength but this ability decline with age, wrinkles develop. There are fibroblast and mast cells in the dermis. Underlying its deepest layer there is areolar tissue and varying amounts of fat. The structures in the dermis are:

- Blood vessels.
- Lymph vessels.
- Sensory (somatic) nerve endings.
- Sweat glands and their ducts.
- Hair roots, hair follicles and hair.
- The arrector pilorum—involuntary muscles are attached to the hair follicles.
- Sebaceous glands (grease glands).
- Fibroblasts.

The dermis also contains certain masses of specialized sensory nerve tissue receptors including Ruffini's corpuscle, Meissner's corpuscle, Krause corpuscle and Pacinian corpuscle.

i. Ruffini's corpuscle is a slowly adapting receptor for sensations of heat and continuous pressure (also called Ruffini's cylinder or ending).

ii. Meissner's corpuscle (*corpusculum tactus*) is a medium-sized nerve ending found in the skin mostly in palms and soles. It is called tactile corpuscle, tactile cell or touch cell.

iii. Krause corpuscle, encapsulated nerves ending in the mucous membrane of mouth, nose, eyes and genitals.

iv. Pacinian corpuscle (*corpusculum lamellosum*) is a type of nerve ending sensitive to pressure and vibration; it is the most complicated of nerve ending and is found throughout the body.

Glands are found in the dermal layer of the skin. Glands are of two types:

1. **Sweat glands:** These secrete a watery fluid (perspiration).
2. **Sebaceous glands:** These secrete a fatty substance called sebum.

APPENDAGE STRUCTURES OF SKIN

It is conventional to consider three important components of skin as appendages of the epidermis, hair follicles and their sebaceous glands, the sweat glands and nails.

Hair Follicle and Sebaceous Glands (Pilosebaceous Units)

The sebaceous glands: The sebaceous glands secrete sebum, which serves as a protective mantle for the outer skin and hair, and keeps the hair shiny and moistured. The sebaceous glands are the bodies built-in-automatic lubrication system. About 100 per square inch of skin is heavily concentrated on the scalp, face and upper torso. Pituitary, thyroid and sex gland hormones primarily regulate them.

Almost the whole skin is punctuated by invagination of epidermis, out of which emerges the keratinized tubes we call hair. The hair follicles are set at an angle into the dermis. The main constituent of the hair shaft is keratin. Each hair bulb contains melanocytes along with the hair follicle which produce pigment.

Sweat Glands

Human skin possesses two kinds of sweat glands—eccrine and apocrine. Eccrine glands are coiled structures lying with a duct opening directly on the surface of the skin. Apocrine glands are coiled structures with a duct which opens into the hair follicle.

The body has about two million sweat glands and they come in two types.

Eccrine sweat glands: Eccrine glands are responsible for recreating the sweat that bathes our skin in moisture, maintains an acid balanced environment that prevents the proliferation of

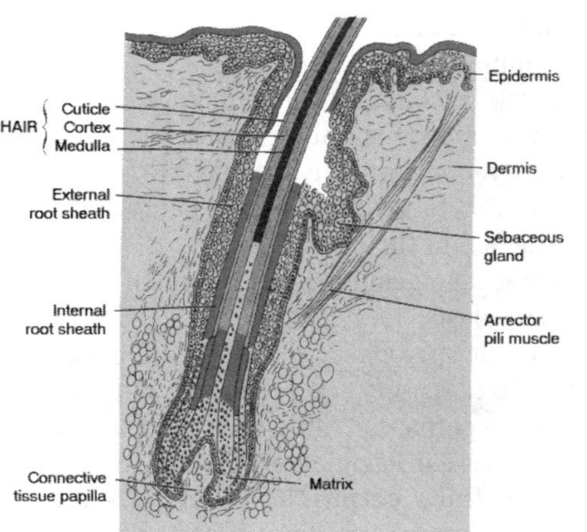

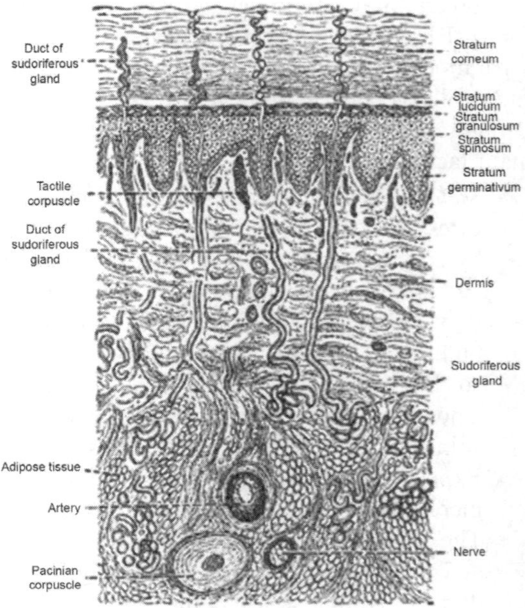

undesirable bacteria or fungi. We normally excrete 800–1230 ml of water during twenty-four hours and this amount can increase ten folds when we perspire heavily. Perspiration from the eccrine glands also regulates our body temperature by evaporating from the skin, cooling us off when we are overheated. Stream and pungent, spicy food will also increase the eccrine glands flow of sweat.

Apocrine sweat glands: Apocrine glands secrete body odour scattered around our body, armpits, face, chest, public and anal areas. These glands empty into the hair follicles near the skin surface. Our personal scents change according to our state of health and mind. The odour is accentuated further by the secretions from sebaceous glands (sebum) which is a fatty, oily substance that the odours from the apocrine glands cling to when deposited on the scalp and hair. Sebum absorbs odours such as smoke and perfume, as well as any of the more pleasant eventual oil fragrances.

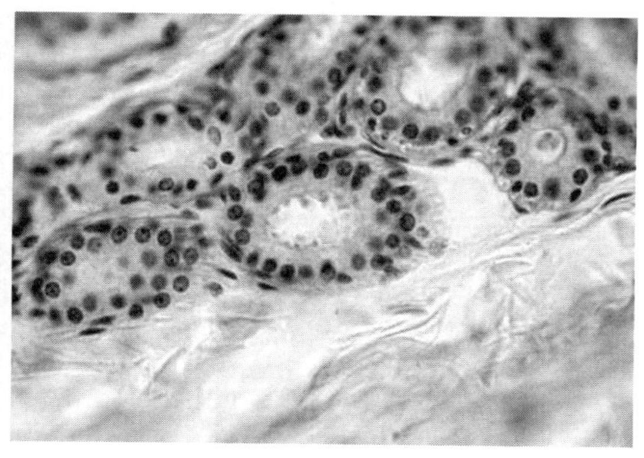

Part way long to hair follicle, a duct delivers sebum, the secretion of the attached sebaceous gland onto the surface of the hair, thus lubricating the surface. Hair is not found everywhere on the body surface. The lips, palms and the soles are hairless (Glabrous). The palms and soles are characterized by remarkable whorled ridge and furrow patterns known as dermatoglyphics. These are unique to each human.

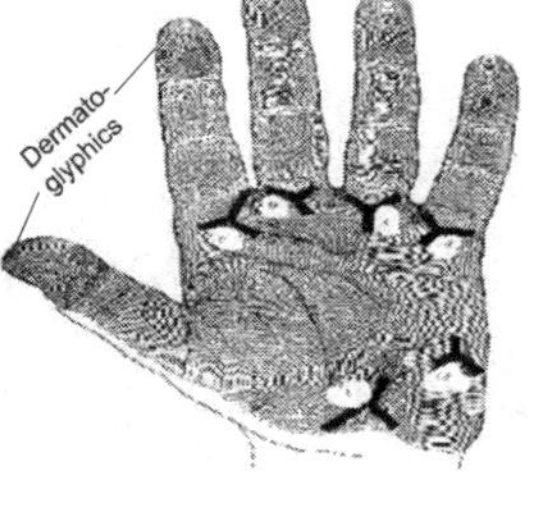

Nails

An epidermal invagination is present at the end of each digit and produces the structure we call the nail. The nail plate is produced from the lower surface of the nail fold and is largely made up of keratin (*more details in Chapter 22—Nail Cosmetics*).

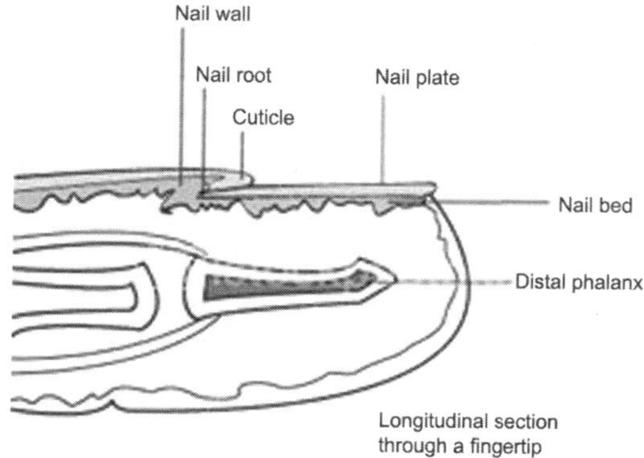

NATURAL ALTERATION OF THE STRUCTURE OF THE SKIN

Age and environment alter the structure of the skin. Some changes are associated with normal maturation and development, e.g. development of facile, axillary and pubic hair and the increased size and activity of apocrine glands, all of which occurs at puberty. Some changes are the result of a process of deterioration, as a part of ageing, e.g. graying of hair, loss of dermoepidermal corrugations. Others are largely due to environmental damage, especially UV radiations, e.g. wrinkling, coarsening and yellowing of facial skin.

Sebaceous glands are lipid-producing structures which, with the exception of the palm, sole and dorsum of the feet, are distributed over the entire body surface. They are most numerous and most productive on the scalp and face and are largest on the forehead, nose and upper part of the back. The forehead may contain as many as 100 glands per square centimeter. The vast majority are connected to hair follicles.

The sebaceous glands are fairly well developed at birth; they soon regress and remain small throughout childhood. They start maturing around

10 years of age and the progress continues through adolescence and remains unchanged until later years. The development of sebaceous glands and the stimulation for sebogenesis are hormone dependent.

The sebaceous glands have two auxiliary actions. The sebum is mildly bacteriostatic and fungistatic, and the gland provides the pro-vitamin D.

DISORDERS OF SEBACEOUS GLANDS

Disorders of sebaceous glands include:

1. Miliaria or prickly heat or strophus
2. Seborrhoea
3. Acne vulgaris.

1. **Miliaria or prickly heat or strophus:** It is heat rash characterized by itchy red spots which develop on the chest, under the armpits and between the thighs in the hot countries, caused by blocked sweat glands.

2. **Seborrhoea:** Seborrhoea is production of a quality of sebum which is excessive for the age and sex of the individual. The patient with seborrhoea complains of oiliness of the face and scalp. In the majority of the cases, the determining factor is genetic. It may also be associated with certain diseases like Parkinsonism, epilepsy, manic depression, Cushing's syndrome, etc.

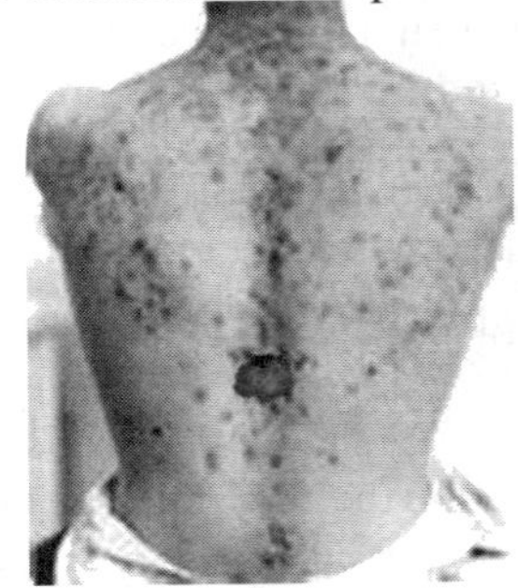

In the common physiologic form, seborrhoea is usually noticed at puberty and tends to reach its peak between 18–25 years. The disease is associated with oily looking skin. The sites of predilection include face, ears, scalp and upper part of the trunk which are particularly rich in sebaceous follicles.

There is no radical treatment for physiologic seborrhoea. Regular washing and frequent use of shampoo may be helpful.

3. **Acne vulgaris:** Acne is self limiting disease of sebaceous glands, manifesting generally in adolescents. It is characterized by lesions like papules, nodules and cysts. Acne lesions begin in and around puberty, the onset is a year or so earlier in girls than in boys. The problem reaches a peak and subsides in the following 3–4 years. Boys tend to have more severe acne than girls.

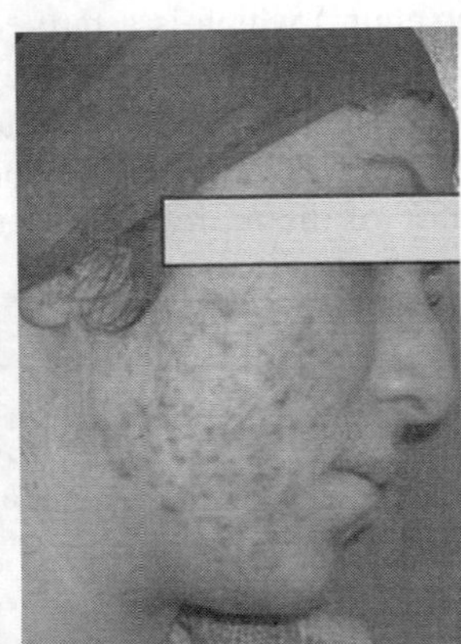

Acne is the classical stigma of adolescence and has been viewed as a normal physiological reaction in the skin. The increase in size of the sebaceous glands and concomitant increases amount of sebum secretion during adolescence are physiologic, but inflammatory changes of acne represent a disease which may be extraordinarily chronic and sometimes produces severe residual physical and physiological scarring which may be lifelong. The disease occurs worldwide in all races affecting 80% of the people sometimes or the other in their life.

PATHOGENESIS OF ACNE

The basic causes of acne are still unknown but five major pathogenic factors are:

1. Circulating sex hormones
2. Increased sebum production
3. Abnormality of microbial flora
4. Hyperkeratinisation of the pilosebaceous duct
5. Inflammation.

The spurt in the levels of hormones during puberty correlates well with the onset of acne vulgaris. The physiological action of male hormones is to increase sebum production and cause sebaceous gland hyperplasia. On the other hand, oestrogens suppress sebaceous gland activity. Acne results from hormonal imbalance—increase in the androgen–estrogen ratio. The levels of circulating hormones do not always correlate with the severity of acne. It is suggested that the sensitivity of the sebaceous glands to circulating androgens is increases in patients with acne vulgaris.

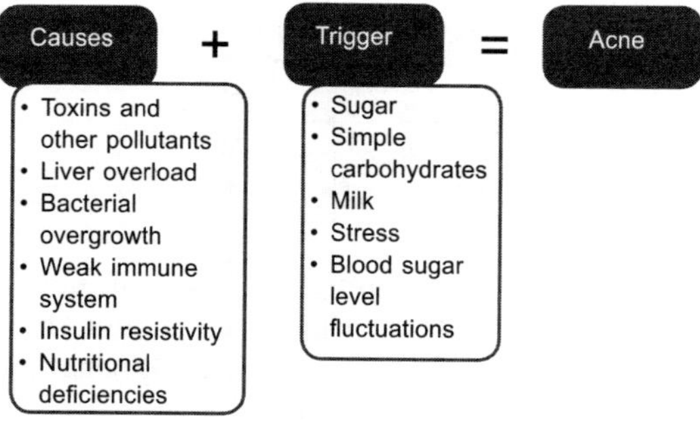

Resident bacterial flora in the sebaceous follicles, namely *Propionibacterium acnes*, *P. granulosum* (both anaerobic pleomorphic diphtheroids) and *Staphylococcus epidermidis* play an important role in the production of acne. The furnish lipases which are responsible for the hydrolysis of sebum triglycerides to the free fatty acids which contribute to follicular hyperkeratosis. They also produce enzymes which play an important role in the inflammatory process. These organisms are more in number in patients with acne. *P. acnes* also produces factors which attract polymorphonuclear leucocytes (PMNL). PMNL ingest *P. acnes* and liberate hydrolytic enzymes that damage the follicular wall further contributing to the inflammation in a hot and humid climate. Emotional stress and use of oily topical comedogenic chemicals may aggravate acne.

The primary changes occur in the pattern of keratinization of the pilosebaceous duct. Normally, keratinous material is loosely arranged in the sebaceous follicles but in acne, it becomes denser. There is increased production of keratinocytes, these initial changes occur in the follicular infundibulum leading to the formation of microcomedones and thereby initiating the process of acne.

Excess sebum secretion and sebaceous gland hypertrophy and hyperplasia usually occur in patients of acne, more so in severe acne.

Sequences of Events

The sequence of events in the development of acne lesions is shown in figure. The blockage of the sebaceous canal due to altered keratinization leads to retention of sebum (comedone formation) and initiation of the inflammatory response. An increase in microbial flora increases inflammation (papule and pustule formation). Further retention of sebum and release of enzymes cause rupture of the

Acne

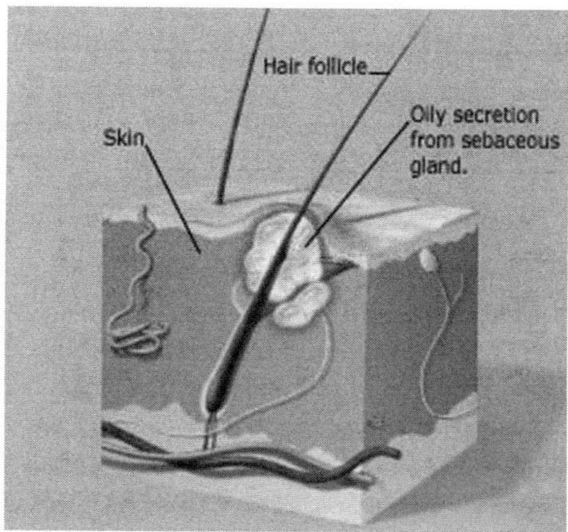

sebaceous gland. The resultant spread of the sebum in the dermis adds to the collection of inflammatory cells in the dermis (nodule formation). The confluence of affected glands results in the accumulation of large areas containing pus, fluid and glandular dermis (cyst formation). Scar formation results when such inflamed lesions and cysts heal after rupture and absorption of fluid. The pattern and size of the scar would depend on the size and pattern of the lesion as well as individual susceptibility. The disfigurement so produced is dreaded by patient and physician alike, and formulates the main aim in treatment of acne vulgaris.

Lesions of acne vulgaris are seen on areas of the body rich in sebaceous glands, mainly the face, mid chest, back, shoulders and upper arm. Comedones can be described as the most characteristic lesions, although they may be seen in other related diseases. A comedo is described as a conical, raised lesion with a broad base and plugged apex. The plug may be black (blackhead) or white (whitehead), and is formed by keratinous material blocking the sebaceous canal. The black colour is due to oxidized melanin seen in open comedones where the blockage is not complete and the plug can be easily expressed. A whitehead, however, is a closed comedo where the plug cannot be removed. This is the initial lesion of acne vulgaris. About 25% of whiteheads resolve within 3–4 days while 75% develop into inflamed lesion.

The inflammatory lesions, namely papules, pustules, nodules and cysts, indicate the severity of the disease. These lesions while healing produce scars.

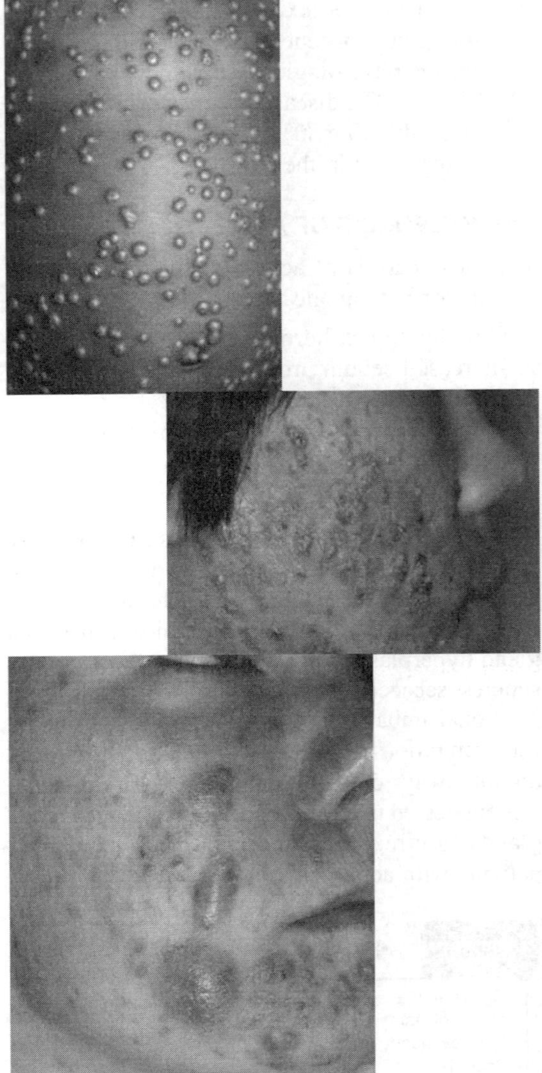

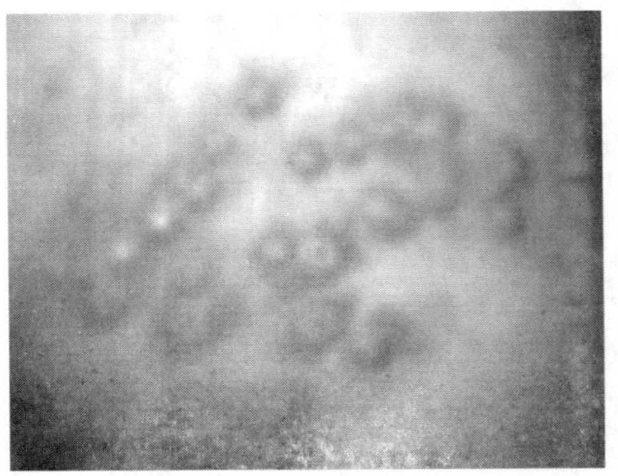

Acne lesions are generally asymptomatic, although inflammatory papules and nodules may be tender. Post-inflammatory hyperpigmentation may develop after resolution of acne.

The appearance of acne in later age, or persistence of acne lesion beyond the thirties, calls for investigation of the endocrine system as acneiform lesions are seen in thyroid dysfunction, adrenal hyperplasia, gonad and pituitary hyper functions.

Genetic and racial factors also play a role in the development of acne.

Sequence of events in acne

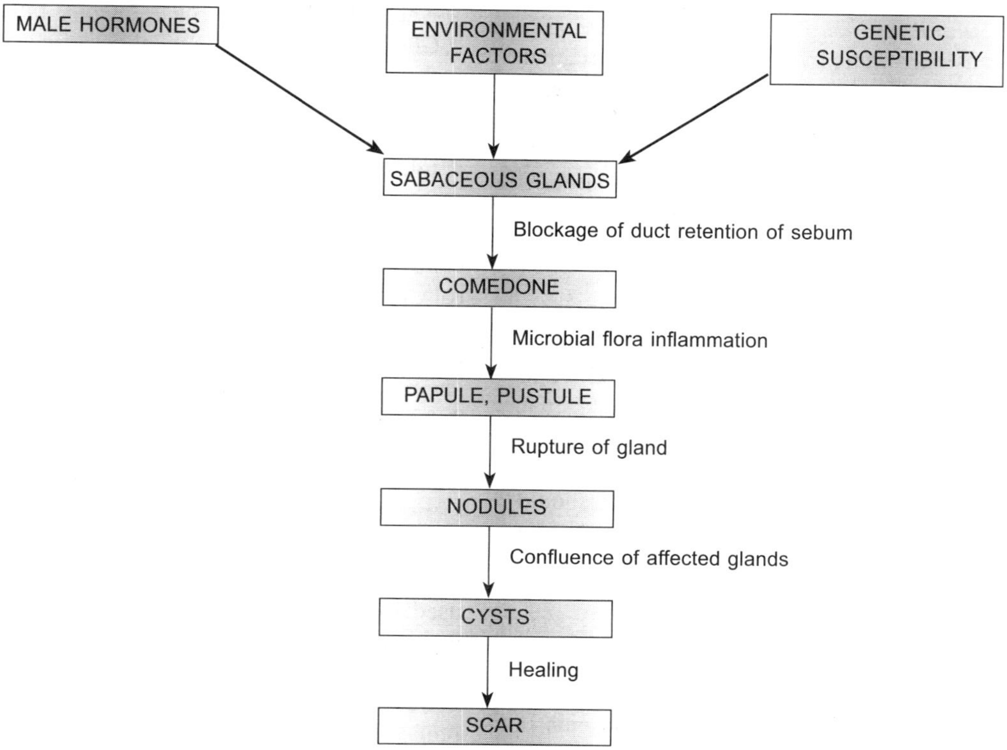

MALE HORMONES → SABACEOUS GLANDS
ENVIRONMENTAL FACTORS → SABACEOUS GLANDS
GENETIC SUSCEPTIBILITY → SABACEOUS GLANDS

SABACEOUS GLANDS
↓ Blockage of duct retention of sebum
COMEDONE
↓ Microbial flora inflammation
PAPULE, PUSTULE
↓ Rupture of gland
NODULES
↓ Confluence of affected glands
CYSTS
↓ Healing
SCAR

Treatment

Acne is mostly a treatable disease except for minority of cases. Mid-acne requires only topical therapy, while moderate and severe acne need both oral and topical therapy. The choice of therapy depends on severity of condition in terms of:

1. The physical extent and symptoms.
2. The degree to which scarring appears to be developing.
3. The psychological effects on the individual.

Since the disease is influenced by diverse factors, various therapeutic modalities exist. Effective therapy is important to prevent dreaded sequelae of scarring, as cosmetic disfigurement is a major problem.

Treatment is directed at:

1. Decrease in sebaceous gland activity.
2. Removing follicular destruction.
3. Reducing follicular bacterial population.
4. Reducing inflammation.

Diet has no role in therapy of acne.

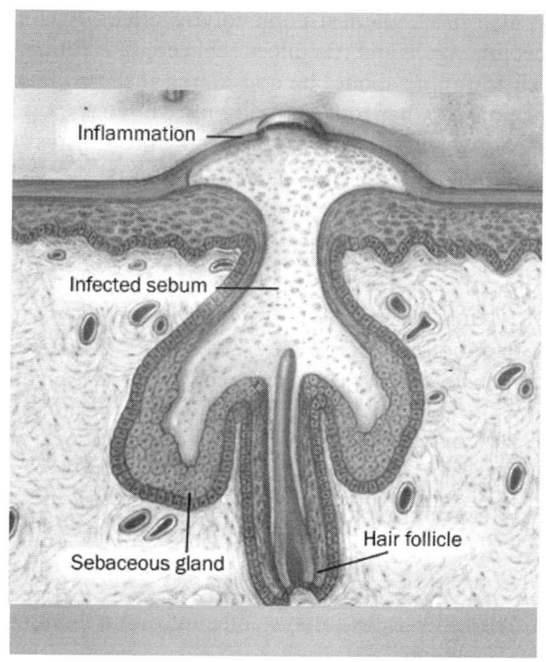

Local Therapy

a. *Frequent washing with soap and water:* Cleansing, simple hygienic measure of regular washing of face with soap and water are recommended to remove keratinized plugs.

b. *Topical agents:* The most widely used are benzoyl peroxide, vitamin A, acid (tretinoin) and antibiotics, e.g. clindamycin, erythromycin, etc.

Retinoic Acid

Retinoic acid is vitamin A analogue. It is used topically. It is used in the strength of 0.05% and is available in a cream or ointment base, it is most effective for comedonal acne. It is a potent comedolytic agent and produces soft keratin by restoring the disturbed keratinization. It should be used with caution as it may produce erythema and exfoliation.

Benzoyl Peroxide

It is most commonly prescribed topical agent for acne. Topical benzoyl peroxide is as effective as retinoic acid. It is used as a lotion and gels in strengths varying from 2.5–10%. It is oxidizing agent with bacteriostatic properties on *P. acnes*. It also has a modest comedolytic effect, it is a peeling agent and it reduces sebaceous secretion. An antibiotic should be considered if there is no response to benzoyl peroxide in two weeks.

Disadvantages: Adverse effects like dryness, irritation and allergic contact dermatitis.

Topical Antibiotics

Topical antibiotics like 2% erythromycin, 2% clindamycin and 5% tetracycline are used in inflammatory acne. They are not as effective as benzoyl peroxide. For severe acne a topical antibiotic must be combined with a systemic one.

Systemic Therapy

a. **Antibiotic and antibacterial agents**

Role of antibiotics in acne

The indications for systemic antibiotic therapy of acne are:

1. Moderately severe or severe grades of acne where topical treatment frequently fails.
2. Chest and back involvement because acne in these areas is less responsive to topical than to oral therapy.
3. Patients with sensitive skin who cannot tolerate irritating topical preparations.

The antibiotics can be very useful in treating acne. They reduce the number of bacteria (*P. acne*) on the skin surface as well as have a role in inhibition of inflammatory process. Antibiotics are bacteriostatic, and they can also inhibit various enzyme activities and affect chemotaxis and lymphocytic function.

The broad spectrum antibiotics are currently used widely in the treatment of acne. Oral antibiotics should be given for at-least 6 months; in case of recurrence repeated courses can be given.

Tetracycline Group: Tetracycline group and macrolide group (erythromycin) are most widely used systemic therapy for acne. Clindamycin is not preferred because of the adverse effect of pseudomembranous colitis.

Tetracyclines, doxycyclines and minocyclines remain the antibiotics of choice. They have a microbially similar spectrum with pharmacokinetic differences.

Doxycycline is preferred over tetracycline or oxytetracycline due to advantage of once daily dosage. Minocycline is reserved for unresponsive acne because of its greater cost and adverse effects of neurotoxicity like dizziness and headache.

Macrolides: The major experience of macrolide therapy in acne is with erythromycin. Erythromycin is given in a dose of 500 mg twice daily.

Disadvantages: *P. acnes* strains resistant to erythromycin become wide-spread limiting its use.

The other systemic modalities that are currently used include estrogenic hormones, glucocorticosteroids and oral synthetic retinoids.

b. **Oestrogens** given in sufficient amounts as anovulatory agents decrease sebum production, hence the use of oral contraceptives for treatment of acne. They are used in female patients. About 100 μg of ethnyl estradiol is required daily for suppression of acne. Side effects include nausea, weight gain, breast tenderness, etc.

c. **Glucocorticoids:** Because of their anti-inflammatory activity and reduction of plasma androgen levels, a short-term use may be advocated in severely affected patients. High doses are required but are resorted to only in severe acne because of their severe side effects. Low dose steroids (2.5–5 mg prednisolone) are effective in woman with excessive androgens.

d. **Isotretinoin** has revolutionized the management of severe and intractable nodulocystic acne. Adequate responses are seen with dose of 1.0 mg/kg for 4–5 months. Residual acne continues to improve after cessation of therapy. It exhibits an anti-inflammatory effect. It is teratogenic. Major side effects include dry skin, itching, dry mouth conjunctivitis, thinning of hair, peeling of palms, soles, etc.

e. **Surgical treatment** is carried out in severe cases.

Other common disorders of the skin:

a. *Pigmentary disorders:* Hyper-and hypo-pigmentations—Hyper pigmentations occurs generally in Caucasians due to imbalance in formation of melanin or also stimulated by exposure to UV or X-ray irradiation. Various conditions are termed viz. lentigens, ephelids, moles, ochronosis, etc. Hypo-pigmentations also known as vitiligo, mostly occurs in non-Caucasians.

b. *Skin scaling disorders:* Dandruff and psoriasis: Dandruff mostly occurs on scalp and characterized by flaking of *Stratum corneum* due to microbial infection or immunological disorders. Psoriasis is scaly red patches on body.

5 Skin Creams and Lotions

Beauty is in the heart of the beholders.

— Bernstein, Al

INTRODUCTION

Cosmetic creams are usually marketed on the basis of their broad claims (e.g. cleansing creams, massage creams, etc.) made on their packaging. Other modes of classification, are based on appearance, texture, subjective feels, ease of spreading, etc.

In skin creams or lotions either fatty phase (oil) can be dispersed in aqueous phase (water) or aqueous phase can be dispersed in fatty phase. If oil droplets are dispersed in aqueous phase (water), emulsion termed as o/w emulsion and if water is dispersed in fatty phase (oil), emulsion is termed as w/o.

These substances possess both hydrophilic (affinity for water) and lipophillic (affinity for lipids) properties.

All the skin care creams are classified according to their different basis:

1. According to function: cleansing cream, foundation cream, massage cream, all-purpose cream.
2. According to characteristic properties: cold cream, vanishing creams, anti-wrinkle cream.
3. According to nature or type of emulsion: o/w cream, bleach cream.

Function-based classification is most widely accepted. According to the function, creams can be classified as follows:

1. Cold creams — Cream for winter
2. Cleansing creams
3. Make-up creams — Foundation cream
   — Vanishing cream
4. Moisturizing creams — Cream for dry skin
5. Night and massage creams
6. Hand and body creams
7. All-purpose and general creams

The four most common skin types are:

1. Normal (no apparent signs of oily or dry areas).
2. Oily (shine appears on skin, no dry areas at all).
3. Dry (flaking can appear, no oily areas at all).
4. Combination (oily and dry or normal areas).

CREAMS

A cream is a topical preparation usually for application to the skin. Creams for application to mucus membranes such as those of the rectum or vagina are also used. Creams may be considered pharmaceutical products as even cosmetic creams are based on techniques developed by pharmacy and unmedicated creams are highly used in a variety of skin conditions (dermatoses).

Creams are semi-solid emulsions that are mixtures of oil and water. They are divided into two types: oil-in-water (o/w) creams which are composed of small droplets of oil dispersed in a continuous aqueous phase, and water-in-oil (w/o) creams which are composed of small droplets of water dispersed in a continuous oily phase. Oil-in-water creams are more comfortable and cosmetically acceptable are as they are less greasy and, more easily washed off using water. Water-in-oil creams are more difficult to handle but many drugs which are incorporated into creams are hydrophobic and will be released more readily from a water-in-oil cream than an oil-in-water cream. Water-in-oil creams are also more moisturizing as they provide an oily barrier which reduces water loss from the *stratum corneum*, the outmost layer of the skin.

Uses of Creams

- The provision of a barrier to protect the skin
- This may be a physical barrier or a chemical barrier as with sunscreens
- To aid in the retention of moisture (especially water-in-oil creams).
- Cleansing.
- Emollient effects.

An emulsion is a mixture of two immiscible substances. One substance (the dispersed phase) is dispersed in the other (the continuous phase). Examples of emulsions include butter and margarine, espresso, mayonnaise. In butter and margarine, a continuous liquid phase surrounds droplets of water (water-in-oil emulsion). Emulsification is the process by which emulsions are prepared.

Emulsions tend to have a cloudy appearance, because the many phase interfaces (the boundary between the phases is called the interface) scatter light that passes through the emulsion. Emulsions are unstable and thus do not form spontaneously. Energy input through shaking, stirring, homogenizers, or spray processes are needed to form an emulsion. Overtime emulsions tend to revert to the stable state of oil separated from water. Surface active substances (surfactants) can increase the kinetic stability of emulsions greatly so that, once formed, the emulsion does not change significantly over years of storage. Home-made oil and vinegar salad dressing is an example of an unstable emulsion that will quickly separate unless shaken continuously. This phenomenon is called coalescence, and happens when small droplets recombine to form bigger ones. Fluid emulsions can also suffer from creaming, the migration of one of the substances to the top of the emulsion under the influence of buoyancy or centripetal force when a centrifuge is used.

Emulsions are part of a more general class of two-phase systems of matter called colloids. Although the terms colloid and emulsion are sometimes used interchangeably, emulsion tends to imply that both the dispersed and the continuous phase are liquid.

There are three types of emulsion instability:

- Flocculation, where the particles form clumps;
- Creaming, where the particles concentrate towards the surface of the mixture while staying separated; and

- Breaking, where the particles coalesce and form a layer of liquid.

Emulsifier

An emulsifier (also known as an emulsifying agent, emulgent or surfactant) is a substance which stabilizes an emulsion. Examples of food emulsifiers are egg yolk (where the main emulsifying chemical is the phospholipid lecithin), and mustard, where a variety of chemicals in the mucilage surrounding the seed hull act as emulsifiers; proteins and low-molecular weight emulsifiers are common as well. In some cases, particles can stabilize emulsions as well through a mechanism called Pickering stabilization. Both mayonnaise and hollandaise sauce are oil-in-water emulsions stabilized with egg yolk lecithin. Detergents are another class of surfactant, and will chemically interact with both oil and water, thus stabilizing the interface between oil or water droplets in suspension. This principle is exploited in soap to remove grease for the purpose of cleaning. A wide variety of emulsifiers are used in pharmacy to prepare emulsions such as creams and lotions.

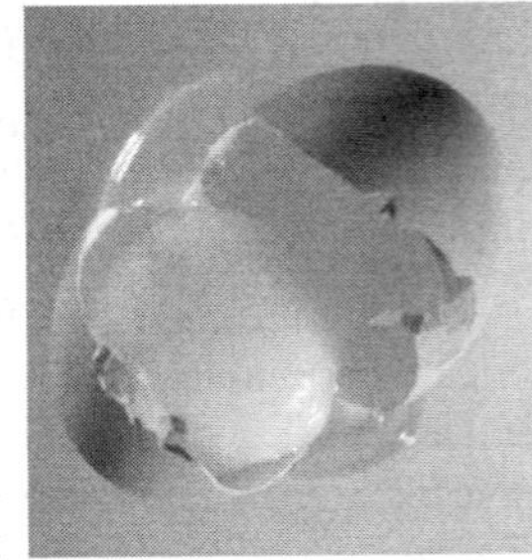

Cosmetic emulsions contain a number of ingredients such as proteins, carbohydrates, sterols, etc. which support growth of microorganisms. Contamination may be introduced by contaminated

Table 5.1: Preservatives for cosmetic emulsions

Group	*Preservatives*	*Action*
Acids and acid salts	Benzoic acid	Antifungal agents
	Dehydroacetic acid	
	Propionic acid	
	Sorbic acid and its salts	
Aldehydes	Formaldehyde	Broad spectrum preservatives
	Glutaraldehyde	
Phenolic compounds	Cresol	Broad spectrum preservatives
	Chlorothymol	
	p-chlorometaxylenol	
	Phenol	
	o-phenyl phenol	
	methyl p-hydroxy benzoate	
	propyl p-hydroxy benzoate	
	butyl p-hydroxy benzoate	
Quaternary ammonium compounds	Benzethonium chloride	Broad spectrum preservatives
	Benzalkonium chloride	
	Cetyl trimethyl ammonium bromide	
Mercurials	Cetyl pyridinium chloride	Broad spectrum preservatives
	Phenyl mercuric acetate	
	Phenyl mercuric nitrate	
Miscellaneous	Sodium ethylmercurithosalicylate	
	Imidizolidinyl urea compound	Synergistic broad spectrum preservatives
	2,4,4-trichloro-2′-hydroxy diphenyl ether	Primarily active against gram positive bacteria

raw materials or poor sanitation during processing. Contamination can also be introduced by the consumer during its use. Therefore, preservatives included in cosmetic emulsion with qualities of low toxicity, stability to heat and storage, chemical compatibility, acceptable odour and colour, reasonable costs, effective against a variety of microorganisms.

LOTIONS

A lotion is a low-to-medium viscosity emulsion medicated or non-medicated topical preparation intended for application to unbroken skin. Most lotions are oil-in-water emulsions but water-in-oil lotions are also formulated. Lotions are usually applied to external skin with bare hands, a clean cloth, cotton wool or gauze. Lotions can be poured under influence of gravity.

The key components of a lotion emulsion are the aqueous and oily phases, an emulgent to prevent separation of these two phases, and, if used, the drug substance or substances. A wide variety of other ingredients such as fragrances, glycerol, petroleum jelly, dyes, preservatives, proteins and stabilizing agents are commonly added to lotions.

Lotions can be used for the delivery to the skin of medications such as:

- Antibiotics
- Antiseptics
- Antifungals
- Corticosteroids
- Anti-acne agents
- Soothing or protective agents (such as calamine)
- Aside from medical use and usage in skin care, lotions are often used as accessories to aid massage.

It is not uncommon for the same drug ingredient to be formulated into a lotion, cream and ointment. Creams are the most convenient of the three but are inappropriate for application to regions of hairy skin such as the scalp, while a lotion is less viscous and may be readily applied to these areas (many medicated shampoos are in fact lotions). Lotions also have an advantage in that they may be spread thinly compared to a cream or ointment and may economically cover a large area of skin. Non-comedogenic lotions are recommended for use on acne prone skin.

LINIMENT

Liniment, word derived from the Latin *linere*, means to anoint, is a medicated topical preparation for application to the skin. Preparations of this type are also called balm. Liniments are of a similar viscosity to lotions (being significantly less viscous than an ointment or cream) but unlike a lotion, a liniment is applied with friction that is a liniment always rubbed in.

FORMULATION OF CREAMS AND LOTIONS

In the formulation of creams and lotions, ingredients of two phases are selected to perform a predetermined function. Formulation consideration of these various creams will be discussed under this section:

Cleansing Creams and Lotions

Cleansing creams and lotions generally used for removal of dirt or soil from skin in different ways like facial make-up residues, pigments, surface grime from atmosphere, oil and grease from sebum, dead cells and crusts. Although water is good and cheap cleansing agent, but it is ineffective against oils. Soaps can emulsify oils, so they also remove too much oil from skin surface that make it rough and dry. Residual alkalinity of soap may cause outermost cells to lift and separate from neighboring cells and these are inconvenient outside bathroom also.

A cleansing cream and lotion can overcome this problem and they have following properties:

- It should have good appearance and stable at room temperature.
- It should soften on application or liquefy at body temperature on skin.
- It should have low viscosity so as to spread easily without dragging but high enough to retain dirt particles.
- It should not have oily or greasy feel during application.
- The cream residue should not become viscous after evaporation of water.
- A thin emollient film should remain on the skin after its use.
- Its physical action should be that of flushing on the skin and pore opening rather than absorption.

Examples of cleansing creams and lotions are:

Beeswax-borax emulsion: It occupies very important place in this category.

Formula I

Beeswax	8.0%
Mineral oil	50.0%
Microcrystalline wax	5.0%
Borax	0.4%
Perfume and preservatives	q.s.
Distilled water	q.s. 100.0%

Formula II

Beeswax	10.0%
Mineral oil	40.0%
Paraffin wax	12.0%
Borax	0.4%
Ozokerite	10.0%
Perfume and preservatives	q.s.
Distilled water q.s.	100.0%

Beeswax and other waxes' quantity influence its stiffness.

Non-ionic emulsifying agents like sorbitan fatty acids ester can be used to supplement beeswax-borax emulsion.

Formula III

Beeswax	10.0%
Mineral oil	48.0%
Lanolin	4.0%
Borax	0.7%
Sorbitan sesquioleate	1.0%
Perfume and preservatives	q.s.
Distilled water q.s.	100.0%

Formula IV

Beeswax	8.0%
Mineral oil	20.0%
Lanolin	3.0%
Borax	0.3%
Hydrogenated vegetable oil	27.0%
Antioxidant	0.5%
Sorbitan stearate	5.0%
Polysorbate 60	2.0%
Perfume and preservatives	q.s.
Distilled water q.s.	100.0%

Dissolve borax in distilled water. Melt all waxes, absorption base and mineral oil together in a separate beaker. Heat both the phases, i.e. aqueous and oily, in a water bath to nearly 70° C and then aqueous phase is added to oily phase with constant stirring. When temperature cools down to 40–45° C, perfume is added with constant stirring to cream.

Liquefying cream: Cream with no water, translucent, formulated with oils and waxes. They liquefy when gently massage. A mixture of oils and waxes is melted and poured while warm. Amorphous materials like ozokerite and petrolatum are added to encounter crusty surface formation. Emollient ingredients like lanolin, cetyl alcohol are added to the formulation which leaves a thin oily film. Melted product should be filtered through filter paper. This cream is used for dry skin.

Formula V

Mineral oil	64.0%
Paraffin	17.0%
Petrolatum	18.0%
Cetyl alcohol	1.0%

Formula VI

Mineral oil	64.0 %
Ozokerite	10.0%
Petrolatum	11.0%
Lanolin (hydrous)	15.0%

Acid cleansing cream: This cream is used for skin return to normal pH. Usually skin is covered with a coating having pH value of 3–5. Certain conditions like perspirations, physical conditions, sebaceous glands secretion, disturbance of protective acid layer of the skin and breakdown natural protection against bacterial attack of skin and allowed microbial growth makes change in skin pH. Acid creams can be prepared by the addition of mild acids like citric, tartaric, lactic, malic and phosphoric acids in suitable formula. Emulsifying agents are used for this purpose, e.g. glycerin stearate cetyl or stearyl alcohol, citric acid, lactic acid and lemon juice are also used as acidic substances.

Formula VII

Beeswax	4.0%
Lanolin	7.0%
Stearyl alcohol	2.0%
Petrolatum	35.0%
Glycerin	5.0%
Lactic acid	1.5
Distilled water q.s.	100.0%

Formula VIII

Wool max	4.0%
Lanolin	7.0%
Stearyl alcohol	2.0%
Petrolatum	35.0%
Glycerin	5.0%
Lactic acid	1.5%
Distilled water	100.0%

Washable cleansing cream: This is with detergent.

Formula IX

Mineral oil	38.0%
Cetyl alcohol	2.0%
Sodium cetyl sulphate	1.0%
Ozokerite	2.0%
Perfume and preservatives	q.s.
Distilled water q.s.	100.0%

Gel like cleansing cream

Formula X

Stearic acid	10.0%
Mineral oil	25.0%
Triethanol amine oleate	5.0%
Terpineol	0.05%
Lanolin (anhydrous)	3.0%
Propylene glycol	6.0%
Perfume and preservatives	q.s.
Distilled water	q.s. 100.0%

Lather cleansing cream: This cream containing surface active agents produces lather on skin when little extra water is used.

Formula XI

Stearic acid	10.0%
Petroleum jelly	1.5%
Mineral oil	5.5%
Cetostearyl alcohol	1.0%
Isopropyl myristate	3.0%
Sorbitan mono oleate	2.0%
Sodium lauryl sulphate	4.5%
Polyoxyethylene sorbitan mono laurate	1.5%
Triethanol amine	1.5%
Glycerin	5.0%
Perfume and preservatives	q.s.
Distilled water	q.s. 100.0%

Cleansing lotions: Cleansing lotions are more popular after introduction of plastic containers. It spreads easily in thin layers. It is more economic but has stability problem. Auxiliary emulsifying agents like glyceryl mono stearate, fatty alcohols, beeswax are also added.

Formula XII

Mineral oil	38.0%
Triethanol amine stearate	8.0%
Beeswax	2.0%
Perfume and preservatives	q.s.
Distilled water	q.s. 100.0%

Triethanol amine stearate is made of one part triethanol amine and two parts of stearic acid; this has to make *in-situ* to save from discolouration of preparation. Another problem occurs is viscosity. If the preparation is more viscous then adjust with water and if it is less viscous then adjust it with triethanol amine stearate.

Cleansing lotion for oily skin: It is based on synthetic detergents. If products are meant for oily skin with acne, substances like salicylic acid, resorcinol and sulphur may be added. Another method for preparing cleansing products for oily skin is to use solvents like ethyl or isopropyl alcohol in concentration up to 60%. More concentration causes dryness and irritation.

Formula XIII

Ethyl alcohol	12.0%
Menthol	0.05%
Camphor	0.1%
Cetyl dimethyl benzyl ammonium chloride	0.2%
Ethoxylated cetyl alcohol	2.0%
Stearyl amine	1.0%
Perfume and preservatives	q.s.
Distilled water	q.s. 100.0%

Transparent weekly acidic lotion: This is prepared with polyoxyethylene esters and ethers.

Formula XIV

Polyoxyethylene (20) sorbitan mono laurate	0.5%
Polyoxyethylene (15) lauryl alcohol ether	1.5%
1,3-butylene glycol	5.0%
Glycerin	5.0%
Oleyl alcohol	0.5%
Ethyl alcohol	10.0%
Perfume and preservatives	q.s.
Distilled water	q.s. 100.0%

Cold Cream

It is found to be one of the most popular and old preparation since long back, Galen is considered to be the founder of this cream. Its popularity is flourishing with time. It must primarily possess an emollient action and is useful for dry skin and quite popular in winter. It should provide a cooling sensation on application because of the evaporation of water separated by breaking of w/o emulsion. The resultant oil film on the skin is non-occlusive since they may remain on the skin for considerable period of time. Cold creams consist of w/o emulsion that is made by using beeswax and alkali, usually borax as the emulsifying agent.

Formula XV

White beeswax	20.0%
Light mineral oil	50.0%
Borax	0.7%
Distilled water	28.8%
Perfume and preservatives	0.5%

Method: Melt white beeswax in a water bath (at 70° C). Add light mineral oil to it. In other beaker heat water to about 70° C and dissolve borax in it. Mix the aqueous phase to the oily phase with constant stirring, till a creamy emulsion is prepared. Add perfume at about 40° C.

Formula XVI

White beeswax	16.0%
Light mineral oil	50.0%
Borax	0.8%
Spermaceti	3.5%
Distilled water	29.2%
Perfume and preservatives	0.5%

Follow the method of previous preparation.

Triethanolamine (trolamine) can be employed in place of borax as an alkali. Cold cream containing glyceryl monostearate are easy to manufacture since it is self-emulsifying agent. Cold creams with borax are superior to those without it, since emulsion is whiter, smoother and more stable.

Night and Massage Cream

This cream is also known as **emollient cream**. This is applied in night and removed in morning. It is formulated with substantial fatty phase and having easy spread. It provides occlusive layer on the skin which slows the rate of transepidermal water loss (moisturizing effect).

So far as skin is concerned, water is the effective emollient, the plasticizer of the *stratum corneum*. Solvents and detergents make skin dry. The dry skin may be made smooth by introducing water into horny layer by following ways:

- By increasing diffusion from living epidermal cells,
- By introducing water to the horny layer from outside, and
- By occluding the skin surface to prevent evaporation of water from it.

Emollient materials are of two types:

i. **Water soluble emollient (Humectants):** This material retards evaporation of water in o/w emulsion and hold water in close contact with skin and thereby introduce water into *stratum corneum* (horny layer), e.g. polyhydric alcohols such as propylene glycol, glycerol, sorbitol, etc.

ii. **Fat soluble emollients:** This material is forming thin film over skin and thereby reducing evaporation from skin surface, e.g.

- *Hydrocarbon oils and waxes:* Mineral oils, petroleum jelly, paraffin, casein, etc.
- *Silicon oils:* Dimethyl polysiloxane, methyl phenyl polysiloxanes.
- *Triglycerides esters:* Vegetable and animal oils and fats.
- *Alkyl esters:* Methyl, isopropyl and butyl esters of fatty acids.
- *Fatty acids:* Lauric acid, palmitic acid, stearic acid, etc.
- *Lanolin and derivatives:* Lanolin, lanolin oil, lanolin wax, etc.

For satisfactory emollient cream, moderate penetration, slip and milder occlusive actions are required.

Emollients have three basic actions:

- *Occlusion:* Providing a layer of oil on the surface of the skin to slow water loss and thus increase the moisture content of the *stratum corneum*.
- *Humectant:* Increasing the water-holding capacity of the *stratum corneum*.
- *Lubrication:* Adding slip or glide across the skin.

Formula XVII

Beeswax	10.0%
Spermaceti	12.0%
Mineral oil	53.0%
Borax	0.5%
Perfume and preservatives	q.s.
Rose water	q.s. 100.0%

This action can be achieved by partially replacing mineral oil with vegetable oils like almond oil, peach/kernel oil. Quantity of waxes could be reduced to reduce drag.

Formula XVIII

Beeswax	9.5%
Spermaceti	4.5%
Mineral oil	35.5%
Almond oil	15.0%
Borax	0.5%
Antioxidant	q.s.
Perfume and preservatives	q.s.
Rose water	q.s. 100.0%

When vegetable oils are used in formulation, antioxidants are invariably used to retard rancidity of vegetable oils. Antioxidants like BHT, BHA, *n*-propyl gallate can be used. When more penetration is required, lanolin can be added.

Formula XIX

Beeswax	12.0%
Mineral oil	30.0%
Palm oil	10.0%
Lanolin	10.0%
Antioxidant	q.s.
Perfume and preservatives	q.s.
Rose water	q.s. 100.0%

Emollient cream with surfactants and emulsifying agents: Most commonly used emulsifying agent include glyceryl mono stearate, cetyl and stearyl alcohol, diethylene glycol mono stearate, propylene glycol mono laurate, poly ethylene glycol esters of fatty acids, poly ethylene glycol ethers of fatty acids. These are used alone or in combination.

Cream prepared with non-ionic emulsifier is stable in acidic pH. Creams prepared with fatty acid esters of polyhydric alcohols are stable in wide range of pH (4.0 to 9.0). Creams prepared with non-ionic ethers are stable in wider ranges.

Formula XX
Cream with non-ionic emulsifier

Beeswax	4.0%
Spermaceti	2.0%
Mineral oil	32.0%
Glyceryl mono stearate	15.0%
Glycerin	7.5%
Perfume and preservatives	q.s.
Distilled water	q.s. 100.0%

Emollient cream with cationic surfactants: Commonly used cationic surfactants are quaternary ammonium compounds, pyridinium and imidazolinium salts. These have conditioning effects on skin and also have antiseptic value. This antiseptic property of cationic surfactants affected by free fatty acids which are usually available in emollient creams.

Formula XXI

Beeswax	12.0%
Mineral oil	22.0%
Hydrogenated vegetable oil	15.0%
Tween 60	2.0%
Arlacel 60	5.0%
Borax	0.6%
Antioxidant	q.s.
Perfume and preservatives	q.s.
Distilled water	q.s. 100.0%

Emollient lotions: They are aesthetically acceptable due to creamy texture, smooth feel and emollient performance. Same materials are used to prepare emollient lotions which are used to prepare emollient creams. Stability is the problem with emollient lotions. Some thickening agents like glyceryl mono stearate, cetyl and stearyl alcohol, sorbitan mono stearate, ethylene glycol mono stearate, polyethylene glycol distearate are used in lotions. These materials act as emulsion stabilizers also and it should be used in low concentration. High concentration may cause thixotropic gel formulation in lotion.

Other thickening agents:

- *Plant hydrocolloids:* Gum Arabic or acacia, tragacanth, Irish moss
- *Cellulose derivatives:* Methyl cellulose, sodium carboxy methyl cellulose, hydroxyl ethyl cellulose, hydroxyl propyl methyl cellulose (HPMC).
- *Synthetic high polymers:* Carboxy vinyl polymer, poly vinyl alcohol.
- *Clays:* Colloidal magnesium aluminum silicate, magnesium aluminum silicate.

Humectants like propylene glycol and glycerin exert profound effect on viscosity and stability of lotions.

Formula XXII

Cetyl alcohol	1.0%
Lanolin	1.5%
Mineral oil	36.5%
Silicon oil	4.0%
Arlacel 80	5.0%
Tween 80	5.0%
Perfume and preservatives	q.s.
Distilled water	q.s. 100.0%

Some formulation containing vitamins A, D, E and K (fat soluble vitamins) and vitamin B_6, pantothenic acid (water soluble vitamins) in creams claiming beneficial effects.

Vanishing and Foundation Creams

Vanishing creams are o/w emulsions spread easily and seem to disappear rapidly when rubbed on the skin. These creams are composed of emollient esters, low percentage of oil, which leave little apparent film on the skin. Traditional formula of vanishing cream is based on stearic acid that imparts attractive appearance to cream. Stearic acid melts above body temperature, and crystallizes in a form so as to be invisible providing a non-greasy film. With stearic acid white creams are produced and sometimes because of this whiteness these creams are also known as '**snow**'.

Vanishing creams are mainly used for providing fairness to skin and as a component of make-up to hold face powder and improve its adhesion. The main ingredients of vanishing creams are excess of stearic acid, soap and purified water. Soap is prepared in situ by chemical action between an alkali and a part (28–30%) of stearic acid added. Water used must be strictly purified otherwise its tendency to convert emulsion and reduce stability of the cream by formation of soaps of divalent ions like calcium and magnesium.

Various alkalis used for preparation of vanishing creams are: potassium hydroxide, sodium hydroxide, sodium carbonate, aqueous ammonia, borax and triethanolamine (trolamine). Out of these potassium hydroxide is most suitable because it makes a cream of fine texture and excellent consistency. Carbonates are not suitable because they liberate carbon dioxide and render the cream too spongy. Aqueous ammonia is objectionable because of its odour and volatility and creams also tend to turn yellow with age.

Since vanishing creams are o/w type of emulsion, there is possibility of drying out the cream due to evaporation of water from external phase of emulsion. To prevent this, 5–10% of glycerin and other polyols are added to creams as humectants.

Formula XXIII

Stearic acid (triple pressed)	17.0%
Potassium hydroxide	0.7%
Glycerin	5.0%
Perfume and preservatives	q.s.
Purified water	q.s. 100.0%

Formula XXIV

Stearic acid	20.0
Isopropyl myristate	2.0
Triethanolamine	1.2
Glycerin	5.0
Water	100.0
Perfume and preservative	q.s.

These creams leave a dry, tacky residual film. This film also has a drying effect on skin. Such creams are suitable for hot climates which cause perspiration on the face. Creams made with sodium hydroxide are harder than with potassium hydroxide. Borax can also be used to produce white cream but it tends to be graining.

Self emulsifying glyceryl mono stearate can also be used as emulsifying agent in preparation of vanishing cream. Advantage with this is that it does not have drying effect on skin as is the case with cream prepared by alkali stearates.

Formula XXV

Mineral oil	2.0%
Cetyl alcohol	3.0%
Glyceryl mono stearate	15.0%
Glycerin	10.0%
Perfume and preservatives	q.s.
Distilled water	q.s. 100.0%

For dry and flaky skin some special oily materials like lanolin, lanolin derivatives are used in vanishing creams.

Formula XXVI

Glyceryl mono stearate	12.0%
Mineral oil	4.0%
Cetyl alcohol	2.0%
Glycerin	6.0%
Lanolin	1.0%
Perfume and preservatives	q.s.
Distilled water	q.s. 100.0%

Foundation creams: These are available in the form of pourable lotions to thicker creams as very widely in viscosity. They are applied on face before starting make-up. They provide a base or foundation surface for better applicability of make-up.

These creams are prepared by incorporating and milling powders like titanium dioxide, kaolin, bentonite and pigments with vanishing creams. The desired shade and spreading properties can be obtained by varying the amount of pigments and titanium dioxide and/or other fillers.

Formula XXVII

Ceto macrogol emulsifying wax	25.0%
Mineral oil	5.0%
Titanium dioxide	10.0%
Pigments	0.5%
Glycerin	8.0%
Perfume and preservatives	q.s.
Distilled water	q.s. 100.0%

Some materials that give porosity to film building materials are called **porositones**, these are ideal for liquid make-up as they do not interfere with insensible respiration of skin.

Formula XXVIII

Glyceryl stearate	3.0%
Triethanol anine stearate	2.5%
Monoglycerides of polyunsaturated acids	0.5%
Isopropyl myristate	2.0%
Propylene glycol	6.0%
Talc	3.0%

Titanium dioxide and pigments	5.0%
Cellulose gum	0.8%
Isopropyl lanolate	2.0%
Hydrated magnesium silicate	0.8%
Branched chain fatty acid/esters	5.0%
Perfume and preservatives	q.s.
Distilled water	q.s. 100.0%

Cream cake foundation: This is used for non-aqueous formulations for heavy make up.

Formula XXIX

Petroleum jelly	75.0%
Lanolin	5.0%
Isopropyl myristate	2.5%
Titanium dioxide and pigments	7.5%
Perfume and preservatives	q.s.

Preparation with vegetable oil and waxes:

Formula XXX

Sesame oil	65.0%
Zinc oxide	10.0%
Oxycholesterol	2.0%
Triglyceryl stearate	1.0%
Titanium dioxide and pigments	7.5%
Perfume and preservatives	q.s.

By incorporation of high melting point waxes this preparation can be made in the form of stick.

Formula XXXI

Mineral oil	24.0%
Isopropyl myristate	20.0%
Petroleum jelly	3.0%
Beeswax	10.0%
Ozokerite	3.0%
Kaolin	25.0%
Titanium dioxide and pigments	7.5%
Perfume and preservatives	q.s.

The method of preparation is same as in case of vanishing cream. Titanium dioxide is not added initially during preparation. It is added after the cream is being prepared. It is milled with the finished cream with the help of roller mill.

Hand Creams and Lotions

It is modified vanishing cream very important to maintain moisture for maintenance of normal skin (soft and flexible) because extensive contact of soap, detergent and even water may damage the cell walls, removal of lipids and other secretions from skin. Chipping and cracking of skin caused by cold and dry winds because it takes out moisture from skin. Though water alone would have been sufficient for treatment of dry skin but water evaporates rapidly leaving behind the skin dry without providing emolliency. An ideal emollient will be a substance which will not only make water available to *stratum corneum* but will also regulate water intake of the *stratum corneum*. Because of this, formulations of creams and lotions meant for hands centre around emollients.

Hand-creams and lotions are used for protecting hands for the persons who are working in such industries where their skin come in contact with hazardous chemicals, or environmental conditions also promote different skin diseases. Occupational dermatitis may be caused in industries of acids, alkalis, soaps, solvents, grease, fats, gases, hypertonic solutions, lye, cement, plasters, metallic oxide, hot water, vapours, etc.

The substances show their harmful effects due to:

Dissolving horny layer, dissolving natural protective grease on the skin, protein precipitates, desiccation, hydrolysis or dissociation in water to form irritating compounds, oxidation or reduction.

The harmful effects depend on: character of substance, concentration and degree of exposure to the irritant.

Protective substances may be classified as:

Group I: Fats and oils, e.g. petrolatum, lanolin. These are very popular but have disadvantage like staining of skin, so require strong detergents to remove, which may turn into dermatitis.

Group II: Glycerin, honey, starch, tragacanth, agar etc., these substances form a barrier film and do not adhere for a long time since they are readily soluble in water and highly absorptive.

Group III: Emulsions formed by the combinations of group I and II, these are easily applied on skin and leave a fatty barrier on it. They also require soaps to remove them.

Group IV: Soap bases, like alkaline salts or stearates, oleates and palmitates. Main drawback of these substances are the chances of skin irritation due to presence of free alkali.

Most commonly used emollients are lanolin and its derivatives, sterols, phospholipids, fatty acids/esters and hydrocarbons.

Desirable characteristics of hand creams and lotions:

- It should soften the hands.
- It should apply easily.
- It should not leave a tacky film.
- It should have a pleasant smell.
- It should not interfere with normal perspiration of skin.
- It should have an appealing colour.
- It should spread easily and must not show any rolling tendency.

Formula XXXII

Cetyl alcohol	2.0%
Lanolin	1.0%
Mineral oil	2.0%
Stearic acid	15.0%
Glycerin	10.0%
Potassium hydroxide	1.0%
Perfume, colour and preservatives	q.s.
Distilled water	q.s. 100.0%

Hand lotion: These are differing from hand creams only in total solid contents. They are prepared to protect hands which receive most of the wear and tear and they are quite useful for persons washing dish, cleaning house and other activities which tends to turn skin rough. Hand lotion's main function is to soothe and soften roughened hands. Hand lotion is non-greasy product and may be applied many times during the day.

Formula XXXIII

Cetyl alcohol	0.5%
Lanolin	1.0%
Stearic acid	5.0%
Glycerin	2.0%
Triethanol amine	1.0%
Perfume, colour and preservatives	q.s.
Distilled water	q.s. 100.0%

Formulation of hand lotions offer two problems:

1. Lotion may exhibit gelling tendencies which make difficult it to pour.
2. Viscosity of lotion may be so low that it may come out of container.

Lotion should also retain its viscosity during shelf life period.

Lotions of stearate type are more susceptible to gelation and are also vulnerable to shear action. Excessive mechanical agitation can make these lotions of very low viscosity.

The following precautions will retard gelation in lotion:

- Excessive amount of polyol fatty acid esters (e.g. glyceryl monostearate) or fatty alcohol (e.g. cetyl alcohol) should not be used in formulation of hand lotions.
- Dispersed wax phase should be plasticized with mineral oils. The concentration of mineral oil that can be used as high as 10%.
- Small amount of alkyl sulphates (0.1–0.5% of sodium lauryl sulphate) should be incorporating in formulations.

All Purpose Creams (Sports Creams)

This term is very popular but it is misnomer. No cream can have the properties which should be possessed by a foundation cream, cold cream, hand cream and protective cream. Any attempted formulation to serve all the functions which the specialized creams have, will be a compromise formulation excelling in none of the functions. These are somewhat oily, but non-greasy. Being w/o type emulsion these spread easily on skin and make a protective film.

Formula XXXIV

Wool alcohol	2.0%
Microcrystalline wax	5.0%
Mineral oil	25.0%
Petroleum jelly	7.0%
Glycerin	5.0%
Magnesium sulphate	0.5%
BHA	0.002%
Perfume and preservatives	q.s.
Distilled water	q.s. 100.0%

Moisturizing Creams and Lotions

When water is lost from *stratum corneum* more rapidly than it is received from lower layers of skin, the skin becomes dehydrated. The dehydrated skin loses its flexibility and appears rough. Water (moisture) plasticizers *stratum corneum*, provides it flexibility makes it soft. Soft skin appears smooth. These creams which restore water to the *stratum corneum* are called moisturizing creams. Water contained in the cream is lost by evaporation when the cream is applied to the body. Therefore, a moisturizing cream must provide a non-volatile residual film capable of retaining the moisture of the skin as well as the moisture which is applied directly by cream. This may be achieved by adding humectants in the aqueous phase of the cream. Commonly used humectants are glycerin, propylene glycol and sorbitol (70% solution).

Formula XXXV

Stearic acid	4.0%
Mineral oil	8.0%
Lanolin	1.0%
Glyceryl mono stearate (self emulsifying)	3.0%
Isopropyl myristate	2.0%
Glycerin	4.0%
Propylene glycol	4.0%
Triethanol amine	0.2%
Perfume and preservatives	q.s.
Distilled water	q.s. 100.0%

Creams with Special Ingredients and Barrier Creams

Vitamin containing creams used in local vitamin deficiencies and alleviated by topical application of vitamins. Fat soluble as well as water soluble are capable of being absorbed through skin. These facts justify the use of vitamins in skin creams. However, these creams should contain sufficient quantities of vitamins in stabilized form.

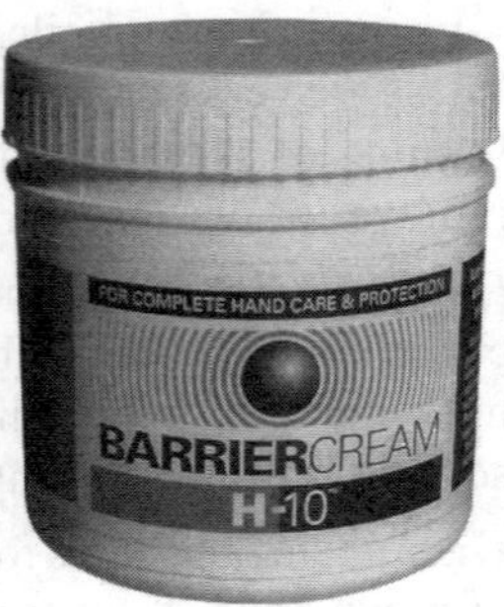

Vitamin B_6 (pyridoxine), vitamin C (ascorbic acid) and pantothenic acid used in toilet preparations. Fat soluble vitamins A, D and E have been used in cosmetics. Vitamin E is said to enhance percutaneous resorption. Emollient cream bases can be used for incorporation of vitamins.

Protein creams containing materials like albumin, amino acids, mucopolysaccharides hydrolyzed protein can be included in moisturizing creams.

Barrier creams protects from skin irritants amongst housewives to garage mechanics. Ideal characteristics for barrier creams:

- It should have good consistency and ease of application.
- It should adhere to skin reasonably and persistently.
- It should have ability to form a coherent, impervious, flexible and non-cracking film.
- It should be free from any tendency to irritate the skin.
- It should be easy to remove when desired and have aesthetic acceptability.

Barrier creams are of two types:

Water repellent cream: It forms a thin film that acts as a barrier to water and water soluble agents having skin irritating properties.

Oil repellent cream: It forms a thin film which acts as barrier to oil and oil soluble irritants.

With the advent of silicones a number of protective cream products can be formulated combination of silicones with other barrier agents, e.g. petroleum jelly, beeswax, paraffin, etc. can produce excellent barrier cream.

Healing agents: This is also associated with hand cream and lotion. Constant use of hands and feet in daily tasks may result in bruises and cracks. Healing agents have ability to stimulate the growth of healthy granulation tissue.

Examples of healing agents: allontoin (0.01–0.1%), urea (3–5%).

Astringent creams and lotions: Astringent creams are used to correct excessive oiliness of skin and hide coarse pores of the skin by reducing the permeability of the cell membrane. The astringent action is accompanied by contraction and wrinkling of tissue and by blanching. They precipitate proteins and cause slight inflammation and reddening of the skin which makes the skin surface openings less prominent.

Examples of astringents:

i. *True astringents:* Precipitate proteins, e.g. salts of aluminium, bismuth, iron and zinc, tannins or related polyphenolic compounds.
ii. *Pseudo astringents:* Produce physiologic action without combining with proteins, e.g. cold water, alcohol.

For proper treatment of oiliness use an astringent cream at night and lotion in morning before the application of face powder.

Formula XXXVI for astringent cream

Alum	0.2%
Zinc sulphate	0.2%
Glycerin	15.0%
Gum acacia (pulverized)	2.0%
White beeswax	15.0%
Theobroma oil	2.0%
Purified water	65.0%
Perfume and preservatives	q.s.

Dissolve alum and zinc sulphate in glycerin, preservative along with equal amount of purified water by means of heat. Dissolve the gum acacia in remaining water, strain and heat the mixture. Melt the white beeswax and theobroma oil, and add acacia mucilage to it with rapid and continuous stirring. Then add alum, zinc sulphate mixture with stirring. Add perfume when it becomes cool.

Formula XXXVII for astringent lotion

Alum	0.75%
Zinc sulphate	0.1%
Glycerin	10.0%
Alcohol	12.0%
Purified water	75.0%
Perfume	q.s.

Dissolve alum in small portion of purified water in one beaker. Add zinc sulphate in the glycerin in another beaker and add remaining purified water in it. Add alum solution and alcohol to the zinc sulphate solution. Allow to stand for 24 hours and then filter.

Hormone creams: Hormone creams containing less than 7500 IU of hormone per ounce produce little effect; above 15,000 IU per ounce, they fall into the realm of therapeutic agents. Their studies with both men and women involved the use of such hormones as estrone, estradiol, testosterone, alpha estradiol and 17-methylandrostenediol, all of which yielded similar results. From a practical stance, an efficacious hormone preparation should contain an estrogenic content of 10,000 IU or 1 mg of festrone per ounce of cosmetic vehicle.

An upper limit of 20,000 IU estrogen per month (based upon the use of oz of cosmetic cream containing 8 mg of estrone per ounce) is generally accepted by the cosmetic industry.

ESTROGEN CREAMS

Formula XXXVIII

Cetyl alcohol	3.0%
Stearyl alcohol	10.0%
Arlacel 83	15.0%
Sorbitan sesquioleate	3.0%
Petrolatum	13.0%
Hormone + Vehicle	1.0%
Amerchol L-101	5.0%

Bleaching creams and lotions: Nowadays they are gaining popularity because they are instantly used to make skin fair for a short period. This is also known as freckle lotion.

Examples of bleaching agents are: acetic acid, lactic acid, citric acid, lemon juice, hydrogen peroxide and other metallic peroxides, potassium chlorate, bismuth sub nitrate, sodium and zinc perborate, etc.

Some skin irritants are also used as bleaching agent like: formaldehyde, salicylic acid, ammoniated mercury and zinc sulphocarbonate.

Formula XXXIX

Hydrogen peroxide (70 vol.)	60%
Distilled water	40%

Mix the two ingredients. Make fresh for better results.

Formula XL

Acetic acid	2.0%
Citric acid	2.0%
Lactic acid	2.0%
Glycerin	10.0%
Alcohol	10.0%
Purified water	74.0%
Perfume	q.s.

Dissolve the acetic acid in glycerin and citric acid in purified water and then add both portion in lactic acid. Dissolve perfume in alcohol and then add into this preparation, keep overnight and filter in morning.

Aftershave lotion: Aftershave lotion is used by men after shaving for emollient effect, astringent effect, softening beard and antiseptic action in case of any cut from razor. They have cooling and refreshing feeling after application, which is obtained by addition of ingredients cantharides, capsicum and high content of alcohol. Perfumes generally used for aftershave lotion are lavender, lilac, bay rum, etc.

Formula XLI

Boric acid	2.0%
Glycerin	4.5%
Menthol	0.05%
Alcohol	7.5%
Witch hazel	85.0%
Perfume	q.s.

Dissolve boric acid in witch hazel and add glycerin to it. Perfume and menthol dissolve in alcohol in a separate beaker. Add alcohol solution slowly to witch hazel solution with constant stirring, filter it.

Gels

Gels and jellies are a type of base which produces a uniform external appearance and give a moist feeling.

Aqueous gels: Aqueous gels are used in summer as under make-up creams. It contains a lot of moisture used for providing moisture and cooling effects. This is also used as the base of a cleanser for light make-up for oily skin. Aqueous gels are prepared by water soluble polymeric substances having gelling abilities used, e.g. carboxy vinyl polymer and methyl cellulose. Besides gelling agents, other ingredients include humectants, surfactants, preservatives, colour, etc.

As gels are usually transparent, care should be taken to uniformly dissolve or disperse other additives in moisturizing gels.

Formula XLII

Carboxy vinyl polymer	0.5%
Methyl cellulose	0.2%
Dipropylene glycol	8.0%
Polyethylene glycol 1500	7.0%
Polyoxy ethylene 15-oleyl alcohol ether	1.0%
Calcium hydroxide	0.1%
Perfume, colour and preservatives	q.s.
Distilled water	q.s. 100.0%

Oily gels: Oily gels are used for supplying oil to the skin and are usually applied together with lotion. These are used in winter to moisturize the dry skin because of their moisturizing properties. Because of their great affinity for thick make-up, they are made into cleansers for removing make-up by combining them with suitable surfactants. These can be rinsed off easily after rubbing thoroughly into the make-up. Because of their moist feeling these gels are liked very much and are becoming popular. They give light feeling after use.

Formula XLIII

Liquid paraffin	10.0%
Glycerol-tri-2-ethyl hexanoate	50.0%
Sorbitol	10.0%
Polyethylene glycol 400	5.0%
Acyl methyl laurate	5.0%
Polyoxyethyl octyl dodecyl alcohol ether	8.0%
Perfume, colour and preservatives	q.s.
Distilled water	q.s. 100.0%

Manufacture of Creams and Lotions

General method: Heat two phases, i.e. oil phase (oils, waxes and other oil soluble ingredients) and aqueous or water phase (water or water soluble ingredients) to 75° C in separate vessels (steam SS jacketed kettles).

In aqueous phase 3–5% excess of water is used to compensate the loss of water due to evaporation during manufacture. Next the inner phase is added to the outer phase very slowly stirring continuously and homogenizing to ensure efficient emulsion. Here flow rate employed is 175–200 g/min. total stirring time, cooling rate are also important. These factors affect lotion viscosity, clean consistency and emulsion stability.

If cooling is started too soon after emulsification, crystallization of higher melting waxes may occur. A gradual cooling assures homogeneous dispersion of waxes.

Temperature at which perfume oil added is also important in manufacturing of creams and lotions. Addition of perfume oil to w/o type emulsion proceeds smoothly due to its solubility in the outer phase, but in case of o/w type emulsion, perfume oil must break through aqueous phase to reach oily phase. It becomes more difficult at room temperature particularly when emulsion is having high water content. Therefore, perfume oil is added very slowly to cream and lotion when it is between 45° C and 50° C.

Flow diagram of manufacturing of creams and lotions

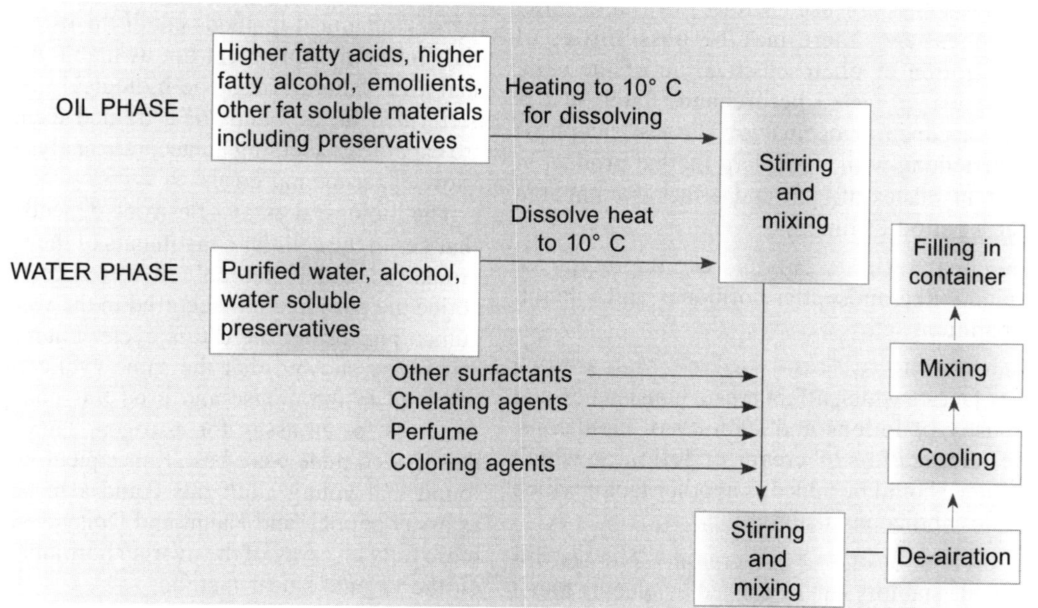

If a cream is to be hot poured, it is stirred to about 5° C above congealing point. Colour solution is added, if coloured cream is to be prepared. Cream is maintained at that temperature throughout filling operation with occasional stirring. If cold filling is to be carried out, the cream is stirred up to 35° C and colour solution is added. If coloured cream is to be prepared cream is then filled into the containers.

EVALUATIONS OF CREAMS AND LOTIONS

Creams and lotions are widely used for various parts of the body, stringent evaluation and quality control is must. General tests like qualitative and quantitative characterization, some other parameters are also important as:

Viscosity: To obtain a consistent quality, viscosity of creams and lotions should be measured during manufacturing process. Viscosity of creams and lotions are non-newtonian in nature that should remain constant throughout their shelf life. A Ford viscosity cup or a Brookfield viscometer can be used to measure viscosity of creams and lotions. If the viscosity is high, it can be corrected adding additional fatty material and emulsifier.

Patch test or sensitivity test: Sensitization of skin is measured with patch test. As various types of ingredients are used with use of antiseptic, hormones, etc., there may be possibilities of sensitization or photosensitization of the skin. This must be tested beforehand. Patch test is normally open or occlusive. The test sample is applied along with a standard market product at different places of skin and effect is compared after a period of time.

Biological testing: This is very important for products like antiseptic, hormones and vitamin preparations, etc.

Cooling time and total stirring time are two other factors which affect cream consistency and viscosity of lotions and stability of emulsions. The temperature of cream or lotion at which perfume should be added is another factor which can destabilize emulsion.

Peroxide stability test in creams: The test for peroxide stability can be carried by placing about 1 gm cream in a test tube and heating it in a constant temperature bath. Upper surface of the cream should be in level with the fluid of the bath. The tube is kept in the constant temperature bath for 24 hours at 95° C. The contents of tube are transferred to a 250 ml flask and peroxide content is determined. Stability of peroxide in cream can be found out by initial concentration of hydrogen peroxide in cream and by final concentration of hydrogen peroxide in cream after above-mentioned treatment from the following formula:

% Stability = (Final H_2O_2 concentration/Initial H_2O_2 concentration) × 100

Stability of peroxide cream should be better than 95%.

Estimation of hydroquinone in creams: About 4 gm accurately weighed cream is dispersed in 10 ml of water in 250 ml flask. To this added 10 ml of 0.1 N sulphuric acid and 10 drops of diphenylamine (1% in ethanol). This mixture is titrated with 0.1 ceric sulphate to a violet end-point.

Content of hormonal substance: In order to verify the declared content of estrogenic substance, accurate assay methods are needed. Both chemical and biological methods are found in the literature. However, because of the small amount of hormone present (of the order of 1 mg/oz), chemical methods entail the use of special (Infrared) equipment not available in many analytical laboratories. The biological methods, performed in the hands of a trained technician are capable of yielding an accurate analysis with a low experimental error.

The biological assay of estrone depends upon changes in the cellular constituents of the vagina. Stockared and Papanicolaou were the first to notice the cornification occurred in the vagina of guinea pig during the estrus cycle. Later, Allen and Doisy showed that the same cyclic change occurred in the mouse and used this change as the basis for an assay for estrogen. They found that spayed mice were best. Subsequent workers found that young adult rats could also be used as assay animal, and Kahnt and Doisy described more fully an assay of the ovarian hormone based on the vaginal smear method.

The estrogen assay depends upon a cellular change in the vagina of the spayed mouse or rat. Following ovariectomy, sinears of the vaginal contents show a predominance of leukocytic cells if a sufficiently high dose of estrone is administered, the vaginal cell picture changes. There is a reduction and finally a disappearance of the leukocytes with an appearance of nucleated epithelial cells. These become mixed with cornified cells, and in some cases a smear containing 100% cornified cells is found. After several hours leukocytes appear, and within 24 hrs they again become the dominant cellular type. This disappearance of a mixture of nucleated cells and cornified cells alone constitute a positive smear.

Marketed Products

- Body Cocoon (body moisturiser from Garnier).
- Body Repair (intensive body moisturiser from Garnier).
- Vaseline Intensive Care Body Lotion.
- Nivea Cream.
- Foot Cocoon (foot moisturiser from Garnier).
- Garnier wrinkle reduction cream.
- Hand Cocoon (hand moisturiser from Garnier).
- Borolin Antiseptic Cream.
- Summer Body (gradual tan from Garnier).
- Everyouth Clear Screen Cream.
- L'Oréal Paris Cream.

6 Face Powders

> *God hath given you one face and you make yourselves another.*
> — William Shakespeare

Face powders have been used from ancient times. The first substance used for face powder was chalk and white lead. Slowly with time more complicated blends of different materials were added to emerge as the modern concept and interpretation of face powder.

A face powder is a cosmetic which is applied on the face with the help of a powder puff that the face powder should be harmless to the skin.

FACE POWDERS

Face

The mirror of self that speaks a language without words. It is a large surface area that embodies the soul of a human being. A face powder is basically a cosmetic product which has as its prime function the ability to complement skin colour by imparting a velvet like finish it would enhance the appearance of the skin by masking the shine due to the secretion of the sebaceous and sweat gland. Face powder becomes very much common among women to improve the personal attractiveness.

Facial powders provide coverage of complexion imperfections, oil control, a matte finish, and tactile smoothness to the skin. Originally, facial powder was applied over a moisturizer to function as a type of powdered foundation. Liquid foundations have largely replaced the powdered foundation; however, for patients who wish sheer coverage with excellent oil control, a powdered foundation performs excellently. An appropriate moisturizer for the patient's skin type is first applied and allowed to set or dry, followed by application of a full coverage translucent powder.

Purpose and Functions

The purpose of face powder is to mask visible imperfections and shine caused by moisture or grease from perspiration. It should impart a smooth finish to skin. The characteristic of appearance and finish given to the skin is called **texture** which is the important characteristic of face powder. Texture is influenced by following factors.

- Powder should have adequate covering power to mask the skin imperfections.

- It should adhere to skin for sufficient time to avoid blow off easily.
- The finish given to skin must complement the skin colour preferably of a matt or peach-like.
- It must be an absorbent without changing its appearance on skin.
- Face powders are sold in two different forms—loose and compact.
- There must be sufficient slip to spread on skin by the puff and without producing a blotchy effect.

Loose Powders

The final outcome of the formulation of face powder depends on quality and characteristic of raw material used. Properties of raw materials like colour, odour density, feel, fineness oil absorption, perfume retention, etc. have direct impact on ultimate quality of the powder.

It is obvious that no single raw material can provide all these characteristics. Therefore, only by using blends of different ingredients a satisfactory product can be obtained. Ingredients are selected on the basis of the characteristic they can provide to the final product.

Covering Power

Covering power is the ability to mask skin imperfections and facial skin blemishes. This can be achieved by using materials like

- Titanium dioxide
- Zinc oxide
- Light Kaolin.

It has been reported that titanium dioxide has about 1.6 times the covering power of zinc oxide in air and about 2.9 times the petroleum jelly. On most greasy skin its covering power is 2.5 times that of zinc oxide covering power. Zinc oxide is preferred to titanium dioxide where less covering power is required. Zinc oxide has a tendency to form balled particles; therefore it must be passed through sieve before mixing with other ingredients. Also excessive use of zinc oxide may result in drying effect.

Adhesion

It is the ability of the face powder to stick to the skin and not dissipate in a short period of time. Adhesiveness is obtained by inclusion of

- Talc (hydrated magnesium silicate)
- Zinc stearate
- Magnesium stearate.

Zinc stearate is most commonly used. Apart from increasing adhesion, metallic stearate also import water repellence properties and make the ultimate product soft. They are used in the range of 3–10% w/w. Other materials which are used to improve adhesion of powder to the skin are

- Stearyl alcohol
- Glyceryl monostearate
- Magnesium myristate
- Petroleum jelly.

Absorbency

The finished powder should have ability to absorb perspiration and sebaceous secretions. Material used should have a high absorptive capacity, but should not change appearance of powder on skin. Components of face powder which have absorptive prosperties are:

- Kaolin (hydrated aluminum silicate)
- Starch
- Calcium carbonate
- Magnesium carbonate.

Kaolin is a hydrated aluminum silicate having the formula $Al_2O_3.2SiO_2.2H_2.2H_2O$. Kaolin has some hygroscopic nature. Therefore, use of kaolin more than 25% w/w is not recommended.

Calcium carbonate is alkaline in nature. It has good absorption characteristic but excessive use (greater than 15% w/w) is not recommended because of its dry feel.

Magnesium carbonate has very high absorption capacity and also it provide a good media for perfume distribution. But it has a greater tendency to dry the skin.

Starches, particularly rice starch have been exclusively used as base of face powder. It has excellent absorptive properties, good covering power and it imparts smoothness to skin. However, starch is a good media for microbial growth when moist also it tends to form cake at higher humidity. Nowadays starch has been replaced by talc.

Applicability

The quality of easy application and spreading of a face powder to produce a smooth feeling is known as applicability. Talc with metatic stearate is commonly used ingredients to impart this characteristic. Talc is hydrated magnesium silicate ($3MgO_2.4SiO_2.H_2O$). Talc is the softest material available. Depending upon its origin, the colour of talc may vary from pure white to dark grey. A suitable talc for face powder should be very white, odourless and feeling somewhat greasy. A good quality of talc is that whose 98% goes though a standard 200-mesh sieve, i.e. 98% of the particles are not larger than 74 microns. There are many ways of reducing particle size of talc. They are:

- Passing though five sieves of 300 mesh or less
- By micronization using fluid energy mill
- By air separation
- By power mill.

The ultimate quality of face powder, depend to a great extent upon the particle size of the talc, as other ingredients are much finer. If larger particles are used the face powder will not adhere to skin. However if very fine particle size of talc will be used it will result in loss of its lamellar structure so also its life.

Also, any change in bulk density of talc may affect quality of final product. Face powder should have constant volume for a given weight; otherwise powder box may be under filled or may overflow, even though the weight may be same in both cases. The talc should not contain more than traces of iron as impurity as its presence can cause odour and colour deterioration.

Blooming

Blooming is to provide peach like finish to skin. The materials which can impart this effect are

- Chalk
- Rice starch
- Powdered silk.

Powdered silk is pulverized fibroin crystallite material extracted from natural silk by removal of sericin and other impurities followed by partial hydrolysis.

The main constituents of powdered silk are: amino acids, glycine, tyrosine, serine, glutamic acid, alanine, leucine, aspartic acid and phenyl alanine. Cosmetic grades are light and bulky and of good colour. Silica may also be used because of its good adhesive property and to increase the bulk as it is very finely divided.

Colour

The colour of the finished face powder is a matter of taste and fashion. Usually a range of shades is produced to enhance the natural complexion.

Several inorganic and organic pigments have been used as colouring agents in face powder.

Inorganic pigments include natural and synthetic iron oxides which vary in colour from yellow, red and brown to black. Naturally occurring iron oxide contain lead and arsenic as impurity. Therefore these pigments are now manufactured synthetically to reduce level of impurities.

CHARACTERISTICS FOR GOOD FACE POWDER

a. **Covering power:** Ability to mask skin imperfections, e.g. skin shine from oily face, pores or minor blemishes and imperfections of the face.
b. **Adhesiveness:** Ability to adhere easily or to be retained to the face and skin for longer period.
c. **Absorbency:** Ability to absorb perspiration and oily secretions.
d. **Bloom and coloring:** Ability to impart a velvety, matt or peach like finish and impart colour according to need.
e. **Slip:** Characteristic smooth feeling.
f. **Perfuming:** To produce a pleasant odour.
g. **Particle size:** It should have very fine particles and should not have any gritty particles.
h. **Irritability:** It should not cause any irritation and toxicity to the skin.
i. **Stability:** It should be physically and chemically stable.

A face powder must be a blend of specific raw materials, if it is to be a product which exhibits the particular characteristics desired. Therefore it would be well to list the basic ingredients normally employed as well as the properties, each may impart to the finished powder formulation.

TYPES OF FACE POWDER

Face powders can be classified into three categories according to their covering ability, nature of skin (dry, normal, oily and high oily).

1. **Light powder:** This is preferably for dry skin, it has low covering power and normally it contains large quantity of talc.
2. **Medium powder:** It having covering power higher than light type, this is for normal or moderately oily skins. It contains talc and zinc oxide.
3. **Heavy powder:** It has more covering power and extensively used for oily skins. It normally contains lower quantity of talc and high quantity of zinc oxide.

FORMULATION

Full coverage powders contain predominantly talc and increased amounts of covering pigments. The covering pigments used in face powder can be listed in order of increasing opaqueness, as follows.

- Titanium dioxide,
- Kaolin,
- Magnesium carbonate,
- Magnesium stearate,
- Zinc stearate,
- Prepared chalk,
- Zinc oxide,
- Rice starch,
- Precipitated chalk, and
- Talc.

It is generally accepted that the optimum opacity is achieved with a particle size of 0.25 microns. Magnesium carbonate also can be used to improve oil blotting, to keep the powder fluffy, and to absorb any added perfume. Kaolin also may function to absorb oil and perspiration. Full coverage face powders usually are packaged in a compact and applied to the face with a puff.

Transparent facial powders are more popular today to add coverage and to improve oil-blotting abilities of a previously applied liquid foundation. Transparent powders have the same formulation as full-coverage powders except they contain less talc, titanium dioxide or zinc oxide because coverage is not a priority. Transparent facial powders commonly have a light shine, produced by nacreous pigments, such as bismuth oxychloride, mica, titanium dioxide–coated mica or crystalline calcium carbonate.

Facial powder usually includes iron oxides as the main pigment, but other inorganic pigments, such as brilliant pink lake, chrome hydrate, ochre, chrome oxide and ultramarine, also may be used. These powders are designed to augment the underlying skin and foundation tones; therefore, transparent powders can be used by patients who have difficulty finding an appropriately tinted facial foundation.

Table 6.1: Raw materials and their characteristics

Raw materials	*Outstanding characteristics*
Talc	Slip
Purified Kaolin (colloidal clay)	Absorbency, adhesion
Precipitated Calcium carbonate	Absorbency, bloom
Magnesium carbonate	Absorbency, fluffiness
Calcium, zinc and magnesium stearates	Adhesion, waterproofness
Rice starch	Absorbency, bloom
Silica and silicates	Absorbency
Zinc oxide	Opacity
Titanium dioxide	Opacity
Frosted-look (guauine bismuth oxychloride, mother-of-pearl, mica, aluminum, bronze)	Sparkle, pearly effect
Metallic soaps	Ease and smooth application
Lithopone	Conceal shine and minor imperfections

Formula I

Talc	75.0%
Kaolin	5.0%
Chalk, precipitated	5.0%

Zinc stearate	5.0%
Zinc oxide	10.0%
Perfume and colour	q.s.

Formula II

Talc	75.0%
Zinc oxide	10.0%
Rice starch	10.0%
Zinc stearate	5.0%
Perfume and colour	q.s.

Method of preparation: Mix the perfume and colour with a part of chalk or magnesium carbonate in a suitable vessel, then pass the mixture through a hand operated sieve with a stiff bristles brush till the perfume and colour is uniformly distributed. Check out by a portion rubbed on a white paper shown on colour flecks. Keep a standard sample for matching successive batches. Weigh out rest of raw materials into a mixer and mix until uniform. Again take the sample and rub it out on a white paper, it should not show any colour flecks on the paper. Match the sample with the previous batch sample. If it matches, then stop mixing, otherwise continue till it is completely matched without any line of demarcation. Finally shift the powder through sieve number 160 or above and fill in a proper container.

Application

Compact facial powders are removed with a puff or dusted loosely from a container with a brush. They impart a matte finish to the face. Patients who desire a shiny or moist semi matte facial appearance should avoid powder because it absorbs the oil in the foundation, thus destroying the dewy appearance. Patients with dry complexions also may wish to avoid facial powder because it can further dry the skin. The oil-absorbing abilities of facial powder are extremely valuable in the patient with an oily complexion prone to develop a facial shine.

Adverse Effects

The incidence of allergic contact dermatitis to facial powder itself is low; however, added fragrances may pose a problem. A more common problem with facial powders is irritant contact dermatitis due to coarse particulate matter, such as nacreous pigments in the formulation. Inhalation of the powders may cause problems in patients with asthma or vasomotor rhinitis. Facial powder may be open or closed, patch tested is same.

COMPACT FACE POWDER

A compact face powder is a dry powder which has been compressed into a cake and is usually applied with the help of powder puff. Compact face powder has become very popular because of its ease of application and storage convenience. The composition of compact face powder is similar to that of loose powder but its effect is slightly differing. The binder contained in compact face powder increasing its adhesion. The materials of compact face powder must compress easily and remain compact, not to break easily in normal conditions of use also must adhere easily to puff. The advantage of compact powder is that large bulk of powder is reduced by compression into cake form.

Compacts Includes

Compact powder, cake make-up powder, cream and liquid powder **(cosmetic stockings, 'night whites')**.

It should have following characteristics:

1. Easily come off with powder puff or moistened applicator or sponge.
2. Stay on skin for a reasonable time and should not come off through rapid drying.
3. Readily washed off with soap and water.
4. Lay on skin uniformly.
5. Repel the moisture caused by perspiration.

The pressure for compaction is very important. The compact face powder must come off easily when rubbed with a puff. Very high pressure will result in a very hard dull cake which will not payoff easily and very low pressure may lead to breaking of the cake during use.

Various binding agents used for compact face powder:

- *Water soluble binders:* Aqueous solutions of gum acacia, tragacanth, gum karaya,

Irish moss, PVP, methyl cellulose, rosin, carboxy methyl cellulose, quince seed, gelatin generally used for making compact powders. A preservative is necessary in gum medium to prevent microbial contamination.

- *Water repellent binders:* Various fatty esters, mineral oil, lanolin derivatives (lanolin dissolved in ether).
- *Dry binders:* talc (to avoid cracking and breaking of cakes it is restricted up to 50% only), colloidal clay, zinc stearate, magnesium stearate.
- *Oily binders:* Light and heavy mineral oil, isopropyl myristate, lanolin, vegetable oils.
- *Emulsion binders:* These prevent loss of moisture and help, in easy manufacturing, prevent the problem of lumping occur when oils alone are incorporated. Examples: soaps like triethanolamine stearate, non-ionic emulsifiers and glycerol monostearate.

Other Agents

Pigments, perfumes, water, humectants (glycerol, glycols), and emulsifying agents.

Methods of Preparation

1. *Kneading or wet method:* Basic materials, binders and colours are kneaded into a paste with water, pressed into moulds and air dried, sometimes it may produce cakes.
2. *Dry compression method:* In this method materials and binders are compressed in special presses under controlled conditions.
3. *Damp method or wet casting method:* In damp method base powder, colour and perfume are mixed uniformly. The mixture is then wetted uniformly with liquid binders till the proper plasticity of mass is obtained. This is commercially acceptable and widely used method.

   Examples of some compact face powders

Formula III

Talc	70.0 gm
Kaolin	20.0 gm
Zinc stearate:	05.0 gm
Titanium dioxide	05.0 gm
Colour, binder and perfume:	q.s.

Formula IV

Talc	60.0 gm
Kaolin	12.0 gm
Chalk, precipitated	12.0 gm
Zinc stearate:	4.0 gm
Titanium dioxide	12.0 gm
Colour, binder and perfume:	q.s.

Mix all the materials of the formula and pass through pulverizer or a ball mill. Moisten the powder with a sufficient quantity of binding solution to make damp mass and then pass through a proper size to form granules. Granules are dried completely in oven and then compressed to form a cake. Care must be taken to keep formula and process standardised. Addition of colour must be carefully done so that uniformity is maintained from lot to lot and batch to batch.

Liquid and cream powders are used for evening wear to offset the glaring effects of lights and applied to face, neck and arms. These have high opacity to cover the colour of skin exposed in evening.

Formula V

Liquid powder

Talc	10.0 gm
Colloidal clay	5.0 gm
Titanium dioxide	5.0 gm
Glycerol	10.0 ml
Rose water	64.0 ml
Ethanol	5.5 ml
Perfume	q.s.

Formula VI

Cream powder

Glycerol monostearate	10.0%
Glycerol	2.0%
Mineral oil (heavy)	5.0%
Stearic acid	2.0%
Spermaceti	5.0%
Caustic potash	0.1%
Titanium dioxide	7.0%
Talc	20.0%
Water	50.0%
Perfume	q.s.

Cosmetic stockings which are the leg make-up close to liquid face powder which provides bare legs the artificial texture-like appearance of women's hose or stockings.

Ideal requirements for cosmetic stockings:

- Colour simulated to particular shades of women's hose.
- Alcohol used for rapid drying.
- Glycerol, sorbitol, propylene glycol are used for providing emollient effect on the skin and wetting agent for even colour deposition.
- Bentonite used to keep the pigments in suspended form for a longer time.
- Talc is to provide the lustre to the mixture, permitting better simulation of silk texture.
- Gum is used for the powder to adhere firmly to skin surface.

Formula VII

Liquid cosmetic stocking:

Talc	6.0%
Precipitated chalk	11.0%
Titanium dioxide	2.5%
Bentonite	2.0%
Glycerol	3.5%
Alcohol	7.0%
Methyl cellulose	0.5%
Wetting agent	0.5%
Water	70.0%
Colour and pigments	q.s.

Mix all the powder ingredients in a powder mixer except colour or pigments. Mix all liquid ingredients in another vessel with a mechanical stirrer. Dissolve the colour or pigments in liquid mixture and add powder mixture in parts with help of stirrer. When all the powder has been added the stirring is continued for at least half an hour for a homogeneous product.

BODY POWDER OR TOILET POWDERS

Body powder or talcum powder is one of the most consumed cosmetic products used extravagantly over lager surface area of body than face powder.

The main purpose of body or talcum powder:

- To absorb moisture or perspiration or body sweat in warmer conditions
- Provide cooling effect,
- Good slip,
- Lubrication and
- Prevent irritation of skin due to chafing.

The very fine particle size of these covers cause a large surface area per unit weight and can cover a large body area which results in strong light dispersion and cooling effect. Body powders consist of mainly of talc with small portions of metallic stearates like zinc or magnesium stearate, etc.

Types of body or toilet powders:

- *Talcum powder:* Talc is main component; it varies from 60 to 90%.
- *Dusting powder:* Usually applied with puff to body of the person.
- *After shave powder:* Consists of talc to which colour and other mineral ingredients has been added to spread smoothly, cling to face with less sheen and more nearly match the colour of men's skin.
- *Baby powder:* Usually less perfumed and without colour than other toilet powder.

Contains of body or dusting powder:

- *Covering material:* Talc, precipitated chalk, light magnesium carbonate
- *Adhesives:* Kaolin, zinc oxide, zinc stearate, magnesium stearate
- *Absorbent material:* Kaolin, light magnesium carbonate, precipitated chalk, starch
- *Slip product and stable filler:* Talc, zinc stearate
- *Antiseptic:* Boric acid, chlorhexidine diacetate, bithional and
- *Perfume:* Lavender, lilac, lime, rose, jasmine, menthol, etc.
- *Fluffiness:* Light calcium or magnesium carbonate.

Talc (Talcum, purified talc, French chalk, soapstone) it is magnesium silicate sometimes containing small proportion of aluminium silicate and must be of high quality and should possess following characteristics: it must be completely white and possess adequate unctuous or slip properties to facilitate spreading over skin; it should be lustrous, very fine and free from grittiness and must be sterilized.

Zinc oxide acts as a mild astringent and relieves the prickling and irritation.

Boric acid (H_3BO_3) colourless scales of a somewhat pearly lustre, or crystals, but more commonly a white powder slightly unctuous to the touch is added in all types of powders for local anti-infective or antiseptic property.

Formula VIII

For Body powder or talcum powder

Talc	70.0 gm
Calcium carbonate	25.0 gm
Zinc stearate	4.0 gm
Boric acid	0.3 gm
Perfume	0.7 gm

Formula IX

For deodorant powder

Talc	79.2 gm
Light calcium carbonate	5.0 gm
Zinc oxide	10.0 gm
Zinc stearate	5.0 gm
Bithional	0.5 gm
Perfume	0.3 gm

Formula X

For baby powder

Talc	84.0 gm
Kaolin	10.0 gm
Satinex	3.0 gm
Boric acid	0.3 gm
Perfume	0.7 gm

Formula XI

For medicated dusting powder

Talc	78.0 gm
Sodium propanedioate	20.0 gm
Boric acid	2.0 gm

Formula–XII

For foot powder

Talc	45.0 gm
Kaolin	45.0 gm
Boric acid	10.0 gm
Menthol	q.s.

Formula XIII

For after shave powder

Talc	70.0%
Titanium dioxide	5.0%
Zinc stearate	5.0%
Precipitated chalk	19.0%
Ochre (golden)	0.5%
Perfume	0.5%

EVALUATIONS OF POWDER

These tests are essential to judge the quality of the finished products of loose powder, compact powder, etc. and done at different levels. In these tests includes like particle size, apparent density, abrasiveness, moisture content, limits for colour, content determination, stability test, etc.

1. **Shade control and lighting:** The first is comparison of the appearance of the body of the powder to a standard when it is spread out and flattened on a white paper background (skin tone). The second manner of evaluating color is comparison of the sample to the standard by skin tone (undertone). It must be stressed that the powder should be applied with the same puff or device that is being employed in the finished package. Skin tone should be the final judgment of powder shade, for it is ultimately the consumer's method of evaluation and usage.
2. **Dispersion of color:** The uniformity of a powder can easily be checked by spreading it on white paper and examining it with a magnifying glass. If any non-uniformity is detected, further processing to achieve maximum color development should result in homogeneity.
3. **Pressure testing:** In a compressed powder, the pressure applied must be of a uniform nature, since the pressure of air pockets will result in a cake which is easily broken. Uniformity and hardness of a cake are checked by means of a penetrometer.
4. **Breakage test:** The most successful manner of checking the tendency of a pressed powder tablet to break or chip is to drop the assembled compact on a wooden surface several times from a height of 8 to 10 inches if the cake remains unbroken, it is an indication that the compact can travel and will undergo normal handling without unsatisfactory results.
5. **Pay-off test:** Pay-off means adhesion characteristic with powder puff, of a compact or pressed powder should be tested on the skin.

   High pay-off due to application of low pressure in formulation, hence soft cake formed that tends to crumble and break easily and more powder adhere on puff.

Low pay-off means high pressure and formation of hard cake so hard that will not rub off the cake enough to adhesion the material to the puff easily.

6. **Particle size:** Particle size determined by stage microscopic method, sieve analysis or by using counter coulter techniques.
7. **Moisture content:** These can be estimated with Karl-fisher apparatus, IR moisture balance and other calorimetric analysis method.
8. **Limits for colour:** By using suitable analytical methods.
9. **Abrasiveness:** Abrasiveness determined with rubbing the powder on a smooth surface and then studying the effect on surface using microscope.
10. **Flow property of powder:** Flow property of powder can be studied by measuring angle of repose of powder by allowing falling on a plate from a funnel and measuring the height, radius of heap formed and time taken. This is important for body powder as they should come out easily from the container for easy application.

Marketed Products

- Shower to Shower
- Ponds
- Persona (Amway)
- Dermicool
- Johnson and Johnson Baby Powder
- Cuticura
- Micderma Dusting Powder
- Tinactin Powder
- Tendrils Body Talk (Kräuter)
- Cinthol Powder
- Nycil
- Smyle
- Touch and Glow (Revlon).

7 Lipsticks

> *Women have face-lifts in a society in which women without them appear to vanish from sight.*
> — Naomi Wolf

INTRODUCTION

Lipstick is a moulded stick composed of colouring material dispersed in a blend of fatty bases (i.e. oils and waxes). Lipstick is used to make appearance of lips (Pouts) attractive by imparting color. Narrow lips can be made to appear wider by applying lipstick above the upper lip line and broad sensual lips can be made to appear narrower by applying lipstick well within natural lip line.

Lipstick is known to have been used around 5000 years ago in ancient Babylon, when semi-precious jewels were crushed and applied to the lips and occasionally around the eyes. Lipstick in some shape or form has been around for a long time and has always been a part of the fashion statement.

Application of colour to lips is practised from ancient time. The Greeks used root called polders to colour their cheek and lips. Ancient Egyptians and Romans extracted purplish-red mercuric plant dye fucius a type of rouge from fucus-algin, 0.01% iodine, and some bromine mannite, which resulted in serious illness and Egyptians know that it was potentially poisonous-talk about the kiss of death. Queen Cleopatra of Alexandria had her lipstick made from crushed carmine beetles, which gave a deep red pigment, and ants for a base. Lipsticks or chapsticks with shimmering effects were initially made using a substance found in fish scales called **pearlescence** or frost.

Although no self-respecting Egyptian would leave home without it, makeup has not always held an accepted place in society. In fact, it has traveled a bumpy road to acceptance.

Lipstick started to gain popularity in the 16th century, during the reign of Queen Elizabeth I who made blood-red lips and stark white faces a fashion statement. By that time, lipstick was made from a blend of beeswax and red stains from plants.

During the Second World War, lipstick gained popularity as a result of its use in the movie industry, and it became commonplace for women to apply makeup, or "put their face on". As with most other types of makeup, lipstick is typically but not exclusively worn by women. It is usually not worn until a female reaches adolescence or adulthood.

Another form of lip color (liquid lipstick), a wax-free semi-permanent liquid formula, was invented in the 1990s by the Lip-Ink International Company. Other companies imitated the idea, putting out their own versions of long-lasting "lip stain" or "liquid lip color", but were not allowed to infringe on the patented wax-free formula. A lip-duo is a combination of lipstick and lip liner. The lips are pools of sensuality.

Various lip colouring preparations including:

- Lipsticks
- Lip gloss (lip paint)
- Rouges
- Lip liner
- Lip salves.

Lipsticks are very popular and frequently used by ladies of any society to brighten the colour of the lips.

Lipsticks are composed of oil-wax base stiff enough to still. In lipstick red staining dye is dispersed in oil and red pigment is incorporated therein perfumed and suitably flavoured and they are moulded and are enclosed in a box. Lipstick are convenient to use so most frequently used in comparison to other cosmetics.

Ideal characteristics of a good lipstick

- It should be smooth and easy to apply on lips and leave a thin film over it.
- It should be non-irritant and non-toxic.
- It should have miraculous wear, moisture, colour and shine.
- It should be free from grittiness and should be non-drying.
- It should have required plasticity.
- It should be innocuous internally as well as externally (due to the fact that a small quantity of lipstick might be ingested).
- It should have a good degree of indelibility.
- It should have pleasant taste, odour and flavour.
- It should not lose its smooth and shiny appearance during storage.
- It should be stable during its shelf life—means free from bloom or sweating during storage.
- It should remain firm (should not melt or harden) within reasonable variation of climatic temperature.

The tubes that hold lipstick range from inexpensive plastic dispensers for lip balms to ornate metal for lipsticks. Sizes are not uniform, but generally lipstick is sold in a tube 3 inches (7.6 centimeters) in length and about 0.50 inch (1.3 centimeters) in diameter. (Lip balms are generally slightly smaller in both length and diameter.) The tube has two parts, a cover and a base. The base is made up of two components, the twisting or sliding of which will push the lipstick up for application. Since the manufacture of the tube involves completely different technologies, we will focus here on the manufacture of lipstick only.

ANATOMY OF LIPS

Skin of the lips (*Cheil-* in Greek) is characterized by an exceptionally thin corneal layer. The *stratum germinativum* is highly developed and corium pushes papillae with high blood content just below the surface. Lips do not content sweat glands and sebaceous glands sparsely, so lips are entirely free from fat. Salivary glands are present in the

1. Raised vermillion border
2. Cupid's Bow
3. Philtrum (Vertical columns)
4. Upper Lip 1/3, Lower Lip is 2/3 size

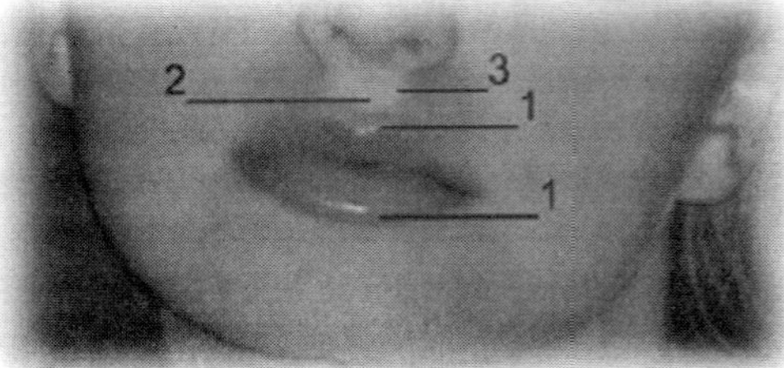

Youth Lip Anatomy

inner portion of lips and this saliva maintains the moisture of the lips. In the extreme conditions of the weather like too much cold or hot corneal layer become dry out and produce cracked lips. Substances applied on lips can easily penetrate to the *stratum germinativum.* Lip contains protective pigment melanin in less quantity.

FORMULATION OF LIPSTICKS

Lipstick is composed of variety of waxes, oils, pigments, and emollients which are adjusted to desired melting point (55–75 °C) and viscosity. Various agents are used to formulate lipsticks:

1. Base
   i. Waxes
   ii. Oils and fats
2. Colour mixtures-colours, staining dyes and pigments
3. Moisturizers
4. Antioxidants
5. Preservatives
6. Flavors.

Bases

For suitable pigment, combination and indelible colour for desired mark and stain, all must be incorporated into a suitable base for uniform dispensability and smooth flow when molten, and which upon congealing form lipstick enough to resist breakage but soft enough to permit easy application.

All desired properties are not fulfilled by one single material so they are used in combination. The basic materials may be divided for purpose of study into oils, fats and waxes.

Vegetable oils: Oils like olive oil due to their propensity for turning rancid and poor solvent power for staining dye has got little use in modern or present time.

Mineral oils: Mineral oils are also used in ancient time as they resist rancidity but due to poor solvent character for colour also they make lipstick smear and run off too readily, make them to use in small proportion to enhance gloss.

Castor oil: It is unique among reg. oils. It is an important ingredient of lipstick for the reason of higher viscosity, presence of hydroxyl group in acid part and for solvent power for bromo acid.

Due to high viscosity of castor oil it delays settling of pigments from molten lipstick mass and to resist flow off and smear applied lipstick. It is disadvantageous as it retards penetration of oil in dry pigment during mixing operation, which in turn causes friction or drug feeling during application of lipstick on lips.

Desirable character for ideal oil as base for lipsticks:

1. Greater solvent power for staining dyes.
2. Lower viscosity, or lower viscosity temperature coefficient.
3. Less odour and greater stability.
4. Dependable supply.

For complete satisfaction, oil must be free from irritation quality or unpleasant taste. Yet no substitute is available which perfectly fits the specification.

BUTYL STEARATE

Butyl stearate wets both bromo acid and pigment more quickly than does the more viscous castor oil, it is very useful in getting the colour into suspension. However, it is not a good solvent for bromoacid. Also butyl stearate dissolves only about 0.2% bromo acid against about 1.7% dissolved by castor oil.

The purer grades are free of disagreeable odour and resistant to turn rancid.

TETRAHYDROFURFINYL ESTER

It has high solvent power for bromo acid. However lower esters as acetate has strong rather disagreeable odour and highly volatile. Higher esters are of somewhat less odour but have low solvent power.

POLYETHYLENE GLYCOL AND ESTERS

Esters with least solubility in water are likely to have low solvent power for bromo acid and are used in lipsticks glycols, are too highly water soluble to be used in large proportion without detracting from the waterproof quality of the lipstick film. Propylene glycol monoester have fair solvent power.

FATTY ESTERS OF LOWER ALCOHOL

Properties are similar to butyl stearate but do not have good solvent power.

However, we come to note that the ideal substitute for castor oil has not been found. Many solvent can dissolve 10% or more of bromo acid but most of them are picked out on account of odour, volatility taste, irritation effect, and immiscibility with waxes, water solubility, tendency to bleed or other effect. So any new material should be tested for above parameter prior to use as lipstick base.

True fat: Seldom used now.

Cocoa butter: Firstly it was though to be ideal base as it melts at body temperature so easy to application but later on it was found that it tends to bloom or comes to the surface in irregular fashion, eventually developing unsightly craters or excrescence so seldom used.

HYDROGENATED VEGETABLE OILS

These are used for shortening; are relatively stable to oxidation and have good texture.

Petrolatum: It is stable, produces good gloss and is used in those formulas where percentage of castor oil is not too high.

Lanolin: Useful in assisting dispersion of colour but in high proportion have tendency to develop a disagreeable odour. Lanolin, also called, Adeps Lanae, wool wax, wool fat, or wool grease, a greasy yellow substance from wool-bearing animals, acts as a skin ointment, water-proofing wax.

Lecithin: In small proportion used to improve smoothness, emolliency and case of application.

Waxes: Waxes are employed for two reasons as to provide, strengthen the stick and to raise the melting point.

Waxes differs greatly in this effect.

Carnauba wax: It is one of the hardest waxes, so it provides good strength and raises melting point.

Candelila wax: Has lower melting point than carnauba but can be used in higher proportion to equable effect.

Bees wax: It is a traditional stiffing agent for lipstick but this result in a rather dull stick with too much drag. Hard waxes provide better gloss.

Ozakerites: Ozakerites along with amorphous hydrocarbon waxes are available in a wide range of melting point and texture.

Paraffins: are too weak and brittle to be used but small quality can improve gloss. As paraffin is immiscible with castor oil so also it is limited in use.

Synthetic waxes: Many synthetic waxes are available and should be judge on individual merits and demerits.

Hydrogenated caster oil in a brittle white wax provides high gloss but little strength.

Ivanovsky has reviewed the theory and use of wax constant 'Paste index' He recognizes three types of semisolid wax preparation.

1. Lattice structure (wax paste)
2. Gel structure
3. Gel paste structure.

Waxes: A mixture of waxes is employed to achieve desired melting point viscosity and other physical properties of the stick. The wax gives lipstick its shape and ease of application. Examples of waxes used include-

- **Bees wax:** 5–20% Among the waxes are bees wax, a substance obtained from bee honeycombs that consists of esters of straight-chain monohydric alcohols with even-numbered carbon chains from C24 to C36 and straight-chain acids also having even numbers of carbon atoms up to C36.

- **Candelia wax:** 5–20% which is obtained from the candelilla plant and is produced in Mexico by immersing the plants in boiling water containing sulfuric acid and skimming off the wax that rises to the surface.
- **Carnuba wax:** 1–3% is an exudates from the pores of leaves of Brazilian wax palm trees
- **Ceresin wax:** 5–20%
- **Cetyl alcohol:** 2–3%
- **Lanolin:** 5–15%
- **Ozokerite wax:** 1–10%
- **Paraffin hard:** 1–5%

Oils and fats: The oils and fats used in lipstick include castor oil, mineral oil, olive oil, cocoa butter, lanolin, and petrolatum. A mixture of oils is used to dispersing eosin dyes and insoluble pigments. It imparts a thin film over the lips. Various oils used in formulating lipsticks are:

- *Isopropyl myristate:* 2–3
- *Castor oil:* 30–40 % More than 50% of lipsticks manufactured in the U.S. contain substantial amounts of castor oil. It forms a tough, shiny film when it dries after application. However, ingestion of large amounts of castor oil may cause frequent rest-room visits.
- Olive oil
- Cocoa butter
- *Liquid paraffin:* 2–3%

Colour Mixtures

Lipstick gets its color from a variety of added pigments.

Colours

In early nineties, carmine was mostly used as lipstick pigment. It is obtained by extracting cochineal insects (dried) with ammonia, carminic acid in solution and is precipitated by alum. Carmine is used in combination with zinc oxide to brighter red colour. This pigment produces deep led lipstick comparable to darker modern shape of present time, but its application results in deficient opacity and intensity. Other colours shades like rosewine, bitter chocolate, brick, crème brulee, cocoa crème, espresso, mocha, red hot, garnet spray, earth, satin ruby, caramel, bridal dreams, moonberry, merlot, jewel pink, poppy, plum pop, tupe, lotus, gold dust, terracotta, nougat, orchid lustre, berrylicious, etc. Indelible lipsticks are made by addition of water soluble dye to carmine. Changeable lipsticks, orange in lipstick and staining skin red were formed by dissolving eosine in stearic acid further melted with wax and alcohol. Other bases include gelatine-glycerol and glycerol-borate.

Opal	Silver	Gold	Green	Blue
Purple	Black	Hot Pink	Coral	Red
Dorothy	Savage	Autumn Rust	Ebony	Nude
Classic Rose	Soft Cherry	Garnet	Plum Wine	Sandal Wood
Mahogany	Spice	Kiss Me	Runway	Fashionista
Couture	Red Glitter	Pink Glitter	Silver Glitter	Gold Glitter

Staining Dyes

Insoluble pigment does not stain lips. Lipsticks coloured with these pigments will rub off results in no or very little colour.

Attempts were made to produce some indelibility along the lines of grinding water soluble dyes in the wax oil base with carmine. For indelibility lips had to be moistened prior to use of stick which enable the dye to dissolve and to stain the lips.

In early nineties, Eosine as a colour for application to lips began. Eosine is tetrabromo derivative of fluoroxein. Acid eosine (Kl was bromo eosine) is orange in colour and insoluble in water, at pH about 4 changes to intense red salt. Hence an indelible stain is thereby formed. Since colour formed is as a result of combination of acid eosine with tissue material, the stain is difficult to remove and long lasting.

In some early lipsticks uses, bromo acid was formed by dissolving stearic acid, further method in mixture of wax and alcohol. Such sticks produced good stain but they dry out and crumble after aging. Later by appropriate means bromo acid is introduced into wax oil base to produce good stain which is stable to aging.

Bromo acid and related dyes are means to produce staining quality of modern lipstick. Without addition of natural pigments they produce natural and changeable lipsticks. Along with pigments these are used to impart stain to all other shade.

Tetrachlorotetrabromofluorescein is used in so called indelible type lipstick, gives strong, brilliant, bluish red stain. Rhodamine gives very blue, intense stain.

In view of the importance of an intense, lasting stain, large work has been done on method improving stain.

Pigments

Carmine was the only pigment which fulfilled requirement in earlier days but became totally inadequate for producing intense and varied colour which were demanded.

Starting with Perkin's synthesis of mauveine, synthetic dyes were made in increasing variety. Uses of dye were delayed due to difficulty of obtaining purified pigment and determining suitable harmless dye for cosmetics.

Dyes for use in lipstick should be converted into insoluble form and dye as well as precipitant or base must be devoid of harmful effects on application to lips. After world-wide processing of such pigments began, soon wide ranges of colours were available.

The red colour of modern lipstick come from iron oxide (rust) or from organic pigments. Typically, the pigment is crushed very finely while being mixed with castor oil and is then mixed with a wax base to form a finished lipstick.

In 1938, Food and drug administration set up standard for cosmetic colour. So all colour are investigated prior to approving them for cosmetic use. Firstly, the particular dye, bases and diluents which were admitted to the approved list had to be free of any toxic or irritation effect. Secondly each batch of dye pigment made by manufacturer must have very low fixed content of arsenic, lead and other impurity. Sample of each batch is sent to Food and Drug Administration for analysis. If sample meets all specification then the batch is certified by that agency as harmless and suitable for cosmetic manufacturing and is then given a certification number. Each shipment of dye or pigment made to manufacture cosmetic must be labelled with the certification number of the batch from which shipment it is taken.

Food and Drug Administration further divides colour into three-classes:

1. Food, Drug and cosmetic colours (FD&C);
2. External drug and cosmetic colour (Ext D&C)
3. Drug and cosmetic colour (D&C).

Out of these three classes only first and third class are used for lipsticks.

Ext D&C colours are used only in cosmetics applied to parts of body where they are not likely to get into the mouth.

FD&C colour incorporate 14 water soluble and four oil-soluble dyes, these are not used in ordinary lipsticks. D&C colours consist of 34 reds, 15 oranges, 4 yellow, 5 green, 7-blues, 2-violet, 1-brown and 1-black. Red and oranges are the only colour much used in lipstick and not all colours.

In addition to purity and harmfulness pigment for lipsticks must be finely divided and should not cause gritty feeling when rubbed on lips.

It must have high intensity and of high opacity. It should be free from rancidifying or drying effect of oils and texture should be smooth and soft.

In spite of certification from Food and Drug Administration, chemist must determine the suitability of each lot offered to him, on basis of physical characteristic. It's compatibility with other ingredient and desirable colour effect. Each pigment has its own peculiarity; two lipsticks look identical in stick but gives different appearance when applied on lips.

Oil soluble dyes are not useful for lipsticks as they do not mask or stain; will also form uneven excrescence on surface of the stick on aging. Oil soluble dyes were the basis of black lipstick which enjoyed a brief vogue year's age. Using these dyes without pigment, dark red stick can be made but gives comparatively lighter red colour upon application.

As skin of lips does not contain enough lipid material to be stained by oil soluble dye neither they become lodged in tiny crevices of the skin as do the insoluble pigments, so when the oil wax base is rubbed off the lips, all oil soluble colour comes off with it.

Titanium dioxide is effectively used to brighten the top note of lipstick or to obtain tints of red (pink).

Some of these dyes are:

- Bromo acid (like tetrabromo fluorescein when a product having high staining properties is desired),
- D&C Red No. 21 and related dyes
- D&C Red No. 27 and
- Insoluble dyes known as lakes, such as D&C Red No. 34, Calcium Lake, and D&C Orange No. 17.
- Pink shades are made by mixing titanium dioxide with various shades of red.

Moisturizers

In recent years, ingredients such as moisturizers, vitamin E, *aloe vera*, collagen, amino acids, and sunscreen have been added to lipstick. The extra components keep lips soft, moist, and protected from the elements.

Antioxidants

As lipstick base contains fats and oils, antioxidants are used to prevent rancidity, e.g. Parabens, Butylated Hydroxy toluene (BHT), Butylated Hydroxy Anisole (BHA). Propyl gallate.

Flavors

Flavors are added in lipstick to have good taste and mask the fatty odour of the base. Perfumes play an important role or key factor in determining consumer acceptance of lipsticks.

As the case of other lipstick ingredient, perfumes also should be selected so they do not produce any irritation effect and have disagreeable taste. Its fragrance should mask fatty odour of base or may be developed after reasonable storage and use. But it should not be so strong to clash with or overpower other perfumes that may be worn concurrently with lipstick.

Floral, lighter spice and fruit flavour type are well accepted since they are used on lips and a suggestion of an edible material might seem natural. However this idea seems to be theoretical and less practical value.

Typical example of lip sticks:

Formula I

Ethyl cellulose	4%
Ethyl alcohol	65%
Petroleum ether	25%
Glycerine	5%
Rhodamine	1%
Preservative, perfume and antioxidant	q.s.

Formula II

Ethyl cellulose	6%
Ethyl alcohol	80%
Castor oil	10%
Bromoacids	0.2%
Lake dyes	3.8%
Preservative, perfume and antioxidant	q.s.

Evaluations of Lip Gloss

Comparisons against a standard should be made by using the flow pour described above. This is a general visual observation. More critical determination of gloss can be made on a Gardeur or other similar instrument.

Abrasion resistance: The standard method of measuring resistance to abrasion is by the use of a Taber or similar abraden. On such a device a film on a panel is subject to wear by abrasive wheels and specified revolutions. Loss in weight after a given number of these revolutions is the measure of comparable resistance to abrasion.

Adhesion: Use of a Holffman scratch adhesion unit, which records the grams loading, needed to scratch a film coating from a substrate glass plate.

Flexibility: Flexibility of the film can be measured on a mandril set in accordance with ASTM Method D-1737–2. Generally, this method determines the flexibility of films. The film coating is then examined for cracks over the area of the bend, and compared against a standard.

Viscosity: The rotational instruments, such as the Brookfield Viscometer, are employed.

ROUGES AND FACIAL BLUSHES

Rouges or facial blushes for cheeks and lips were very popular among women in late nineteenth and in early twentieth century. Red pigment from hybrid saffron is employed for formation rouges. Aluminium Lake of brasileine obtained from Brazil wood is used for the same.

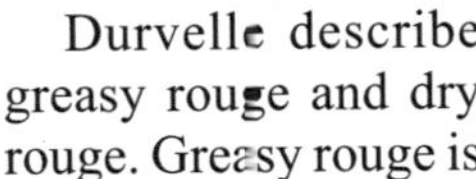

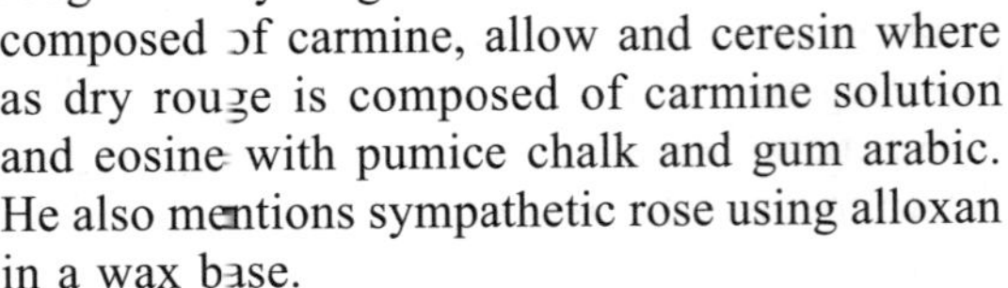

Rouges or facial blushes are designed to enhance rosy cheek color. In many cases, rosy cheeks simply indicate vasomotor instability or fine telangiectatic mats from actinic damage; however, cheek color remains fashionable.

Durvelle describe greasy rouge and dry rouge. Greasy rouge is composed of carmine, allow and ceresin where as dry rouge is composed of carmine solution and eosine with pumice chalk and gum arabic. He also mentions sympathetic rose using alloxan in a wax base.

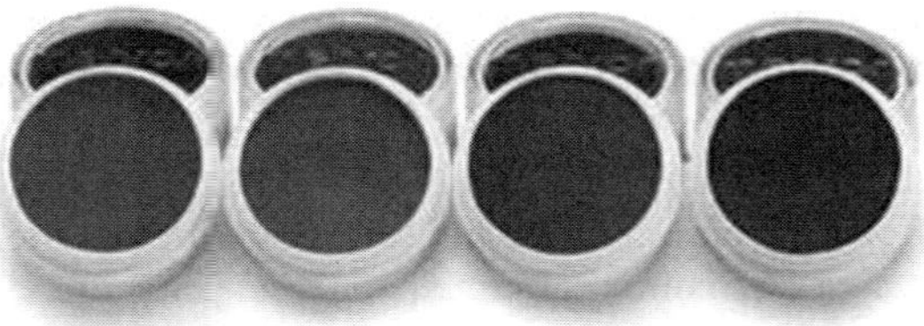

Formulation

Blush and rouge are actually synonyms for a cosmetic designed to add color to the cheeks; however, to many consumers, blush denotes a powdered product, while rouge denotes a cream product. Powdered blushes are more popular and are formulated identically to compact face powder, except more vivid pigments are added. Because color rather than coverage is desired, powdered blushes do not contain much zinc oxide. Cream rouges are formulated like anhydrous foundations that contain light esters, waxes, mineral oil, titanium dioxide, and pigments.

Application

For a natural appearance, cheek color should be applied beginning at a point directly beneath the pupil on the fleshy part of the cheek, sweeping upward beyond the lateral eye. This placement is designed to create or accentuate high cheekbones, which are a desired quality among women.

Adverse Effects

The adverse reaction concerns with blushes and rouges are identical to that for facial powders (see Adverse effects of facial powders). The products can be open or closed patch tested.

Contour Irregularity Camouflaging

The correction of irregular facial surface contours is based on the principle that dark colors make protuberances appear to recede, while light colors make surface depressions appear shallower. Creating an even-appearing surface on a scarred face is achieved through artist shading. Powdered blush-type products are best suited for this purpose. Areas of the face that need to be lightened should be brushed with a light pink or peach pearled blush or buffer. Areas of the face that need to be darkened should be brushed with a deep plum or bronze matte finish blush or highlighter.

These same principles can be used to optimize the shape of the face, the size of the forehead and chin, or the contour of the nose, based on established ideal facial proportions. The perfect facial shape is considered to be oval and symmetrical about the midline. An oval face is one and one-half times as long as it is wide and should taper gradually from its widest dimension at the forehead to its smallest dimension at the chin. The face should be divided into equal thirds from superior to inferior, as follows: forehead to the glabella, glabella to the subnasale, and subnasale to the base of the chin. The face should divide equally into fifths from ear to ear, with each fifth being the width of one eye.

A round face can be camouflaged to appear more oval by shading the lateral margins with a darker colored blush to deemphasize the increased width. An oblong face is shaded with a darkly colored blush along the forehead and chin to deemphasize the increased length. A square face is darkened bilaterally at the jaws.

This same shading technique also can be used to correct forehead and chin dimensions. To achieve the effect, low-set foreheads should have a light blush applied beneath the hairline, while high foreheads should have a dark blush applied at this location. A receding chin should have a light blush applied at the tip and sides. A double chin should be shaded with a dark blush under the entire jawbone.

The artistic application of facial cosmetics is no substitute for surgical revision, which may be recommended in certain cases.

LIP SALVES

Lip salves are devoid of colours and form an adherent, moisture resistant film on lips. Lip salves are used to protect lips from cracking, chapping and drying during winters. It is homogeneous suspension of ingredients which are petroleum or vegetable oil base. Lip salve is available in **stick** and **unctuous** (oily) forms.

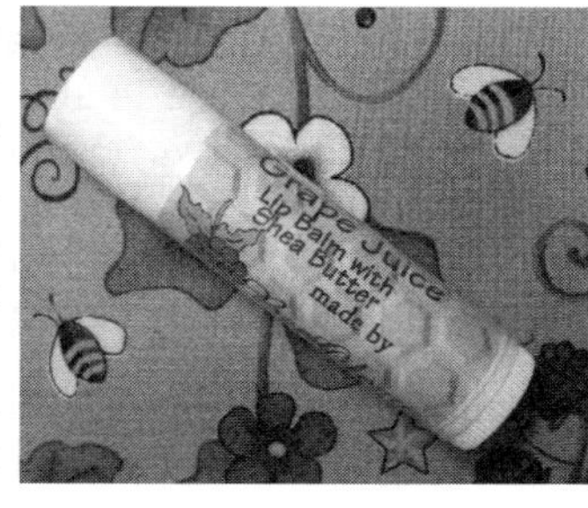

Typical example of lip salve:

Formula III

Castor oil	40%
Paraffin wax	20%
Bees wax	20%
Anhydrous lanolin	20%
Preservative, perfume and antioxidant	q.s.

Lip salves are standardized during manufacture with respect to:

- *Consistency:* Consistency may be determined with the help of penetratometer.
- *Melting range:* Lip salve of stick form has melting range of 45°–60° C and unctuous form 35°–50° C and it can be found out with simple thermometer.
- *Rancidity:* Lip salve contains vegetable oils as a base hence it is liable to rancid. Rancidity test can be determined by shaking 10.0gm of conc. HCl and 10.0 ml of phloroglucinol solution. If it is rancid then the solution will develop pink colour.

LIP JELLY

A liquid lipstick is prepared by perfuming and tinting castor oil with colour and prepared with petroleum jelly, lanolin, etc. A simple liquid lip gloss is prepared by perfuming and tinting castor oil with colour.

Formula IV

Petroleum jelly	45.0 %
Anhydrous lanolin	55.0%
Perfume and color	q.s.

MANUFACTURE

Colour-Grinding and Dispersion

First step in lipstick manufacture is dispersion of colour in base or oil to obtain smooth texture throughout mass. The pigments used are supplied as power with small particle size. Many times these particles form aggregates of various size and many times these are difficult to break up so that single particles can be dispersed in base. While preparation of dry pigment during process of precipitation filtration, drying and grinding, particles are firmly connected together. For preparation of smooth lipstick these particle should be separated and should be surrounded by oil.

To determine, the firmer of the pigment suspension obtained will depend upon nature of oil used for wetting and dispersion of pigment and also on mechanical work done on mixture penetration and wetting of agglomerate pigment will be more for liquid with low viscosity and low surface tension such as butyl sulphate in comparison to heavier liquid as castor oil.

So thrive liquid are must for wetting but heavier liquid are also recommended to reduce setting.

Grinding operation are performed not to reduce particle size but are performed to break up agglomerated. To break up agglomerated, an efficient rubbing or shearing force of specific magnitude is required which is normally supplied by roller or colloidal mill.

In roller mill, the pigment in oil suspension is passed between cylinder at different speeds and the clearance is being too small to allow agglomerates to pass unbroken.

In colloidal mill, the mixture is forced between two closely spaced plates or cones, one of which rotates at high speed.

Mixing

For manufacturing of lipstick, the equipment should be simple enough. Basic equipment consists of:

Jacketed Kettle constructed by inert material, which can be made of aluminium, stainless steel, tinned copper or glass lined steel. It can be water or steam jacketed with water heating by electricity or gas.

Overheating should not be there. Kettle with hemispherical base is easy to empty and clean.

In mixing a batch, slow agitation first sufficient to ensure thorough blending is preferable in order to present introduction of air. Rapid stirring should be avoided.

After mixture is melted or blended, perfume is mixed in and resulting pomade is strained through a fine mesh screen into storage container or moulding reservoir. Lipstick pomade should be stored in air tight container of inert material in dark room at low temperature for long duration storage time.

For short duration storage shortcut may be justified such as casting the pomade in baking pans and wrapping the resulting blocks in paper for shelf storage.

Moulding

For moulding, the pomade is kept hot in a small jacket kettle provided with a slow speed agitation and a bottom outlet. To prevent setting of pigments agitator blade should clear size and bottom by small distance

Pomade is removed from storage container with attention to cutting from top to bottom if entire content is not used at one time. In molten condition in moulding kettle it should stand 20–30 minutes with slow agitation to allow rising to the surface. Stirring should be continued throughout operation.

It is important that pomade should be free from air bubble before moulding. If the pomade cannot otherwise be freed of air it may be subjected to heating under vacuum in order to remove tightly held air.

Due care must be taken to avoid this source of unsightly holes in surface of lipsticks because every time the pomade is poured or agitated it is possible to introduces air bubbles

Usually moulds are formed of strips of aluminium or brass. Cavity shaped like the lipsticks to be moulded are bored in rows centering along the function of each two faces, so that each stripe of metal contains a row of half cavities on each face. Generally moulds are made with 5–7 strips of metal containing a total of 60–144 cavities. The top is milled down, excluding the edges and ends to form a reservoir to hold surplus lipstick mass above the cavities. To avoid splash, there should be little vertical spacing between mould and faucet. The molten pomade is run into the seat of mould not directly into cavity at a uniform rate which is slow enough to avoid splash but enough to prevent cooling before last mould fills. Excess is run into the space over the cavities to develop central hole or well due to shrinkage incident upon congealing. Then mould in stand without motion until surplus pomade on top has congealed, after scraping off the surplus, the mould is set on a chilled plate of metal which rapidly withdraws heat.

Flaming

Sticks are inserted in plastic container which is available in a variety of types. The free end of sticks is reheated for a very short time so the surface becomes smooth and glossy. This process is done by passing the lipstick through gas flame.

ALLERGIC REACTIONS CAUSED BY LIPSTICK

The incident of cheilitis or dermatitis of the lips, attributable to the use of lipstick is rare percentage wise in comparison to number of users. Each manufacturer should be aware in avoiding the use of any ingredient which might provoke allergic manifestations in more than the smallest possible number of subjects. Lip are sensitive than any other body parts as they lack horny layer of cells.

Ingredients like perfume oils usually have complex composition. So, to make sure they are appropriate to use is only by test. Though patch test is useful in preliminary testing an actual test should be made before perfume is adopted for large scale distribution.

Chemically active odorants such as aldehyde, ketones and unsaturated compound in general are expected to be more likely to arouse allergic action or even genuine irritation in comparison to more inert saturated alcohol and esters. This does not mean that the former class is entirely avoided. Some compound of that class can be used for many years with complete satisfaction.

Colours for lipsticks which are commonly used in lipsticks are derived from about 20 days. So problem is much simpler than that of perfumes. Further the colours in use have all been certified by Food and Drug Administration as suitable for cosmetic use. Before admitting any colour to the approved list FDA gives it a rather sever test to prove it free of toxic or irritation quality.

Few cases of dermatitis has been traced out to individual dyes but it impossible that some cases attributed to dyes have actually been caused by the presence of intermediate as impurity.

Betanapthol and resorcinol have both irritating and sensitising qualities and are used as intermediates in preparation of some of certified colours. Low limit has been set by FDA for such intermediates permitted in certified colour, a precaution which should further decrease the incidence of cosmetic dermatitis.

Interesting sidelight on strictness of FDA standard for colour is provided by comparison of arsenic tolerance for colour with those for fruit.

The limit for Arsenic oxide (As_2O_3) in colour is 0.0002% and that for fruit is 0.0003%.

Average lipstick contains about 0.4 gm of colour so a person would have to eat 200 lipsticks to adjust the amount of arsenic allowed on one 2-ounce apple.

Bromo acid has come in for more attention that it probably deserves in connection with cheilitis. Many cases have been reported by bromoacid, other have been loosely described as probably due to bromo. Presence of untreated resorcinol may have caused some difficulty before standard of purity reached their present excellence. At any rate tens of million of lipsticks containing bromo acid that are sold annually attest to the extremely low level of its allergenic effect.

Bromo acid along with other dyes used in lipsticks has photosensitizing properties. So person ordinarily not allergic to given lipstick may experience allergic reaction to it under the effect of strong sunlight.

Recent introduction of high staining or inducible type increased the complaints of dryness of lips after use of lipsticks.

Factors responsible for it are:

1. Larger amount of staining dye are used.
2. D&C Red No. 27 is used more.
3. Thin oil used as stain solvent.
4. More perfume of many type are used to make solvent odour.
5. Blotting the lips is advocated in order to prevent smear.

These factor singly or in combination result in dryness of lips. Care should be taken to remove these effects.

EVALUATIONS OF LIPSTICKS AND ROUGES

Quality control procedures are strict, since the product must meet Food and Drug Administration (FDA) standards. Lipstick is the only cosmetic ingested, and because of this strict controls on ingredients, as well as the manufacturing processes, are imposed. Lipstick is mixed and processed in a controlled environment so it will be free of contamination. Incoming material is tested to ensure that it meets required specifications. Samples of every batch produced are saved and stored at room temperature for the life of the product (and often beyond that) to maintain a control on the batch.

As noted above, appearance of lipstick as a final product is very important. For this reason everyone involved in the manufacture becomes an inspector, and non-standard product is either reworked or scrapped. Final inspection of every tube is performed by the consumer, and if not satisfactory, will be rejected at the retail level. Since the retailer and manufacturer are often times not the same, quality problems at the consumer level have a major impact on the manufacturer.

Color control of lipstick is critical, and one only has to see the range of colors available from a manufacturer to be aware of this. The dispersion of the pigment is checked stringently when a new batch is manufactured, and the color must be carefully controlled when the lipstick mass is reheated. The color of the lipstick mass will bleed over time, and each time a batch is reheated, the color may be altered. Colorimetric equipment is used to provide some numerical way to control the shades of lipstick. This equipment gives a numerical reading of the shade, when mixed, so it can identically match previous batches. Matching of reheated batches is done visually, so careful time and environment controls are placed on lipstick mass when it is not immediately used.

There are two special tests for lipstick:

- The Heat Test and
- The Rupture Test.

In the Heat Test, the lipstick is placed in the extended position in a holder and left in a constant temperature oven of over 130° F (54° C) for 24 hours. There should be no drooping or distortion of the lipstick.

In the Rupture Test, the lipstick is placed in two holders, in the extended position. Weight is added to the holder on the lipstick portion at 30-second intervals until the lipstick ruptures. The pressure required to rupture the lipstick is then checked against the manufacturer's standards. Since there are no industry standards for these tests, each manufacturer sets its own parameters.

In manufacturing lipsticks, following in-process checks can be performed:

- Colour match
- Texture
- Softening point

Lipsticks are evaluated by means of following tests:

1. Color matching test
2. Breaking point test
3. Softening point test
4. Droop point test
5. Test for penetrability
6. Test for force of application
7. Perfume stability test
8. Stability test
9. Microbiological examination.

Colour matching test: Every time, for a particular shade, the same shade is produced as that of standard shade. Therefore, color match should be done on the skin. The color matching should also be done on the skin. The other important criterion is texture. Texture should not be gritty, the product should be smooth.

Breaking point test: This determines the strength of lipstick. To carry out this test, the lipstick is held horizontally in a socket fitting over about ½ inch of its base and weight is applied at measured distance from the edge of support. The weight applied is increased gradually every 30 seconds by predetermined increment (say 10 gms) until the lipstick breaks. On a given lot, at least four readings should be made. The test should be made at a given temperature (25° C or 30° C).

Softening point test: The softening point is an important test as it can give indication whether lipsticks will be able to withstand variation in climate or not. Since India is tropical country and temperature in some regions goes as high as 48° C, the softening point of lipsticks should be sufficiently high so that the lipstick can withstand such high temperatures.

Droop point test: The temperature at which the lipsticks started oozing out oil or flatten from within the case is known as droop point or yield point. Droop point should be above 50° C for safe handling and storage.

Test for penetrability: This test indicates the rheological property of the lipstick with the help of penetrometer. A needle of specific diameter is allowed to penetrate the lipstick, and depth of penetration is noted.

Test for force of application: Lipstick is applied on the piece of the paper (kept on a balance) at an angle of 45° C, and the force required for applying is read from the balance.

Perfume stability test: Perfume stability can also be assessed by storing lipsticks in oven at 40° C and by making periodic comparison of perfume with fresh lipsticks.

Stability test: The stability of lipsticks can be determined by means of accelerated stability test in which lipstick is kept at various temperatures (say 0°, 45°, 70° C) and assessing the lipstick for surface defects, perfume, colour and application characteristics.

Ageing stability may be determined by storing the lipsticks in oven at 40 °C and by making periodic observation of:

- Application characteristics
- Crystallization of wax on surface
- Oil bleeds.

Microbiological examination: Lipstick should not contain more than 100 microorganisms per gram; raw materials should also be subjected to microbiological examination.

Lipstick is the least expensive and most popular cosmetic in the world today. There are no accurate figures for current sales of lip balm, since the market is expanding. Manufacturers continue to introduce new types and shades of lipstick, and there is a tremendous variety of product available at moderate cost. As long as cosmetics remain in fashion (and there is no indication that they will not) the market for lipstick will continue to be strong, adding markets in other countries as well as diversifying currently identified markets.

EVALUATION TESTS FOR ROUGES

Melting point test: It is determined for cream rouges by capillary tube method and it should be below 50° C for good application and storage point of view.

Colour dispersion: This test is done under microscope. Colour particles above size 50μ, may cause agglomeration so particles should be below this size range for better rouge.

Sedimentation characteristics: This property is important for liquid rouges. Viscosity should be adjusted in such a manner that has low sedimentation.

Accelerated stability study: For liquid or cream rouge it is important that it is compatible with the container in which it is packed. This can be studied at higher temperature to predict its normal shelf life.

Aging stability: This is performed by keeping product at 40° C for one month and compare with fresh one after one month, discarded if it shows difference.

Available Marketed Products

- Bloom Lipstick
- Candy Colour (By Colour Bar)
- Lakme (Nine To Five Lipcolor) and Colour Bar
- Lancôme Lipstick
- Lip Stain or Liquid Lip Color (Lip-ink International Company)
- May Be Line Water Shine Lipsticks
- Mylipmiracle (Lipcolor by Avon Beauty Products)
- New Chic Lipsticks
- Princessa Lipstick
- Revlon Lipstick
- Sassy Lipstick
- Sephora Lipstick
- Sumptuous Lipstick
- Tangee Lipstick
- Blue and Blues
- Lip Balm by Forest Essentials
- Lip Liner in Plum And Mocha
- Lip Pencil (Nina Ricci)
- Lip Spa Balm (Oriflame)
- Rouge Sensation (Piere Cardin)
- Shinepops (Elizabath Arden)
- Lumieres Magiques De Chanel Rouge
- Perfect Glows (Colour Bar)
- Loreal Colour Juice Lip Gloss
- Mac Carbon Pencil
- Maybeline Sweet Nothings Lip Colour.

Sunscreen Preparations

Trust not too much to an enchanting face.

— Virgil

INTRODUCTION

Sunlight is vital for all living beings; it is required for formation of vitamin D, circulation of blood, formation of haemoglobin, etc. exposure of sun light on the human body may exhibit beneficial and harmful effects depending upon length and frequency of exposure.

The purpose of sunscreen **(anti-sunburn and suntan)** preparations is to assist the skin painful effects and the purpose of the anti burn preparations is to minimize the harmful effects of sunburn. The materials which are used for the above purpose are known as sun tanning agents and sun burn preventive agents respectively. In combination these are known as **sunscreens** (also known as **sunblock, suntan lotion**).

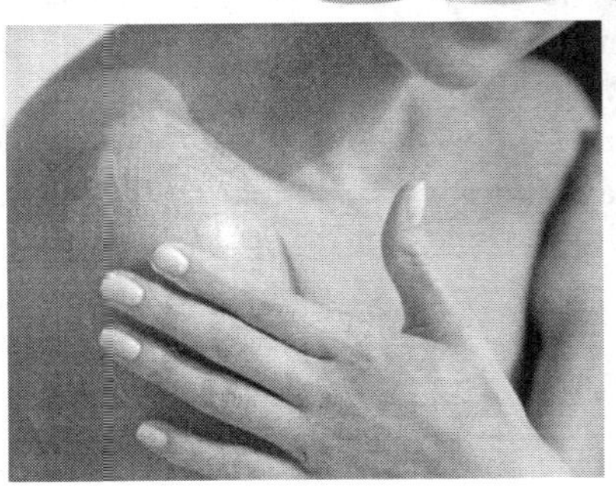

Sunrays reaching to the surface of the earth contain **visible rays (wavelength between 400–760 nm), ultraviolet rays** with shorter wavelength **(between 220–400 nm)** and **infrared rays** with longer wavelength **(between 760–5100 nm)**. Lower the wave length more the energy of the rays. Ultraviolet rays particularly below 320nm are responsible for most of the therapeutic as well as most of the detrimental effects of sun rays on human body that depends on length and frequency of exposure, intensity of the sunlight and individual sensitivity also concerned.

The best sunscreens protect against both **UVB** (ultraviolet radiation with **wavelength between 290 and 320 nm**), which can cause sunburn, and **UVA** (**between 320 and 400 nm**), which damages the skin with more long-term effects, such as premature skin aging and skin cancer. Most sunscreens work by containing either an organic chemical compound that absorbs ultraviolet light (such as oxybenzone) or an opaque material that reflects light (such as titanium dioxide, zinc oxide), or a combination of both. Typically, absorptive materials are referred to as chemical blocks, whereas opaque materials are mineral or physical blocks.

Prior to the 1920's, skin color was an important indicator of one's social status in European and eastern Asian countries. Pale skin was a widely accepted indicator of upper-class status. The first important link between skin cancer and exposure to the sun was made in 1918 by Norman Paul of Sydney, Australia. He outlined the dangers of extended exposure to the sun. Discussing the management of *xeroderma pigmentosum*, Paul wrote, "Protection from sunlight is essential and can be aided by the wearing of red or brown veils or the application of similarly coloured ointments". Although by this time the practice of applying "zinc cream" to the nose, lips, and cheeks for outdoor activities was becoming more commonplace, it was done so for the prevention of sunburn. It was not until the 1920's that suntanned skin emerged as a fashionable trend. Thought to be serendipitous, a well known designer named Coco Chanel developed a suntan during a cruise from Paris to Cannes aboard a yacht owned by the Duke of Westminster. The fashion world was inadvertently given a new trend. Adding to the popularity of a darkened complexion at the time, a caramel-complexioned American born singer named Josephine Baker was becoming increasingly popular in Paris.

In the 1930's, a south Australian chemist by the name of H.A. Milton Blake formulated a protective agent containing 10% salol (phenyl salicylate). By 1935, protective lotions containing quinine oleate and quinine bisulfate had appeared in the United States. The first commercially available sunscreen product was introduced by L'Oreal founder E. Schueller in 1936. During World War II (1939-1945) large numbers of US servicemen in the Pacific theater were exposed to the adverse effects of the tropical sun. Severe sunburn imposed a serious problem. At the time, the military used red petrolatum as a sunscreen.

In the 1940's, dermatologists began prescribing 2-5% p-aminobenzoic acid (PABA) in aqueous cream or in 70% alcohol. Patients were instructed to rub the product into the skin well, because it binds to protein and will remain in the *stratum corneum* for hours. Most of these early sunscreens were directed toward UVB. The first commercial self-tanning product, Man-tan®, was introduced to the market during the fifties. Deeply tanned skin continued to gain popularity as a status symbol and suntanning was recognized as a cosmetic attribute.

In 1972, the FDA re-classified sunscreens from cosmetics to over-the counter (OTC) drugs, and applied more stringent labeling requirements. Johnson & Johnson, Inc. developed the first waterproof sunscreen in 1977, with Coppertone, inc. quickly following with a product containing polyanhydride resin, PA-18, as the agent imparting water resistance. In 1978 a Federal Register published the established guidelines for the formulation and evaluation of sunscreens marketed in the United States. These guidelines were reevaluated in 1988 and further revised in 1993 and in 1999. By the late 1970's, the FDA declared sunscreens as safe and effective in helping to prevent skin cancer, alleviate premature aging of the skin, and prevent sunburn. **The sun protective factor (SPF)** numbering system (2-15) was also introduced. The cosmetic industry began to add sunscreens to various skin care cosmetics, carefully avoiding claims that would place the products clearly in the drug category Coppertone developed the first UVA/UVB sunscreen. Sunscreens continued to be incorporated into more and more cosmetic products. Due to the growing incidence of skin cancer, the American

Academy of Dermatology (AAD) became the first medical society to start a public education campaign on skin cancer prevention.

Beginning in the 1990's, the sunscreen industry offered products that provided protection against UVA and UVB radiation. Sunscreens were incorporated into an even greater range of consumer products, including daily-use cosmetics. Foundation make-up with sunscreen offered full-spectrum UVA protection, achieved through high pigment content and inorganic particulates. Coppertone premiered a new sun care specialty line for outdoor sports enthusiasts called Coppertone Sport Ultra Sweatproof Dry Lotion[R]. The sport-type sunscreen products were designed to hold the sunscreen actives in place to prevent spreading. Shade UVA Guard Lotion[R], SPF 15, was introduced in 1993, delivering the broadest UV protection available in the U.S. It was also the first PABA-free sunscreen formulation in the U.S. to use the UVA blocker Parsol 1789. On May 12th, 1993 the Tentative Final Sunscreen Monograph (TFM) was published, outlawing claims such as those relating to aging and wrinkling. On May 21, 1999, the FDA published its Final Sunscreen Monograph. The FDA proposed SPF 30 as the upper limit for SPF labeling. Products with SPF values over 30 may be labeled as "**30 plus**" or "30+". Permissible labeling claims are limited to the prevention of sunburn.

Sun tanning agents are those sunscreens which absorb a minimum of 85% ultra-violet radiations of the wave-lengths of 290–320 nm, but which transmit ultra-violet radiations of wave-lengths of longer than 320 nm and produce a light transient tan. The sun tanning depends on one's capacity to produce pigment, melanin. There are three types of skin tanning:

- *Immediate tanning:* It is elicited by radiations of wavelength of 300–660 nm at one to ten hours of sun exposure and effects fade within 2–3 hours of exposure.
- *Delayed tanning:* It is stimulated by radiations of wavelengths of 290–320 nm at long exposure and fade within 100–200 hours.
- *True tanning:* It is also known as melanogenesis and starts about two days after exposure and reaches maximum in about 2–3 weeks.

Sunburn and suntan preparations are classified as: sunscreen preparation, palliative preparations and simulative preparations. **Sunscreens preparations** absorbs erythemal portion of the sun's radiant energy and scatter the incident light effectively. Opaque powders such as zinc oxide and titanium dioxide, either in dry state or in a vehicle, will serve to scatter the ultraviolet light effectively falling upon it.

Palliative preparations are used for the relief of the irritation and other problems causing from sunburn. This generally contain antiseptic because in sunburn bacterial infection chances are higher than steam burn case, and because damage of skin cells are involved in sunburn.

Simulative preparations are artificial suntan preparations. The purpose is enhanced colour to prevent the skin damage by absorption of the erythmal radiations. An artificial suntan is obtained by staining of the skin. Several natural materials like henna, walnut juice, olive oil extracts, etc. now mostly synthetic staining materials are used.

Sunburn preventive agents are those sunscreens which absorb more than 95% or more of ultra-violet radiations of the wavelengths of 290–320 nm. There is another type of sunburn preventive agents which scatter the sunlight. These include titanium dioxide, kaolin, talc, zinc oxide, calcium carbonate and magnesium oxide.

There are two factors which are responsible for natural protection of skin against sunburn.

a. Thickness of the *stratum corneum.*
b. *Pigmentation of the skin:* The increase in melanin content of the epidermis increases the protection of skin.

SYMPTOMS OF SUNBURN

- Damage or destruction of cells in the prickle cell layer of epidermis
- Histamine is released
- Dilation of blood vessels
- Erythema (redness of skin, caused by hyperaemia of the blood vessels near the surface)
- Swelling of skin and proliferation of basal cells.

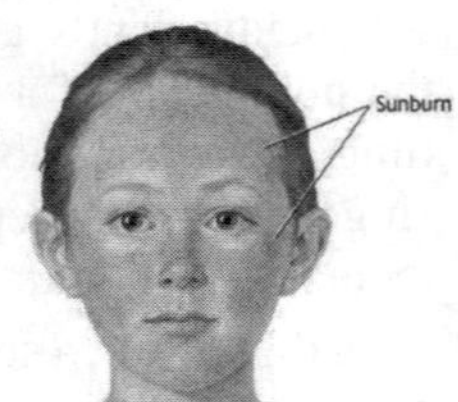

CLASSIFICATION OF SUNBURNS

It depends on degrees of burn by sun exposure:
Degree 1: Minimum erythema
Degree 2: Vivid erythema
Degree 3: Painful burn
Degree 4: Blistering burn.

Ideal sun screening agent should have following characteristics:

- It should absorb erythmogenic radiations in the range of 290–320 nm without its break down.
- It should be resistant to water and perspiration
- It should allow full transmission of radiations in the range of 300–400 nm for tanning effect.
- It should be non-volatile.
- It should have suitable solubility characteristics in suitable vehicle
- It should be stable to heat, light and perspiration
- It should be odourless, or mild odour acceptable to user.
- It should be non-toxic, non-irritant and non-sensitizing.
- It should be capable of retaining its sun screening property for several hours.
- It should not stain on body and cloths.
- It should be neutral.

LIST OF SUNSCREEN AGENTS

- Anthranilates (o-aminobenzoates, linalyl, terpenyl and cyclohexenyl esters);
- Azoles (2-acetyl-3-bromoindazole, methyl naphthoxazole, phenyl benzoxazole, various aryl benzo-thiazole);
- Cinnamic acid derivatives (phenyl cinnamonitrile, butyl cinnamoyl pyruvate);
- Coumarin derivatives
- Diazoles and triazoles
- Dibenzalacetone and benzal-acetophenone;
- Dihydroxy cinnamic acid derivatives (umbelliferone, methyl umbelliferone, methylaceto-umbelliferone);
- Dihydroxy-naphthoic acid and its salts;
- Hydrocarbons (diphenyl butadiene, stilbene);
- Hydroquinone;
- Naphthosulphonates (sodium salts of 2-naphthol-3,6-disulphonic and of 2-naphthol-6, 8-disulphonic acids);
- p-aminobenzoic acid and derivatives;
- Quinine salts;
- Quinoline derivatives (8-hydroxy quinoline salts, 2-phenylquinoline);
- Salicylates;
- Tannic acid and its derivatives (e.g. hexaethyl ether);
- Uric and violuric acids.

SUN PROTECTION FACTOR (SPF)

This term is defined by Plough Corporation to determine the relative effectiveness of sunscreen agents to protect the skin. USFDA also recommended SPF as a means of numerically identifying the effects of products and guide customers about suitability of their type of skin. This is the ratio between UV exposure required to produce a minimally perceptible erythema on protected skin and the exposure that will produce the same erythema for unprotected skin. The larger the SPF the greater is the sun protection. SPF rating numbers are in the range of 2–8. Sensitive individuals should use products of ultra protection category, i.e. SPF of 15 or more, e.g. sweating, swimming, playing, humidity conditions, etc.

SPF value = Minimum erythemal dose for protected skin (DPS)/Minimum erythemal dose for unprotected skin (DUS)

Dose required for DPS is 2 mg/cm^2 or 2 µg/cm^2, no dose required for DUS

SUNTAN LOTION

Formula I

Glyceryl p-aminobenzoate	3.0%
Propylylene glycol ricinolate	8.0%
Glycerin	10.0%
Alcohol	60.0%
Purified water to make	100.0%
Perfume and color	q.s.

EVALUATIONS OF SUNSCREEN PREPARATIONS

Spectrophotometric Evaluation

This is to evaluate the UV radiation absorption ability of the sunscreen preparations with UV

Table 8.1: Skin types: based on first 30–60 min sun exposure after winter season or no sun exposure

Skin types	*Sunburn and tanning history*	*Examples*	*SPF*
I (sensitive)	Always burns easily, never tans	People with fair skin, blue eyes, freckles, unexposed skin is white	8 or more
II (sensitive)	Always burns easily, tans minimally	People with fair skin, red or blonde hair, hazel, blue or even brown eyes, unexposed skin is white	6–7
III (normal)	Burns moderately, tans gradually and uniformly (light brown)	Normal average white person, unexposed skin is white	4–5
IV (normal)	Burns minimally, always tans well (moderate brown)	People with white or light brown skin, dark hair, dark eyes (e.g. Mediterranean, Mongoloid, Orientals Hispanics), unexposed skin is white or light brown	2–3
V (insensitive)	Rarely burns, tans Profusely (dark brown)	Brown skinned persons (e.g. American Indians, East Indians, Hispanics), unexposed skin is brown	2
VI (insensitive)	Never burns, deeply pigmented	Blacks (e.g. African and American negroes, Australian and south Indian aborigines), unexposed skin is black.	Not indicated

spectrophotometer. Other parameters can be calculated like concentration of the substance in the preparation, molar extinction coefficient or absorbency and compared with other substances.

The intensity (I) of radiation of a given wavelength is transmitted through a layer of absorbing material of thickness (t) and concentration (c) depends upon the intensity of incident radiation (I_0). This relationship is known as the Bouguer-Beer's Law, is given by the equation:

$$I = I_C v^{-kte} \qquad ...(1)$$

Where k is a constant which depends upon the intensity of absorbing material. In logarithmic form, the equation may be written as:

$$\ln I^{-}/I_0 = -kte \qquad ...(2)$$

Where e is the optical density. The optical density of a film 1 cm thick ($t = 1$) and containing 1g more of absorber per 1000ml ($c = 1$) is the standard value of the absorption coefficient k.

The ratio I^{-}/I_0 give the fraction of incident relation transmitted through the film of absorber. Equation (3) is more conveniently used with Briggsian logarithms, and becomes-

$$2.203\, e = \log I_0/I = (2.203k)\, tc$$

$$A = \log I_0/I = \text{etc.} \qquad ...(3)$$

Where A is the molar absorbency and E is the molar extinction coefficient. The molar extinction coefficient is a characteristic, intensive property of a pure compound and may be used to identify an unknown screen in a suntan preparation.

Erythemal Dosage

Primary interest of the cosmetic chemist is the estimation of erythemally effective radiation, or E-vitons/cm^2, transmitted by a suntan preparation. The erythemal energy is the product of the solar energy transmitted through the film I_3 and the effective factor at that wavelength. This can be computed by the relationship:

$$\text{uW/cm}^2 = I_s = I_{s0} \times I/I_0$$

The quantity I_{s0} is a fraction of the wavelength, which is given in the last column of the table. The summation of the uW/cm^2 for 50 Å wide bands centered from 2925 to 3375 Å gives the erythemal energy transmitted by the preparation.

$$\text{E-vitons /cm}^2 = 0.1\ (\text{uW/cm}^2)_2 = 0.1$$

And the erythemal dosage for a given solar exposure would be given by:

$$\text{E-vitons /cm}^2 \times \text{sec exposure}$$

The physiological effects that should be observed with such a dosage, provides a first approximation of the effectiveness of the screen. However, it must be remembered that many other factors enter into the operation of the screen; stability, film thickness, film continuity, percutaneous absorption, and the effects of the skin milieu on the transmittance of the screen all can contribute to the variations in the actual behavior of the preparation. For comparative purposes, the term provides a preliminary figure of merit for different sunscreens or suntan preparations, the highest protection is afforded by the preparation with a minimum.

The Sunscreen Index

Kumler proposed a method for evaluating the relative screening activity of sunscreen compounds. He proposes to compute $E_{0.1}^{0}$ cm at 3080 Å. to obtain a figure of merit he calls the sunscreen index or S.I. This particular wavelength was chosen as the peak of the sunburn curve, and the S.I. at this wavelength is taken to indicate the total efficacy of the screens.

Dose of Sunscreen Preparation

Dosing for sunscreen can be calculated using the formula for body surface area and subsequently subtracting the area covered by clothing that provides effective UV protection. The dose used in FDA sunscreen testing is 2 mg/cm². From a sample calculation in a FDA monograph, if one assumes an "average" adult build of height 5 ft 4 in (163 cm) and weight 150 lb (68 kg) with a 32 in (82 cm) waist, that adult wearing a bathing suit covering the groin area should apply 29 g (approximately 1 oz) evenly to the uncovered body area. Considering only the face, this translates to about 1/4 to 1/3 of a teaspoon for the average adult face.

Sensitivity Test

This test is performed directly on the rabbit's backside or abdomen. Suntan preparations are directly applied on these sites because these are very sensitive sites and these are kept for exposure to radiation along with control unprotected site for specific period of time. The effects are observed at the end of the period. Several factors or variables like radiation source, site of the test field etc. to be taken care during the test as they may influence the results.

Available Marketed Products

- Sun Defiance for Max Sun Defence (Fair & Lovely)
- Protect & Lighten Sun defence moisturiser.
- Garnier Synergie Sun Control
- Lakme Sun Expert Sun Screen Souffle'.
- Lotus Sun Screen Gel.
- H_2O + Body Soother Solar Relief Gel
- Elovera SPF
- Copper Tone
- Ever Youth Sun Screen (Cadila Health Care)
- Perfect Radiance Safe Sun (Lotous)
- Terra a Bella Sun Powder (Christin Dior)
- Nycil Prickly Heat Power
- Shower to Shower
- Olay Total Effects Sun Block Mosturizer

Antiperspirants and Deodorants

> *Beauty always promises, but never gives anything.*
> — Weil Simone

DEODORANTS

Everybody wants to get rid of the objectionable body odour. This problem is caused by perspiration or sweat or sudor that is deposited slowly on skin surface, which may become malodorous in a period of five to six hours after bath. Perspiration is a phenomenon which the body has for regulation of body temperature and protection of skin from dryness. This problem cannot be eliminated by taking frequent baths. Perspiration takes place with help of sweating glands or sudoriferous glands, i.e. eccrine (occur over almost all body surfaces) and apocrine glands (occur over areas like armpits, pelvic region, around nipples etc.).

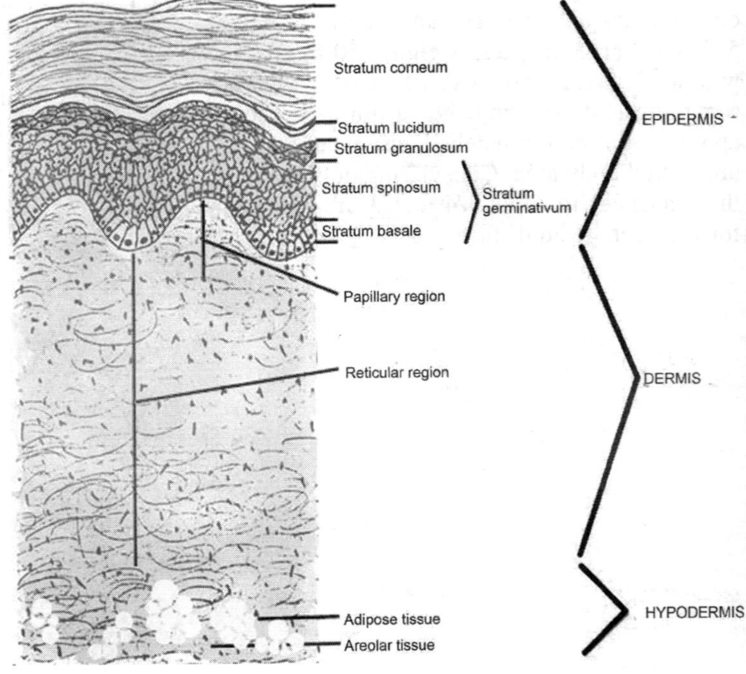

Eccrine sweat glands are distributed over the entire body surface but are particularly abundant on the palms of hands, soles of feet, and on the forehead. These produce sweat that is composed chiefly of water with various salts. These glands are used for body temperature regulation.

Eccrine sweat glands are coiled tubular glands derived from the outer layer of skin but extending into the inner layer. They are distributed over almost the entire surface of the body in humans and many other species, but are lacking in some marine and fur-bearing species. The sweat glands are controlled by sympathetic cholinergic nerves which are controlled by a centre in the hypothalamus. The hypothalamus senses core temperature directly, and also has input from temperature receptors in the skin and modifies the sweat output, along with other thermoregulatory processes.

Human eccrine sweat is composed chiefly of water with various salts and organic compounds in solution. It contains minute amounts of fatty materials, urea, and other wastes. The concentration of sodium varies from 35–65 mmol/l and is lower in people acclimatized to a hot environment. The sweat of other species generally differs in composition.

Apocrine sweat glands develop during the early to mid puberty ages approximately around the age of 15 and release more than normal amounts of sweat for approximately a month and subsequently regulate and release normal amounts of sweat after a certain period of time.

These glands produce sweat that contains fatty materials. Mainly present in the armpits and around the genital area, their activity is the main cause of sweat odor, due to the bacteria that break down the organic compounds in the sweat.

Emotional stress increases the production of sweat from the apocrine glands, or more precisely: the sweat already present in the tubule is squeezed out. Apocrine sweat glands essentially serve as scent glands.

Note that the name apocrine sweat gland is archaic; these glands are no longer believed to secrete their products by an apocrine mechanism in which the apical portion of the cell is sloughed off with secretory products inside. Rather, the apocrine sweat glands secrete in an eccrine fashion: membrane-bound vesicles bind to the plasma membrane of secretory cells and release products by exocytosis with no net loss of the plasma membrane. These glands are still called apocrine sweat glands to distinguish them from the eccrine sweat glands.

Deodorants are cosmetic substances applied to the body, most frequently the armpits, to reduce the body odor caused by the bacterial breakdown of perspiration. A subgroup of deodorants is **"antiperspirants"**, which prevent odor and reduce sweat produced by parts of the body. Antiperspirants are typically applied to the underarms, while deodorants can also be used on feet and other areas in the form of body sprays.

Human sweat itself is largely odorless until it is fermented by bacteria that thrive in hot, humid environments such as the human underarm. The

armpits are among the consistently warmest areas on the surface of the human body, and sweat glands provide moisture. Underarm hair adds to the odor by providing increased surface area on which these bacteria thrive. Body odor is controlled by reducing moisture, killing bacteria or over powering the bacteria's smell with perfume.

Bromhidrosis or bad body odour is a very common problem both in male and female. Bromhidrosis or body odor (also called bromidrosis, osmidrosis and ozochrotia) is the smell of bacteria growing on the body. These bacteria multiply considerably in the presence of sweat, but sweat itself is almost totally odorless. Body odor is associated with the hair, feet, crotch (upper medial thigh), and anus, skin in general, breasts, armpits, genitals, pubic hair, and mouth.

Body odor is specific to the individual, and can be used to identify people, though this is more often done by dogs than by humans. An individual's bodily odor is also influenced by diet, gender, genetics, health, medication, occupation, and mood.

Sweating (also called perspiration or sometimes transpiration) is the production and evaporation of a watery fluid, consisting mainly of sodium chloride in solution that is excreted by the sweat glands in the skin of mammals. Sweat also contains the chemicals or odorants 2-methylphenol (o-cresol) and 4-methylphenol (p-cresol). In humans, sweating is primarily a means of temperature regulation. Evaporation of sweat from the skin surface has a cooling effect due to the latent heat of evaporation of water. Hence, in hot weather, or when the individual's muscles heat up due to exertion, more sweat is produced. Sweating is increased by nervousness and nausea and decreased by cold.

COMPOSITION OF SWEAT OR PERSPIRATION

- Several fatty substances like lower fatty acids (e.g. lactic acid from muscles)
- Steroids
- Lactones
- Residual bacterial substances (responsible for malodorous or foul smell)
- Moisture

To correct body odour, there are several methods

1. Use of deodorants,
2. Use of antiperspirants (these are rather drugs than cosmetics to check bacterial growth),
3. Bathing timely to remove secretions of both types of sweat glands.
4. Use of sprays, deosprays and perfumes to absorb body odours.

Two types of deodorants:

1. To deodorize perspiration without restricting its flow
2. To deodorize and prevent decomposition through bacteria inhibiting action.

Deodorants may be liquids, creams, pastes, powders, compacts and sticks.

Usually deodorants are applied in arm pits (axillae, oxter) to reduce apocrine sweating, sometimes used for deodorizing sanitary napkins. Deodorants may be antiseptic or astringent in action, but remember sanitary napkins should not be an astringent. Deodorants are usually alcohol-based, which kill bacteria effectively. Deodorants can be formulated with other, more persistent antimicrobials such as triclosan, or with metal chelant compounds that slow bacterial growth. Deodorants also often contain perfume fragrances intended to mask the odor of perspiration.

Materials used for deodorant actions are boric acid, benzoic acid, oxyquinoline sulphate, formaldehyde, alum (aluminum sulfate), aluminum acetate, aluminum chloride, aluminum chlorhydrate, aluminum zirconium chlorhydrate (this product is banned since 1977 by UNFDA), zinc salicylate, zinc sulphocarbonates, etc. since no one material possess all the property so for effective deodorant it becomes necessary to use two or more product to produce its desired effects.

Formula I

For deodorant liquid

Oxyquinoline sulphate	3.0%
Aluminum chloride	15.0%
Magnesium sulphate	5.0%
Alcohol	5.0%
Purified water	72.0%
Perfume	q.s.

Mix all the salts in purified water and then add perfume dissolved I alcohol.

Theoretically any compound which has antibacterial action can be used as deodorant.

Many of the quaternary ammonium compounds have been found non toxic and sufficiently non-irritating. Such compounds can be used in deodorant liquids.

Formula II

Chlorhexidine diacetate	0.5%
Propylene glycol	0.2%
Denatured spirit (50%) q.s.	100.0%
Perfume	q.s.

Formula III

For deodorant powder

Purified zinc peroxide	40.0%
Boric acid	20.0%
Talc	40.0%
Perfume	q.s.

Mix all ingredients in ascending order of their weight, sift and perfume.

Formula IV

For deodorant sticks

Sodium stearte	10.0%
Sobritol solution	5.0%
Isopropyl myristate	2.5%
Triclosan	0.5%
Ethanol q.s.	100.0%
Perfume	q.s.

Dissolve sodium stearate, Isopropyl myristate and triclosan in ethanol and then dissolve sorbitol and perfume in it. Finally pour the solution into moulds to obtain sticks.

1. Antimicrobial substances used in antiperspirant and deodorant formulations such as:
   i. Hexachlorophene,
   ii. Aluminum phenolsulfonate, and
   iii. Zinc peroxide,
2. Quaternaries compounds such as:
   i. Cetyltrimethylammonium bromide,
   ii. Alkyldimethylbenzylammonium chloride, and
   iii. Di isobutyl phenoxy ethyl dimethyl benzyl ammonium chloride

Tyrothricin and neomycin have also been used in deodorant formulations. Studies have shown that neomycin alone or in combination with aluminum chlorhydroxide produces effective odor suppression, along with a significant increase in gram-negative organisms. Its effectiveness as a deodorant is essentially attributed to its exceptional suppression of gram-positive organisms.

Trichlorocarbanilide is a high-duty, wide-spectrum bacteriostatic agent also used in deodorants. It has lasting safe and stable bactericidal effect. It also has high effects on inhibiting and killing gram-positive, gram-negative bacteria, fungus, *saccharomycetes*, and virus.

ANTIPERSPIRANTS

A variety of substances which have astringent action inhibit the flow of perspiration. The mechanism by which such antiperspirants act has not been clearly defined. The narrowing of the openings of the sweat ducts and the formation of a keratotic plug in the sweat duct orifice to obstruct the flow of sweat were suggested by dermatologists as the possible causes of anhydrosis. It was found that the sweat suppression by formalin is due to a high-level obstruction of the eccrine duct, but that aluminum chloride anhydrosis results from increased transductal absorption of sweat.

A simple type of antiperspirant can be formulated as represented in formula.

Formula V Liquid Antiperspirants:

Aluminum chlorhydrate (50% solution)	25.0%
Glycerin	5.0%
Alcohol	45.0%
Borax	2.0%
Water (q.s.)	100.0%
Perfume	q.s.

EVALUATION OF ANTIPERSPIRANTS AND DEODORANTS

The most widely used method, subjects with shaved axillae do not use any antiperspirant or deodorant for one week which acts as conditioning period. The next week at 9.30 a.m., tare absorbent pads are placed in dried axillae and are held in place by keeping the arm close to the body. These pads are retained for about half an hour or until a minimum of 100 mg of perspiration is obtained from the least perspiring axilla. The above mentioned procedure is repeated 3 to 6 times to obtain control data. The number of repetition of the procedure and collection of data depends on the reproducibility of the ratios of the weight of perspiration from the right axilla to that from the left.

This value is known as PR (perspiration ratio). In the afternoon of the same day the product to be tested is applied to the axilla which perspire the most. After one hour, pre-weighed pads are applied to both the axillae and 3 to 6 collections are made in this manner. This method is repeated for next three days applying the product at 9.00 a.m. and collecting the perspiration on weighed pads from 10.00 a.m. The PR is then calculated from these collections. The collected data is tabulated indicating control readings and readings after 1, 2, 3 or more application. The effectiveness is calculated by comparing the data obtained each day of the test with control data. The increased perspiration flow subjects may be placed in room having controlled humidity and temperatures. Buffered aluminum chloride solution could be used as a benchmark preparation for use in checking the test procedure.

The test developed by Daley's is a back test method which is more rapid and convenient. However, this method is not reliable in case of certain cream type products which form a coating on the skin resulting in large reduction of perspiration.

Fredell and Longfellow is widely used. On the first day of the test, odour of both axillae is recorded. A scale of 0 to 3 is used for recording and direct sniffing is used for judging the odour. The product to be tested is applied to one axilla and nothing is applied to the control. After 6 hours, both axillae are again sniffed and the odour is recorded. The test may be repeated on succeeding day.

Available Marketed Products

- Axe
- Rexona Deostick
- Dynamite
- Fa
- Eva
- Cinthol
- Spice, Good Mornings (Parkavenue)
- Spinz.

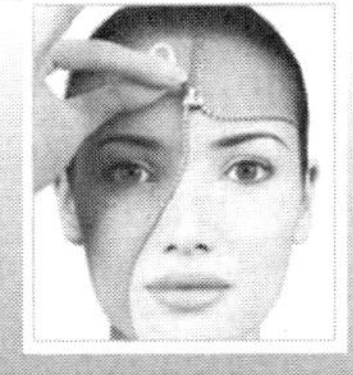

Depigmenting Agents

> *Beauty is worse than wine, it intoxicates both the holder and beholder.*
>
> — Immermann

DEPIGMENTING AGENTS

Hyperpigmentation is the result of an increased amount of melanin in the epidermis, the dermis, or both. This pigmentary change can be divided into two pathophysiologic processes:

1. Melanocytosis (increased number of melanocytes) and
2. Melanosis (increased amount of melanin).

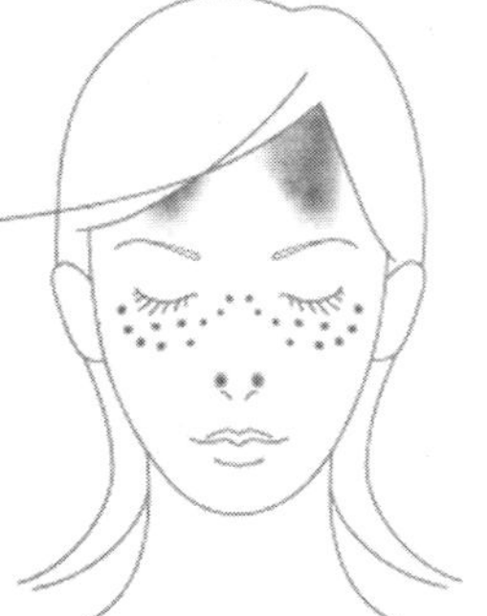

Depigmenting agents work best when melanosis or melanocytosis is restricted to the epidermis. Other methods of depigmentation being used are chemical peels.

Depigmenting agents can be divided into several groups:

1. Phenolic compounds include the following:
   - Hydroquinone
   - Monobenzyl ether of hydroquinone
   - 4-methoxyphenol
   - 4-isopropylcatechol
   - 4-hydroxyanisol
   - N-acetyl-4-S-cysteaminylphenol

2. Nonphenolic compounds include the following:
   - Corticosteroids
   - Tretinoin
   - Azelaic acid
   - N-acetylcystein (NAC)
   - L-ascorbyl-2-phosphate
   - Kojic acid
3. Combination formulas include the following:
   - *Kligman's formula:* Consists of 5% hydroquinone (HQ), 0.1% tretinoin, and 0.1% dexamethasone in hydrophilic ointment.
   - *Pathak's formula:* Consists of 2% HQ and 0.05–0.1% tretinoin.
   - *Westerhof's formula:* Consists of 4.7% N-acetylcysteine (NAC), 2% HQ, and 0.1% triamcinolonacetonide.
   - *N-acetyl-4-S-cysteanimylphenol:* A phenolic thioether, N-acetyl-4-S-cysteaminylphenol is a new type of depigmenting agent. It claims to be more stable and less irritating to the skin than HQ and is specific to melanin-synthesizing cells.

A monophenol compound tert-butyl-4-hydroxyanisol (mequinol) has been studied in the treatment of solar lentigines and related hyperpigmented lesions. The topical combination product containing 2% 4-hydroxyanisole/0.01% tretinoin solution (Solage) is well tolerated and superior to either active component.

ARBUTIN

Arbutin, or hydroquinone-beta-D-glucopyranoside, consists of HQ bound to glucose; arbutin is a naturally occurring beta-D-glucopyranoside derivative of HQ. Arbutin can inhibit melanogenesis by affecting the activity of tyrosinase rather than by killing melanocytes and decreasing the synthesis of melanin. Arbutin executes its activity by mimicking the amino acid tyrosine, the usual substrate of tyrosinase.

AZALEIC ACID

Azaleic acid (AZA) is a dicarboxylic acid originally isolated from *Plasmodium ovale*. It has been reported to have depigmenting effects, while showing no significant activity on normal skin. AZA is believed to selectively inhibit tyrosinase in hyperactive melanocytes.

CHEMICAL COMPOUNDS

Chemical Peeling Agents

Chemical peeling has become an established technique in treating cutaneous hyperpigmentation. Chemical peeling agents include glycolic acid, resorcinol, and salicylic acid.

Kojic Acid

Kojic acid (5-hydroxy-2-[hydroxymethyl]-4-pyrone), a fungal metabolic product, has been increasingly used as a skin-depigmenting agent in skin care products.

Paper-mulberry Compound

The compound 5-(3-2,4-[dihydroxyphenyl]propyl)-3,4-bis (3-methyl-2-butenyl)-1,2-benzenediol from paper-mulberry root bark has been shown to inhibit mushroom tyrosinase, to scavenge free radicals, and to depigment UV-induced hyperpigmentation in guinea pigs. Studies in humans show no irritation or sensitization.

Vitamin C

Vitamin C (L-ascorbic acid) and its derivatives are believed to act as reducing agents on melanin intermediates. They block the oxidative chain reaction from tyrosine/dihydroxyphenylalanine (DOPA) to melanin at various points.

ANTICELLULITES

Lipolysis is mediated, in part, by beta-adrenergic receptors, which induces fat breakdown, and alpha2-adrenergic receptors, which inhibits fat breakdown. Agents that bind to these receptors may hypothetically serve a therapeutic effect on cellulite. Beta-adrenergic stimulators include theobromine, theophylline, aminophylline, caffeine, isopropylarterenol hydrochloride, and epinephrine. Alpha2-adrenergic inhibitors include yohimbine, piperoxan, phentolamine, and dihydroergotamine.

ENZYMES

Papain

Papain, an enzyme found in the papaya fruit, chemically digests intercellular bonds. Papain has been studied in the treatment of hypertrophic scars, and it can be used to exfoliate keratotic skin. Deoxyribonucleic acid (DNA) repair enzymes.

GROWTH FACTORS

Epidermal Growth Factor

Epidermal growth factor (EGF) is found in plasma, sweat, urine, saliva, and semen. When EDG is bound to the epidermal growth factor receptor (EGFR), it stimulates epidermal growth and differentiation. It has been used in the treatment of burns and excision wounds, where it accelerates re-epithelization.

Transforming Growth Factor

Transforming growth factor (TGF) stimulates normal skin growth and cellular growth and repair. TGF exerts positive regulatory effects on the accumulation of the body's extracellular matrix proteins. TGF is also a mediator of fibrosis (repair tissue formation) and angiogenesis (development of new blood cells), and it promotes the healing of wounds.

Hormones

Hormonal creams claim to be the most effective means to stop or slow the aging process by reversing the loss in tone and elasticity of the skin. Confirmatory evidence of this claim is yet to be published.

Estrogens

Some studies have shown anti-aging effects of estrogens. One investigation found that after 6 months of applying 0.01% estradiol and 0.3% estriol compounds, the elasticity, firmness, wrinkle depth, and pore sizes of the skin were markedly improved. On immuno-histochemical analysis, significant increases of type III collagen labeling were combined with increased numbers of collagen fibers at the end of the treatment period. Also, no systemic adverse effects were noted. However, better studies are needed before these agents are routinely used for their anti-aging effects.

Progesterone

Progesterone creams are being marketed as formulations that reverse the chemical changes that occur in collagen with aging and that normalize the immune system. Some manufacturers' claims also state that progesterone cream heals skin conditions, such as acne, psoriasis, rosacea, seborrhea, and keratoses. Other manufacturers claim that progesterone creams are a topical supplement for women who experience symptoms relating to premenstrual syndrome (PMS), menopause, or osteoporosis. None of these claims is supported by well-designed studies.

Testosterone

Topical application is quickly becoming the preferred method of testosterone administration. Theoretically, when topically applied, testosterone bypasses the stomach and the liver and does not cause an undesirable rise in estrogen. Manufacturers claim that testosterone creams have many benefits, such as memory enhancement, antidepressant effects, increased resistance to stress, and the ability to treat disorders associated with hypogonadism (e.g. increased storage of fat, shrinkage of muscle mass). Recent studies have shown that testosterone gel applied to the skin once daily restored blood levels of the hormone in men with hypogonadism.

Growth Hormone

Growth hormone (GH) and its mediator insulin like growth factor-1 (IGF-1) are responsible for many effects on growth, development, immunity, and metabolism. Produced and secreted by the anterior pituitary gland in the brain, GH is released in pulses in response to signals from the hypothalamus. GH exerts anabolic effects throughout the body, favoring the growth of tissues, bones, and muscles. Studies have shown that the aging population has lowering levels of GH in the body, with resultant decreased lean body mass, fat deposits, immunity, and overall energy.

Botulinum A Exotoxin

Botulinum A exotoxin is a neurotoxin produced by the bacterium *Clostridium botulinum.* This toxin is now being used by cosmetically oriented specialists for the treatment of a large variety of movement-associated wrinkles on the face and neck.

As a simple and effective nonsurgical procedure, the injection of botulinum toxin A seems to be an effective method of eliminating crow's feet and wrinkle lines that are on the upper third of the face and producing temporary browlift. This form of temporary chemical denervation compliments the cosmetic practitioner's armamentarium. In addition, the use of botulinum toxin to block sympathetic innervation of eccrine sweat glands is proving to be valuable in treating hyperhidrosis of the axillae, palms, and soles.

PEPTIDES

Microcollagen Pentapeptides

Fibroblasts in aged tissue produce less collagen than those in younger skin, but their capability to produce collagen is still present. Fibroblast collagen production has been reported to be stimulated by a pentapeptide fragment of the collagen molecule.

At the carboxyl-terminal end of the collagen molecule is a fragment that has been identified as a participant in the regulation of its own synthesis. Lys-Thr-Thr-Lys-Ser pentapeptide is a potent stimulator of collagen and fibronectin synthesis, which are both important components of the interstitial matrix.

Copper Peptides

The copper-dependent lysyl oxidase (LO) plays a critical role in the biogenesis of connective tissue matrices by cross-linking the extracellular matrix proteins, collagen and elastin. Levels of LO increase in many fibrotic diseases, while expression of the enzyme is decreased in certain diseases involving impaired copper metabolism.

Within the past decade, the gene encoding LO has been cloned, facilitating investigations of the regulation of expression of the enzyme in response to diverse stimuli and in numerous disease states. Transforming growth factor-beta, platelet-derived growth factor, angiotensin II, retinoic acid, fibroblast growth factor, altered serum conditions, and shear stress are among the effectors or conditions that regulate LO expression. Since the production of both collagen and elastin is reduced in aging skin and in skin exposed to ultraviolet light, copper peptides may be able to help produce new collagen and hence repair aged skin.

Marketed Products

- Nature's Essence
- Joy
- Lakme
- ISPA
- Eraser
- Aroma magic
- Ponds
- Mederma
- Revoln
- Herbal Facia
- L'orel.

Part

3

Hair Cosmetics

11. Hair Care Products
12. Hair Conditioners
13. Shampoo
14. Hair Sprays
15. Hairdressings
16. Permanent Waving Preparations
17. Hair Straighteners
18. Hair Removal Agents
19. Hair Colouring Products
20. Shaving Preparations

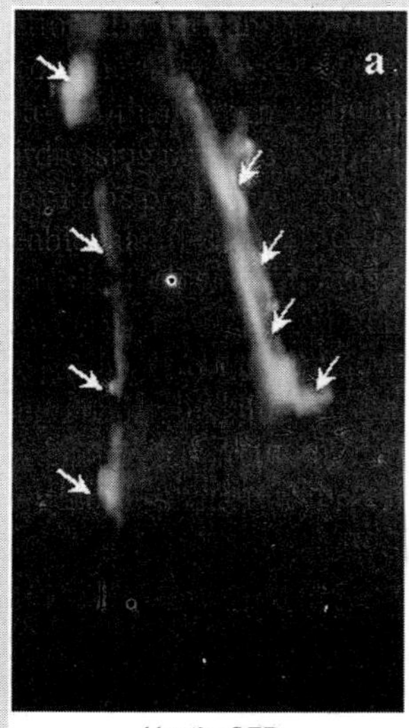

Nestin-GFP

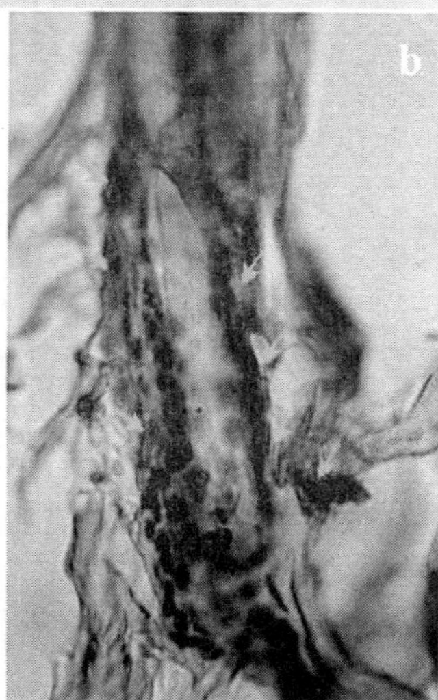

Keratin-15

11

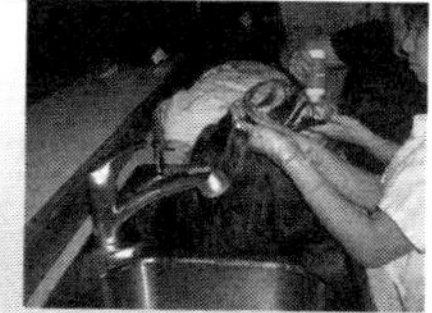

Hair Care Products

> *Everything has beauty, but not everyone sees it.*
> — Confucius

INTRODUCTION

Hair follicle originates from an interaction between epidermal and dermal layers of the skin. The hair follicle is a sheath of epidermal cells and connective tissue that encloses the root of the hair.

Head hair is a type of hair that is grown on the head (sometimes referring directly to the scalp). The most noticeable part of human hair is the hair on the head, which can grow longer than on most mammals and is more dense than most hair found elsewhere on the body. The average human head has about 100,000 hair follicles.

A thickened plate of epidermis, over an assembly of dermal cells, grows forward to form a 'peg', which in time grows round the dermal papilla to create the hair bulb. The epidermal cells adjacent to the dermal papilla then multiply and push towards the scalp surface a column of keratinizing cells. (Keratinisation or 'cornification' is the process by which a cell accumulates the fibrous protein keratin. This displaces the cytoplasm of the cell.) At a certain stage in the keratinisation of the cell, it is essentially 'dead'. The column of keratinizing cells becomes the hair shaft, invested in an inner layer root sheaf, and a hair canal is formed in this process.

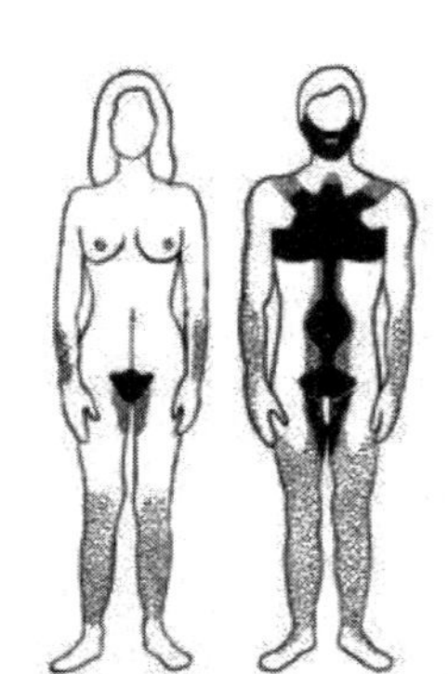

Black or brown colour of hair is due to pigment **melanin.** Melanin is formed in epithelium cells of the matrix and is carried with them when these are pushed up and are keratinized. Due to change in metabolism, melanin is not formed and therefore not carried with the cells and

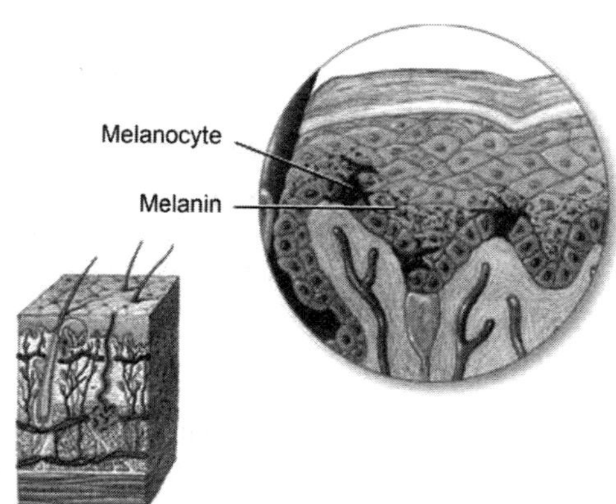

hair becomes grey. The average human scalp measures approximately 120 square inches (770 cm^2). Hair density depends upon race and hair color. Caucasians have the highest hair density, the growth rate is average. Blacks have medium density and slowest growing hair. Asians have the lowest density but the fastest growing hair.

Melanin is also of two types:

Eumelanin (black and brown pigment) and

Phaeomelanin (yellow or reddish)

The study of hair, mane, locks, curls or tresses and its related science is known as **Trichology**.

STRUCTURE OF HAIR

In this context it is the nature of the hair-shaft which is of primary interest. Its structure may be described as follows. The majority of the shaft is made up of elongated keratinized cells connected together to form the cortex of the hair fibre. The cortex is enclosed by a cuticle derived from a single strand of cells in the bulb of the root, which becomes a surface structure of the hair fibre five to ten cell layers thick.

The bulk of the hair is made up of the cortex, the cells of which are toughly high in Indians. The straight hair of the Chinese and Japanese has virtually no orthocortex, while the crimped hair of Negroes has an easily recognized band of orthocortex.

The cuticle cells on the surface of hair have an overlapping, roof-tile like formation. This means that the friction of the fibre is considerably less when it is measured from root to tip than when measured the reverse way. The hairdressing practice of back-combing takes advantage of this property. Backcombed hair is more manageable than if it had been conventionally combed or brushed. In back-combing, the hair surface is, to a certain extent, damaged through the detachment of some of the cuticle cells. The loss of these cells roughens the hair surface and thus increases interfibre friction. This means that individual hairs are less likely to separate from their neighbours.

TYPES OF HAIR

Humans have three different types of hair:

- **Lanugo**, the fine hair that covers nearly the entire body of fetuses.

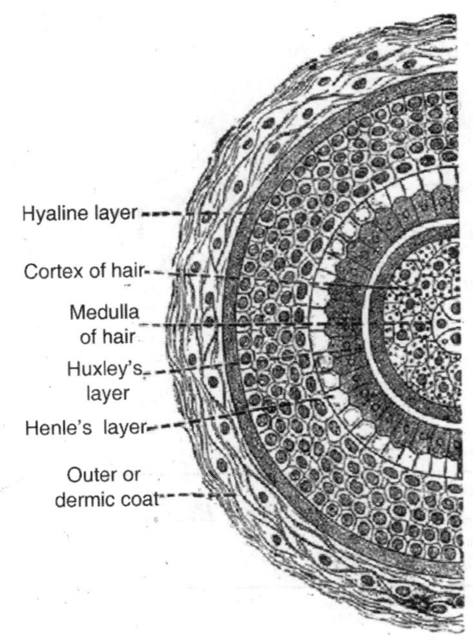

- **Vellus hair**, the short, fine, "peach fuzz" body hair that grows in most places on the human body in both sexes.
- **Terminal hair**, the fully developed hair, which is generally longer, coarser, thicker, and darker than vellus hair.

Different parts of the human body feature different types of hair. From childhood onward, vellus hair covers the entire human body regardless of sex or race except in the following locations: the lips, the nipples, the palms of hands, the soles of feet, certain external genital areas, the navel and scar tissue. The density of the hairs (in hair follicles per square centimeter) varies from one person to another.

The diameter of human hair ranges from 17 to 181 μm. There are usually four major types of hair texture: fine, medium, coarse and wiry. Within the four texture ranges hair can also be thin, medium or thick density and it can be straight, curly, wavy or kinky.

GROWTH OF HAIR

A typical growth rate for human scalp hair is 0.3–0.5 millimetres (mm) per day. A healthy scalp (the top of the head extending from one inch above the ear.) supports something of the order of 100000 hair follicles. Few workers have considered it as high as 200000 hair follicles.

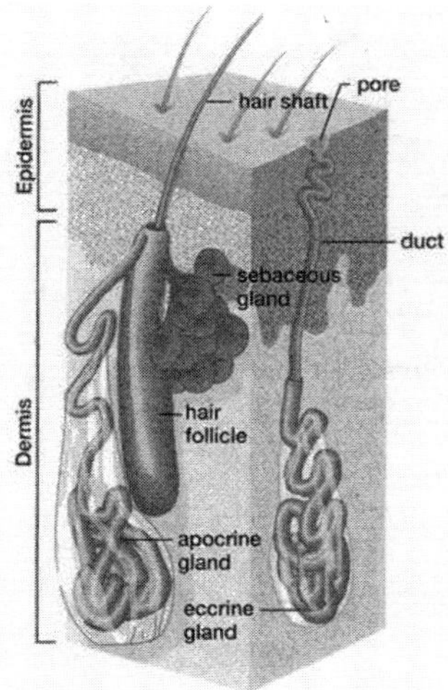

The Cyclic Nature of Hair Growth

Practically all of the follicles participate in a cyclic activity. The active growth or anagen phase in which the hair is produced alternates with a resting period known as the telogen phase. In the latter period the fully formed or club hair remains fixed in the follicle by its expanded base and the dermal papilla is free from the epidermal matrix. The latter is reduced to a small secondary growth. Between the anagen and telogen stages there is a short transitional period, known as the catagen, in which the newly formed club hair moves towards the skin surface. All of the hair fibres reach a terminal length which is determined by the duration of the anagen phase. At any one time some 85% of the scalp hairs are said to be in the anagen phase and 12% and 3% are in the telogen and atagen phases, respectively. These proportions are very similar to these found in a study of the mechanical properties of scalp hairs front on particular head. A scalp anagen phase may be as long as 3 years. Normal hair or strands loss is around 100 to 200 per day wherein a single hair can last up to two to five years.

Sometime, the rate of hair loss becomes more than the growth of hair or new hair do not grow. Such a situation results in baldness or hair loss. The factors responsible for hair loss are:

- Reductions in hair follicle function due to male hormones: conversion of testosterone to 5α-dehydrotestosterone (DHT) which is thought to be responsible for male pattern baldness or *androgen alopecia* or *alopecia aerata*.
- Reductions in metabolic functions of hair follicle and hair bulbs: supply of nutrients to dermal papilla may not be sufficient due to reduction in blood flow in capillaries surrounding hair follicles.
- Excessive build up of dandruff flakes at pores of scalp: Dandruff (pieces of dead skin which form on scalp also known as scurf or *pityriasis capitis*) causes blockage of hair exit at epidermis and is decomposed by bacteria which adversely affect the condition of hair and causes irritation to scalp. If such condition remain as such for a long time it results in hair loss called pityriasis type hair loss. Excessive secretion of sebum at hair follicle also causes seborrhoeic alopecia due to decomposition of sebum which results in irritation and redness.
- Hair loss due to loss in flexibility of scalp results in poor blood circulation in peripheral blood vessels in subcutaneous tissue of scalp.
- Other factors like stress, malnutrition, genetic reasons, drug induced hair loss due to side effects.

Hair Composition

Hair contains 97% Protein and 3% moisture. This strong fibrous protein is called keratin. Each hair is an incredibly strong and resistant fiber, having a complete structure, which can be altered by our daily activities like washing, drying, blow drying, and chemically changing color and texture. Good quality hair is one which is quite elastic, can stretch upto 30% beyond its normal length and spring back and having ability to absorb and hold moisture up to 50% of its weight called porosity.

Each hair comes out of the scalp through the hair follicle. The angle at which the hair comes out of the skin is directly related to the shape of the hair. Straight hair is round in cross section and shoots straight out from the scalp. Wavy hair is oval in shape and comes out of the scalp at a straight angle, curly hair is flattened and comes out of the scalp at an extreme angle. Each hair has

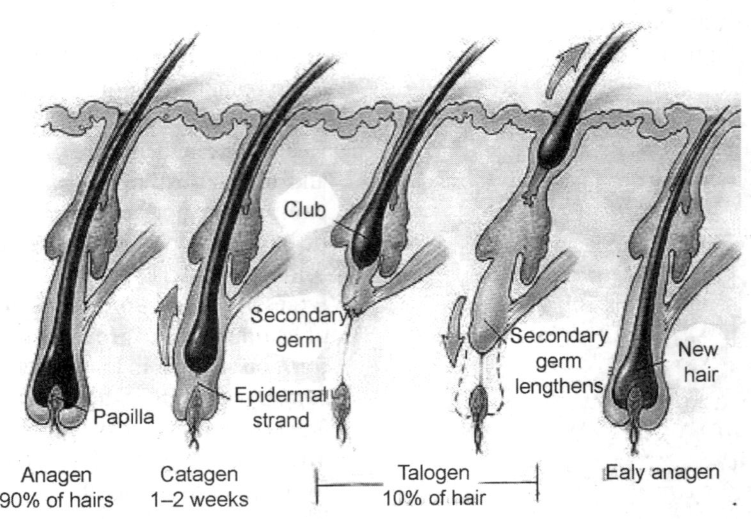

its own blood, nerve, and muscle supply, found within the dermal layer of the skin. At the base of the hair follicle, embedded in the dermis, is the papilla, the root through which a rich supply of oxygenated blood feeds hair growth via blood capillaries. Blood is the communication link between the body and the hair imbalances or toxicity in the body are interpreted and transferred to the hair through the blood supply, such as in sluggish or blocked flow of blood. Stress, pollutants taken into the body, hormonal fluctuations, illnesses, Pharmaceutical or illicit drugs, and poor diet all come into play in this integral part of nourishing hair growth. The hair bulb envelops the hair papilla. Within rich fertile soil, so does the hair germinate in and sprout from the hair bulb, growing prolifically with papilla's rich source of nutrients.

Our hair grows approximately 13 mm a month and sometimes slightly faster. When we are highly active, growth usually slows. The active, growing phase of an individual hair is called anlagen phase and it could grow 45–75 cm through this period. The hair then goes into an intermediate or transitional phase called cartage phase just before entering the tillage phase, in which it rests for about a hundred days and then falls out. About 85–90% of all hair on the head is in an active growing phase while 10–15% is in resting phase. Loss of hair is known as alopecia which is due to hormonal imbalances, anemia, mineral imbalances, exposure to poisons, X-rays, liver and kidney disease, autoimmune disease, stress, poor diet, genetic, hereditary thinning or balding.

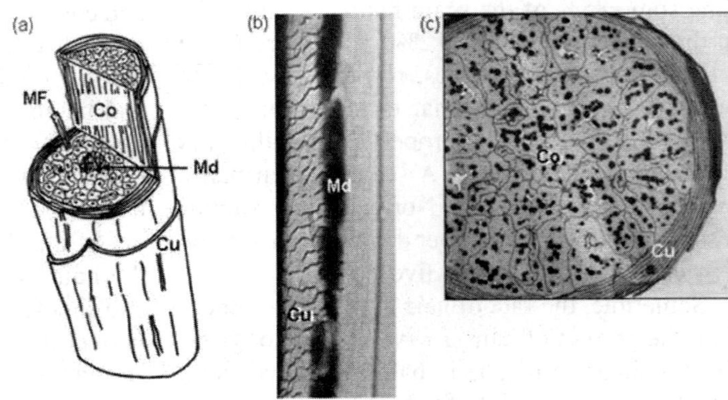

HAIR CHEMISTRY

Hair fibres are composed of approximately 85% of the complex protein keratin, with some 7% of associated water. The other principal ingredients are lipids 3% and pigment 2%. This last component is melanin, derived in biosynthesis from the amino acid tyrosine. Also present are trace amounts of many metals such as aluminium (Al), chromium (Cr), calcium (Ca), copper (Cu), iron (Fe), manganese (Mn), magnesium (Mg) and zinc (Zn), the latter in a fairly high concentration of 22 mg per 100 g of hair. Phosphorus compounds are also abundant (80 mg per 100 g hair), mainly derived from degraded nuclei of the cortex cells. The most important material is keratin which has been formed biochemically from the condensation of some 18 types of amino acids (Table 11.1).

Table 11.1: Amino acid composition of hair keratin

Amino acid	mol %
Alanine	5.6
Arginine	7.0
Aspartic acid	7.0
Cystine*, cysteine*, methionine	12.3
Glutamic acid	12.9
Glycine	6.0
Histamine	0.8
Isoleucine	3.2
Leucine	8.0
Lysine	2.9
Ornithine	0.2
Phenylalanine	2.0
Proline	7.0
Serine	10.2
Threonine	6.7
Tyrosine	2.0
Valine	6.2

* Sulfur acids

Eumelanin is formed from tyrosine by oxidation with help of an enzyme tyrosinase and phaeomelanin from tyrosine in presence of o-amino-phenol derived from tryptophane.

AMINO ACIDS IN HAIR

Amino acids may be represented by the generic structural formula below:

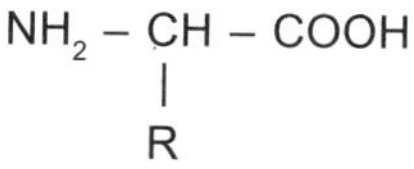

$$NH_2-\underset{\displaystyle R}{\underset{|}{CH}}-COOH$$

Amino acids may be classified according to the following convention. Other systems of classification are often used, but none of which is entirely satisfactory.

Hydrophobic: Examples: valine and phenylalanine

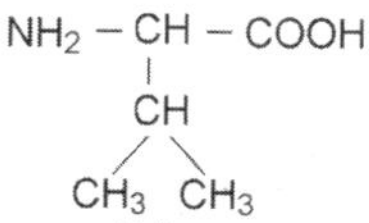

$$NH_2-CH(CH_3)_2\text{-}CH-COOH \quad \text{i.e. } NH_2-\underset{\displaystyle CH_3\ CH_3}{\underset{/\ \backslash}{\underset{\displaystyle CH}{\underset{|}{CH}}}}-COOH$$

Valine

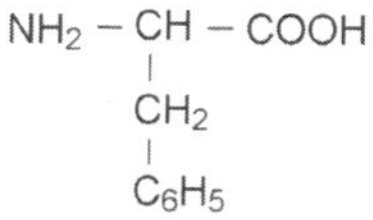

$$NH_2-\underset{\displaystyle C_6H_5}{\underset{|}{\underset{\displaystyle CH_2}{\underset{|}{CH}}}}-COOH$$

Phenylalanine

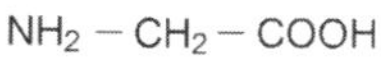

$$NH_2-CH_2-COOH$$

Glycine

$$NH_2-\underset{\displaystyle SH}{\underset{|}{\underset{\displaystyle CH_2}{\underset{|}{CH}}}}-COOH$$

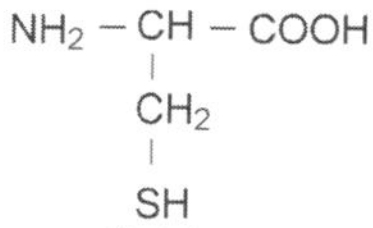

Cystein

Cysteine is the fission product of cystine (below). Later these two amino acids will be shown to be technologically important in the context of hair-waving products.

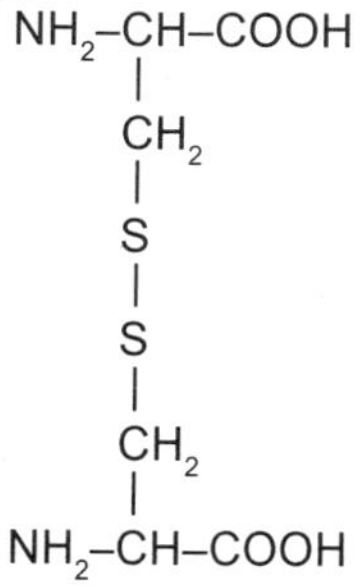

$$\begin{array}{c} NH_2\text{–}CH\text{–}COOH \\ | \\ CH_2 \\ | \\ S \\ | \\ S \\ | \\ CH_2 \\ | \\ NH_2\text{–}CH\text{–}COOH \end{array}$$

Cystine

Acidic: Examples: aspartic and glutamic acids

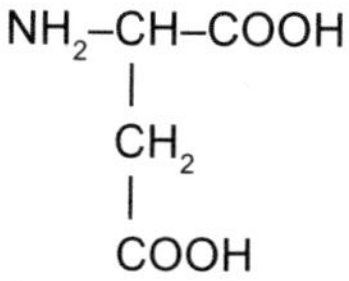

$$\begin{array}{c} NH_2\text{–}CH\text{–}COOH \\ | \\ CH_2 \\ | \\ COOH \end{array}$$

Spartic acid

$$\begin{array}{c} NH_2\text{–}CH\text{–}COOH \\ | \\ CH_2 \\ | \\ CH_2 \\ | \\ COOH \end{array}$$

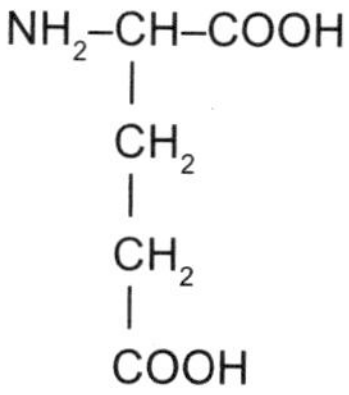

Glutamic acid

Basic: Examples: lysine and arginine

$NH_2 — CH_2 — COOH$
|
$(CH_2)_4$
|
NH_2
Lysine

$NH_2 — CH_2 — COOH$
|
$(CH_2)_3$
|
NH
|
C
NH_2 NH
Arginine

PEPTIDE LINKAGE

In the protein chain, individual amino acids are joined linearly by the group formed from the condensation of a carboxyl group with an amino group to form what is known as a peptide linkage; i.e. —CONH— Thus repeat unit in the keratin chain is:

–NH–CH–CO–
|
R

Where R may represent the same or different amino acid characterizing groups, e.g. for glycine R = H, for lysine R = — $(CH_2)_4$—NH_2.

In the hair fibre keratin chains run roughly parallel to each other and to the fibre axis. These chains are bonded to each other in a lateral sense through physicochemical interactions between the R groups of the amino acid 'residues'. Thus where hydrophobic residues are adjacent, hydrophobic links form the bonds in that particular segment of the protein chain:

↑ CO
NH
$CHCH_2$ (CH_3 CH CH_3) (CH_3 CH CH_3) — CH
NH
NH ↓

↑ NH
CO
CH
NH
CO ↓

Similarly, where an acidic amino acid (e.g. glutamic acid) on one keratin chain is opposite a basic R group on an adjacent chain a salt linkage can be formed:

↑ CO ↑ NH
NH CO
CH $(CH_2)_4NH_3OOC \cdot (CH_2)_2$ — CH
CO NH
NH ↓ CO ↓

In hair keratin there is an abundance of sulphur amino acids (cysteine and cystine), and in the biosynthesis of keratin, cysteines in neighbouring chains can combine via a disulfide bond, and this provides a strong covalent bond between the chains:

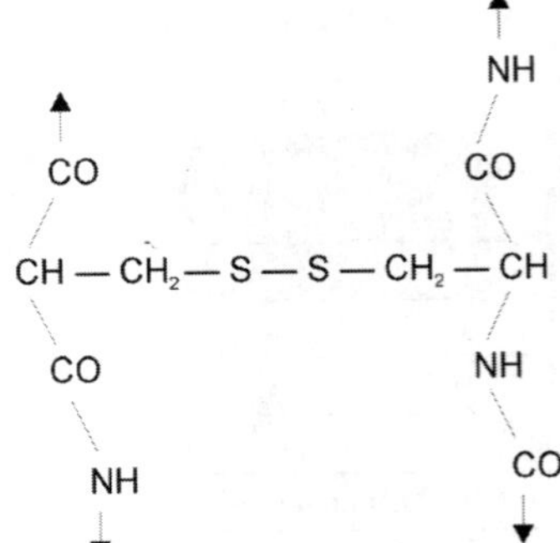

Quantitatively, hydrogen bonds contribute considerably to the cohesion between keratin chains. These may arise in a number of ways; one of which is illustrated below.

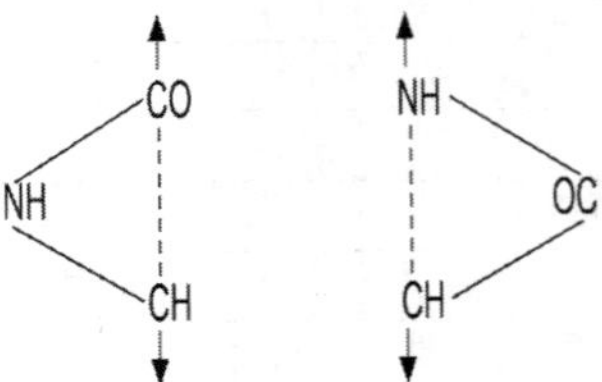

Here the hydrogen bonding arises from the interaction between the peptide linkages of adjacent protein chains. These several mechanisms by which keratin chains can be bonded laterally account for the unusually high values for hardness, density, optical refraction, etc., which characterize hair keratin.

LIMITATION OF AMINO ACID ANALYSIS

Amino acid content determination is inadequate basis for the correlation of chemical composition with the morphology, physical properties and functionality of the keratins. This approach could be termed 'chemical affinity' classification. It involves chemically separating the various protein types liberated after the disulfide bonds have been broken by bisulfite reduction (aided by urea acting as a swelling agent), followed by sequential precipitation into fractions. Following fractions can be recognized:

1. A low sulfur component which contains a high level of extensible alpha-helix protein. This is the component of hair keratin which is responsible for the phenomenon of 'set'. It is the main constituent of the protofibrils of the cortex cells.
2. A high sulfur component, which is present in the matrix, or cement, between the cortex cells.
3. A third fraction which is rich in the amino acid tyrosine.

Consider two contrasting forms of mammalian keratin, human scalp hair and porcupine quill. The hair is very rich in the high sulfur protein fraction, about 40% and contains only a nominal amount of the high tyrosine fraction. At the other extreme the porcupine quill is low in high sulfur protein (about 7%), but very rich in the tyrosine fraction. Other examples along these lines may be quoted.

CHEMICAL REACTIVITY OF HAIR KERATIN

Alkali Hydrolysis

Hydrolysis by strong alkali is less complete than acid hydrolysis, in the sense that as well as liberating individual amino acids, some of the keratin is broken down only so far as the peptide state. This is demonstrated by use of a colorimetric test, the biuret test, which works positively upon the alkali hydrolysate. This test is applied only where peptide linkages, –CONH–, are present.

Acid Hydrolysis

The acid hydrolysis of keratin, e.g. with moderately strong hydrochloric acid, breaks down the protein into its constituent amino acids almost entirely. Thus, upon completion, virtually no material which incorporates the peptide unit is present. However, through the mechanism of deamination of the amino acids with the reagent ninhydrin a purple colour (Ruhamann's Purple) can be developed. This reaction can be used to detect and quantify proteins in their hydrolysates. It is useful in measuring the extent of the damage to hair cuticle and the alleviation of the damage by certain hair-care treatments.

Reactions of Sulfur Amino Acids

The most interesting sulfur amino acid, cystine, may be reduced to generate thiol groups (—SH) by a number of reducing agents. These agents include mercaptans such a thioglycollic acid, alkali bisulphite (e.g. $NaHSO_3$) and certain phosphorus derivatives such as tertrakis (methylol) phosphonium chloride (TMPC). Commercially the mercaptan, ammonium thioglycollate is the most important as it forms the basis of most permanent waving processes and also many hair-straightening products.

HAIR CARE PRODUCTS

In all cases when the degree of reduction is sufficient for the hair to be made plastic enough to be 'set' it can then be fixed (neutralized) by application of an oxidizing agent. Alkali bromates and hydrogen peroxide are practical examples of oxidants.

Alkylation of thiol (SH) groups

Alkylation can readily be accomplished with a wide range of alkylating agents; some of the better known ones are described below.

Reaction of thiols with iodoacetic acid: This has an obvious application in the analytical determination of the degree of reduction of hair which has been treated with reducing agents.

Reaction of the thiol group with maleimides: This includes the difunctional compound

$$\underset{|}{\overset{|}{CH}} - CH_2 - SH + ICH_2COOH \longrightarrow \underset{|}{\overset{|}{CH}} - CH_2 - SCH_2COOH + HI$$

Cysteine Iodoacetic acid

phenylenedimaleimide (PDMI). In theory the reagent should also work as a cross-linking agent, establishing a covalent bond between two keratin chains. Sterically, this would appear to be an unlike linkage. It has not been verified in terms of changes in the physical properties of hair which has been reacted with PDMI. The reaction with the simpler N-ethylmaleimide proceeds as follows:

$$\overset{|}{\underset{|}{CH}} - CH_2SH + C_2H_5N\begin{matrix} CO-CH \\ \| \\ CO-CH \end{matrix} \longrightarrow \overset{|}{\underset{|}{CH}} - CH_2 - SC_2H_5 + \begin{matrix} CH-CO \\ \| \\ CH-CO \end{matrix} NH$$

N–ethylmaleimide

Reactivity of 'free' Amino Groups

'Free' amino groups are those which do not participate in either the continuous protein chain, or in the cross-linking bonds formed between adjacent protein chains. The deamination of amino acids is an example of amino reactivity, which has already been discussed. Amino groups can be made to enter into the formation of interprotein chain bonds by reaction with aldehydes, including formaldehyde.

$$\underset{\text{N-terminal amino acid}}{-NH_2} + \underset{\text{Formaldehyde}}{HCHO} + -NH_2 \longrightarrow -NH + CH_2 - NH - + H_2O$$

A physiologically safe way of producing such bonds are to employ a N-methylol compound as a formaldehyde donor. Examples of such compounds are the N-methylol ureas, thioureas and melamines. These are readily prepared by reacting the parent compound quantitatively with the aldehyde. The latter is released ready for the reaction with amino groups by acidification.

Some useful alkylation reactions can be performed upon the free amino groups of hair keratin. In particular one such reagent in this connection is 1-fluoro-2, 4 dinitrobenzene (FDNB)

$$-NH_2 + \underset{\text{FDNB}}{F - C_6H_3(NO_2) - NO_2} \longrightarrow -NH - C_6H_3(NO_2) - NO_2 + HF$$

These derivatives are not decomposed in hydrolysis and in consequence can be used to identify the individual amino acids in their hydrolysates.

PHYSICAL PROPERTIES OF HAIR KERATIN

Hair keratin is highly complex both in respect of chemical composition and molecular architecture. Much information on the structure of hair can be gained from the study of its physical properties. Hair has an unusually high specific gravity for an organic substance of the order of 1.32 and a very low thermal conductivity comparable to that of asbestos.

Mechanical Properties

Human scalp hair fibres have diameters within the range 30–100 ml depending upon age and racial group. They have hardness similar to that of stainless steel. (Hair fibres pressed between the jaws of a stainless steel vice leave a score mark.) The breaking load of a typical hair is of the order of 100 g substantially lower than its root strength. Young's modulus for hair in the stretching mode is of the order of 2×10^{10} dynes/cm^2. The corresponding value for a typical polymer as used in hairspray formulation is typically 10^7 dynes/cm^2.

Electrical Properties

The resistivity of hair is of the order of 800 mega ohms, measured after equilibrium in an atmosphere of 40% relative humidity. It also has a particularly high dielectric constant of 8.3 similar to that of calcite. This measurement was carried out at 25°C and 60% relative

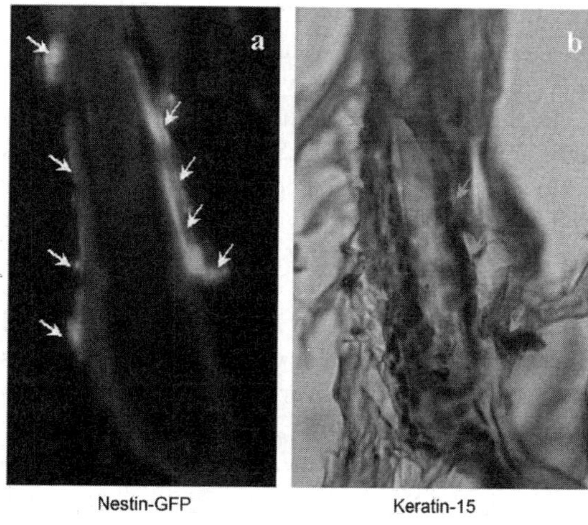

Nestin-GFP Keratin-15

humidity. The iso-electric point (pH value at which the substance has no net electric charge) is approximately 5.

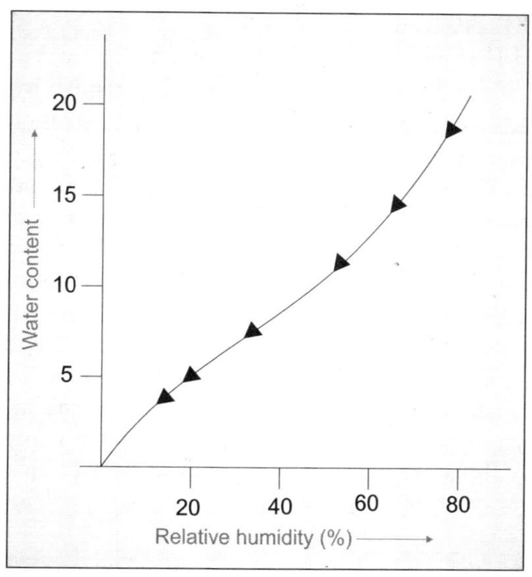

Influence of Water

The water content of hair varies according to the relative humidity of the surrounding atmosphere. **Figure ahead** represents the relationship between water content and relative humidity at constant temperature, this type of figure is generally described as the regain isotherm. At a humidity of 85.87% the water content and consequently the physical properties of hair undergo large changes. Water uptake at much less than 85% humidity occurs in the adsorption part of the regain isotherm. The corresponding regime at humidity greater than 85% is termed the solution region.

The thermal effects following the treatment of hair with water are surprisingly large. This observation applies to both the heat of surface wetting and to the heat of water absorption. The latter effect is easily detected by the senses. In water saturated atmospheres hair fibres will swell to the following extent radially 16% longitudinally 1.2% and in volume, 37%.

Optical Properties

The refractive indices of hair measured at a wavelength of 578 nm are respectively 1.557 and 1.510. These values are high and similar to those of crown glasses.

PHYSICAL PROPERTIES OF HAIR

1. The extreme anisotropy of the hair fibre is demonstrated by a number of its properties, c.p. the contrasting values for radial and longitudinal swelling.
2. Hardness measurements also give unexpectedly large values, indicating the closeness of packing and strength of the lateral bonding operative in the keratin molecule.
3. The equally large values for the mechanical moduli, e.g. Young's modulus and rigidity modulus can be attributed to the factors described in (2) above.
4. That strong thermal effects are met when hair is allowed to interact with water reflects the highly polar nature of the keratin chain segments and the generous surface area available for adsorption of water.
5. The magnitude of the volume, swelling of hair, low dielectric constant, in practical norms these properties relate well to the fact that it is easy to generate electrostatic charges by brushing and combing. These charges leak away only very slowly giving rise to the hairdressing phenomenon of flyaway hair.
6. The refractive index of hair is high, making it difficult to enhance its gloss by depositing films of organic polymers upon its surface.

PHYSICAL PROPERTY OF KERATIN FIBRES

Keratin fibres (hair, wool) may be extended by a pulling action to exactly twice their initial length. The unstretched fibre give rise to a spacing on the X-ray diagram corresponding to 5.2 angstroms (Å); the corresponding spacing for a fully extended fibre is 3.4 × 3 = 10.2Å. The 3.4 Å spacing was deduced to be the length of a single amino acid residue. The above observations allowed a molecular model to be developed; unstretched keratin chains are folded (corrugated), whilst in the fully stretched condition keratin exists in a planar, sheet-like form. These states are known as alpha (half-length) keratin and beta (1 in 1) keratin, respectively. Astbury viewed the keratin chains as molecular springs, which has already served as a good analogy.

In the stretched condition the fibre is said to have set quantitatively. Set may be defined as follows. If the original length of the fibre is L and the extended length is $L + SL$ then the set produced in the fibre is L/I. This set, like the beta state of keratin is intrinsically unstable, thus the set fibre will try to revert to its original length, i.e. L diminished with time. This process is accelerated by exposure to liquid water or its vapour thus after a time has elapsed the excess length will be $/L$ where L is < 01.

The set retained of L expressed in percentage terms will be given by expression.

$$\text{Set retained} = \frac{100\,(\text{set at } t)}{(\text{Initial Set})} = \frac{100\,(\delta L_t)}{(\delta L/L)} = \frac{100\,(\delta L_t)}{\delta_L}$$

$$\begin{aligned}
\text{Let the length of the unwound trees} &= l \\
\text{Initial length of spiral wound form } (L_t) &= L_o \\
\text{Initial set} &= L - L_0 \\
\text{Length of spiral at time } t &= L_t \\
\text{The set at time } t &= L - L_t \\
\text{The percentage set at time } t &= \\
&= \frac{100\,(L - L_t)}{(L - L_0)}
\end{aligned}$$

HAIR CARE PRODUCTS

In the 1960s and early 1970s hair care products, in common with other personal products, enjoyed considerable commercial success. However, in the middle of the 1970s the industry was to meet new and unfamiliar challenges. These affected the profitability and indeed, the survival of some hitherto highly successful products. The constraints may be classified under the headings of health hazards, environmental concerns and product functionality (in relation to claims). An early example of the find was identification of the carcinogenicity of paraphenylene diamine, oxidation dyes; the second was the reputed effect of refrigerant type aerosol propellants upon the Earth's protective layer; and the third constraint is purely a consumer requirement that performance claims should be capable of being confirmed in objective evaluation tests. The net effect of the above and the one of most concern in this chapter is that in the 1980 the hair care industry undertook the radical reformulation of some important products.

CLASSIFICATION OF HAIR CARE PRODUCTS

This section describes post shampoo products with the exception of hair colorants. Frequent reference to shampoos must none the less be made, since they represent the beginning and end of the wash cycle. In almost all treatments, the hair is shampooed to being with. Then for example, it may be towel-dried and a rinse conditioner applied. After the final rinse there is a wide choice of subsequent treatments, for example;

a. A wave set which because of its relatively short term effect may be reinforced by

b. A daily application of hairspray. However, after a few days of this regime it may be necessary to complete the wash cycle, and shampoo again.

Hair-care Products May Be Classified into Two Groups

1. Those which work by purely physical mechanisms, of which shampoos, conditioners and hairsprays are typical examples;

2. Those which bring about chemical changes in or on the hair. This category includes permanent waving and hair straightening preparations and the oxidation dyes which form the basis of permanent hair colorants.

An alternative classification, equally valid can be based upon the extended (functionality of the product, i.e. whether it sets or increases the body of the hair or has a conditioning effect).

Hair-care products are intended to promote certain favourable conditions of hair and to reduce or eliminate properties of hair which are regarded as undesirable. Recent advances in consumer research enable as to improve the definition of these properties. The reason for stressing the contemporary nature of the research is that the preferences of the consumer change over the years; hence the research must be repeated at (say) 5 year intervals. The geographical and cultural situation of the user must also be taken into account. The properties of hair which the product user sees as important are given in Table 11.2. The information was gathered in surveys conducted in the British Isles and the findings appear to apply also to North America, Australia and New Zealand, but would not necessarily apply to neighbouring countries in Western Europe. The terms employed in the following lists are in the vocabulary of the user of hair-care products, they have not been prompted by the scientific staff that designed and carried out the survey.

HAIR-CARE PRODUCTS: RELEVANT HAIR CHARACTERISTICS

Characteristic properties of hair can be measured objectively in the laboratory or in psychometric tests. The latter involve the use of human observers in a statistically designed experiment. The data obtained in the objective tests must be able to be correlated with consumer's perception.

CURRENT POST SHAMPOO HAIR CARE PRODUCTS

One strategy in developing new products is to relate the ability of existing hair products to produce the desirable properties listed in Table 11.2 inefficiencies can be made up and negative

Table 11.2: Hair characteristics descriptive terms

Original hair condition	*Set/Control*	*After treatment condition*	*Set/control*
Favourable			
Gloss	Body	Gloss	Curl strength
Shine	Springiness	Shine	Springiness
Lusture		Lusture	Body
Softness		Soft	Ease of styling
Ease of combining		Ease of combing (wet and dry)	Ease of setting
Unfavourable			
Dry	Fine	Dry	Fine
Greasy	Limp	Greasy	Limp
Coarse	Unmanageable	Coarse	Unmanageable
Damaged	Bushy	Damaged	Bushy
Flyaway		Flyaway	

characteristics can be reduced or eliminated. Products asigned an asterisk(*) in Table 11.3 utilize a good deal of the technology of film forming polymers. Similarly those marked with a cross (+) involve a detailed knowledge of pressurized packaging technology. Instead of quoting highly detailed formulae, a building blood formula will initially be examined for each type of product. This will be followed by an investigation aimed at finding how the simplified formula may be modified to achieve the properties required by the products listed in Table 11.3.

The formulae quoted in the following sections do not include detailed information and proportions of product builders such as perfumes, preservatives and pigments. These important ingredients must, however be included early on in the development of a product so that they may be exanimate for long-term stability and physiological safety.

Some formulae contain special additive which are regarded as being effective in relatively low concentrations. The term effective means bringing about significant changes in product performance or introducing novel properties, i.e. properties not formerly associated with that type of product. The ranges of silicone compounds now available are examples of such additives, and the use of many of them in hair-care products. These raw materials are well defined chemically unlike others proposed such as protein derivatives.

Table 11.3: Important post-shampoo hair products

Physical mechanisms
Hair conditioners also available as aerosol foam (Mousses)
Cream rinse
Clear rinse
Wave sets
Lotions
Aerosol lotions
Aerosol foams (mousses)
Hairsprays
Aerosols
Non-aerosol (pump) sprays
Hair dressings
Brilliantine gels and liquids
Emulsion creams (O/W and W/O)
Non-greasy gels
Aerosol foams (mousses)
Tonics
Chemical mechanisms
Permanent waving preparations
Roller or pin emulsion types
Hair straighteners (mainly for Negroid hair)
Thioglycollate creams
Caustic creams

12

Hair Conditioners

> *Beauty is the promise of happiness.*
> — Stendhal

INTRODUCTION

Conditioners which are applied to hair after shampooing are intended to promote the following properties:

1. Smooth, easy combing in both wet and dry hair.
2. A reduction in the static electricity caused by combing and brushing dry hair, resulting in flyaway hair.
3. The enhancement of the gloss or lustre of hair.
4. The addition of body or volume to hair.

The rinse conditioner is the first treatment to be applied after shampooing and rinsing away the shampoo. The hair is, then towel-dried until it is just damp and the conditioner is distributed, usually with gloved fingers and some optional combing. A few minutes after application, the treated hair is again rinsed and combed prior to drying thoroughly.

Hair conditioner will also alter the ultimate equation and can be healthy, normal, oily, dry, and damaged or a combination.

FORMULATION OF RINSE CONDITIONERS

Formula I represents a traditional composition for this category of product. The key ingredient is the cationic wetting agent cetyl trimethyl ammonium bromide (CTAB). This is known to be strongly adsorbed on the hair surface resulting in a diminished comb-to-hair friction under wet and dry conditions. This leads to a lessening of tribo-electrical charging off the hair surface and in consequence, a reduction in flyaway. This benefit is co-existent with an improvement in ease of dry combing. Yet other useful properties result from the adsorption of CTAB, most particularly in increase in the surface electrical conductivity which further reduces any tendency towards flyaway hair. CTAB may also increase the hydrophobic nature of the hair surface which can lead to a reduction in resistance to combing in the wet state. The other function of CTAB is to emulsify (in oil in water emulsions) the cetyl alcohol which gives the finished product a high viscosity, making it easier to manipulate on a

Table 12.1: Common hair conditioners			
Category	*Primary ingredients*	*Main Advantages*	*Hair-Grooming Benefit*
Cationic detergent	Quaternary ammonium compounds (Quats)	Smooth cuticle, decrease static electricity	Excellent for restoring damaged, chemically processed hair
Film former	Polymers (PVP)	Fill defects in hair shaft, decrease static electricity, improve shine	Improve appearance of dry hair; improve grooming of coarse, kinky hair
Protein containing	Hydrolyzed proteins	Penetrate hair shaft to minimally increase strength	Temporarily mend split ends
Silicones	Dimethicone, Cyclomethicone, Amodimethicone	Place thin coating on hair shaft, decrease static electricity	Decrease combing friction, add shine

head of hair. This avoids the lack of control in use of a product which is too thin.

The polypeptide additive is assumed to make the product effective on damaged hair, i.e. it is intended to be absorbed on sites where the cuticle is damaged or incomplete.

Silicones

Silicones are the newest ingredients used in hair conditioners because silicone is water resistant. Some of it remains on hair shaft after it is rinsed with water. Silicones influences manageability of hair by reducing static electricity, minimize tangles by decreasing frictions, also imparting shine by smoothing roughened scales of cuticles. No cases of silicone toxicity have been found yet. Silicone impartes a thin, non greasy film on hair shaft, therefore, it does not create the limp appearance as other hair conditioners do.

Protein Conditioning Agents

There agents strengthen the hair shaft by penetrating damaging hair shaft and increase fructose strength. The diffusion of protein depends on contact time. The longer the protein conditioner left is contact with the hair shaft. The greater the amount of protein that diffuses into the shaft.

These proteins are derived from keratin, animal collagen placenta and other sources.

Formula I

Hair conditioner

Cetyl trimethylammonium bromide (CTAB)	1.3% w/w
Cetyl alcohol	1.75% w/w
Polypeptide additive	0.20% w/w
Potassium bromide	0.04% w/w
Perfume	0.15% w/w
Colour	0.003% w/w
Water (demineralised) to	100.000 w/w

Procedure

Part A

1. Dissolve the CTAB and additive, in 80% of the water.
2. Heat to 70° C and maintain at this temperature

Part B

3. Melt the cetyl alcohol in a water jacketed vessel and maintain temperature at 70° C
4. Add B to A slowly with constant stirring
5. Cool to 30–50° C
6. While maintaining this temperature, add the potassium bromide little by little, while observing its thickening effect.
7. Cool to ambient temperature and add preservative and colour dissolved in the remainder of the water.
8. Lastly add the perfume and stir for three minute.
9. Homogenisation may improve the consistency and stability of the product. The potassium bromide will reduce the solubility of the CTAB and will also improve the colloidal stability of the emulsion below a critical concentration.

Formula II is a more modern version of a conditioner. Like the traditional one it contains an additive which could make it especially suitable for use on damaged hair. In this case DC929 emulsion is exemplified which contains an amino active silicone.

Below: Structural formula for a reactive (amino) silicone.

$$CH_3-\underset{CH_3}{\overset{CH_3}{Si}}-O-\left[\underset{CH_3}{\overset{CH_3}{Si}}-O\right]\left[\underset{\underset{CH_2-NH-CH_2NH_2}{R}}{\overset{CH_3}{Si}}-O\right]\underset{CH_3}{\overset{CH_3}{Si}}-CH_3$$

Amodimethicone

The side chain amino group on the above amido-dimethicone would be expected to become substantive to hair surface by means of the ionic bond it could form with hair surface carboxyl groups.

Other newer additives which benefit hair condition include Poly-quaternium II and Dimethiocone Copolyol.

$$\left[CH_2-\underset{\underset{\text{(pyrrolidone ring)}=O}{N}}{CH}\right]\left[CH_2-\underset{\underset{\underset{\underset{\underset{C_2H_5}{CH_3-\overset{+}{N}-CH_3}}{(CH_2)_2}}{O}}{C=O}}{\overset{CH_3}{C}}\right] \quad C_2H_5HOSO_3^-$$

Polyquaternium. 11

Counter ion

General formula for silicone glycol surfactants

$$CH_3-\underset{CH_3}{\overset{CH_3}{Si}}-O-\left[\underset{CH_3}{\overset{CH_3}{Si}}-O\right]\left[\overset{CH_3}{Si}-O\right]\underset{CH_3}{\overset{CH_3}{Si}}-CH_3$$

$$C_3H_5O\,(C_2H_4O)_m\,(CH_3C_2H_3O)_nH$$

Dimethiocone copolyol

They will be discussed as they appear in examples of hair care formulae.

Formula II

Rinse conditioner

DC 929 emulsion	5.0% w/w
Cetylalcohol	2.0% w/w
Natrosol 250 M	1.5% w/w
Water (deionised) to	100% w/w

Cationic emulsion of amidodimethione (Dow corning), cellulose (HEC) derived from hydro-cellulose (Hercules).

Procedure

1. Little by little add the Natrosol to the water (which is being maintained at 70°C) while constantly stirring.
2. Melt the cetyl alcohol in a vessel with a water jacket and maintain the temperature at 70°C.
3. Add the heated cetyl alcohol to the Natrosol solution with constant stirring.
4. Cool to 30°C
5. Mix in the DC 929 at high speed.

The introduction of acrosol mousse products has in formulation terms, tended to narrow the division between conditioning and control (set) products. Thus Formula III deserves comparison with the setting mousse of Formula VI. Formula III represents a fairly recent proposal for a hair conditioning cream.

Formula III

Hair conditioning mousse

Polymer VC 713*	5.0% w/w
Mulgofen ON 870[#1]	0.5% w/w
Cationic emulsion 929[#2]	0.15% w/w
Ethanol	12.00% w/w
Water (deionised)	82.35% w/w

* Use 90% of the solution with 10% propane/butane mixture in an aerosol pack.
Polymer VC 713- terpolymer of vinyl pyrrolidone, vinyl caprolactam and an aminomethacrylate-50% active (GAF)

#1 Mulgefen ON 870- ethoxylated fatty alcohol (Rhone Poulenc).

#2 Cationic emulsion 929- emulsion of an amido-methicone silicone. Dow corning 929 (Dow corning)

Formula III represents essentially a styling product. It is generally agreed that styling encompasses element of both setting and conditioning. Specialist co polymers have been developed by SP which can improve the condition of hair. By varying the ratios of the constituent monomers, film-formers with either optimized conditioning or setting properties can be obtained. Or, indeed, versatile materials with a balanced combination of both attributes.

The current popularity of mousse products may be due partly to the case of application. The correct dosage can be selected merely by filling the palm of the hand to varying degrees; for example, the size of an egg, the size of a golf ball, etc. directly from the container. Owing to the opacity of the foam the distribution on the hair can be easily observed. The mousse foam breaks down fairly rapidly on wet hair, but not too quickly so as to prevent the user from being able to control its distribution.

Evaluation of Conditioners

Conditioning products are well served by objective methods to establish their various properties. Their effect on the combining of hair may be assessed *in-vitro*, i.e. on test switches of hair, by use of a comb with a spring gauge attachment to measure the resistance. The method can be used without modification in-vivo on actual heads of hair. Similar comment may be made concerning the conditioner's effect on static electrification of dry hair. Here the actual charge can be measured using a charge locator, as employed by factory safety engineers. This instrument is basically a valve voltmeter in which the grid is connected to the probe. The greater the bias of the grid the larger the charges affecting the probe. Measurements can be made with equal facility *in-vitro* or *in-vivo*

The gloss or lustre of conditioner-treated hair may be quantified by scientific instruments in vitro. Human perception of gloss correlates only with an expression which takes into account both the reflected light and the scattered light. In the absence of instrumentation the psychological methods which employ large panels of human observers can confidently be used *in-vitro*. Gloss measurements *in-vivo* are difficult (but not impossible) as so many extraneous factors affect the outcome of these tests.

Wave Sets

The purpose of a setting lotion is to prolong the life of a water wave which has been conveniently made at the end of the shampooing stage. Traditional setting lotions are basically polymer solutions where the solvent is a mixture of ethanol and water Jeffries. These products are applied to hair which is just wet (towel dried) and spread by combining. The hair is then put into curling rollers and dried in a current of warm air (hair dryer). Finally the rollers are removed and the hair gently combed through. Note that the combing at the end of the process will destroy a significant proportion of the hair fibre-polymer bonds. Fundamentally, setting lotions do not work by sticking hair fibres together with the polymer acting as the glue. Their effect is mainly to increase inter fibre friction, thereby giving the hair mass an extra degree of cohesion and control.

Formula IV

Wave set lotion

Luviskol K 30*	2.50%w/w
Ethoxylated castor oil	0.50 %w/w
Ethanol	50.00 %w/w
Water (deionised) perfume, etc. q.s.	47.00 %w/w

Formula V

Wave set lotion

Luviskol VA 371*	2.50% w/w
Ethoxylated castor oil	0.30% w/w
Ethanol	50.0% w/w
Water (deionized)	47.2% w/w
Perfume etc.	q.s.

* Copolymer of vinylpyrrolidone and vinyl acetate (Supplied as solution in ethanol -E) (BASF).

Formulas IV and V illustrates the point made earlier, that traditional setting lotions are dilute polymer solutions. Ideally the solvent is a mixture of ethanol and water. From a health point of view, Formula IV uses the ultra-safe polymer Luviskol K30, whose chemical nature is given above. It has two drawbacks; it is rigid and brittle when dry, but also takes up moisture to become tacky. The substitution in Formula V, largely overcomes these objections, and gives a more acceptable product.

In both formulae the function of the ethoxylated castor oil is as a plasticizer for these relatively brittle polymers. The plasticizer may be substituted with advantage by more modern synthetic materials, e.g. the Carbowaxes (Goodrich) or the Mulgofens (ISP) (Rhone Poulene).

Formula VI

Styling mousse

PVP/VA E 735[#1]	2.00% w/w
Mulgofen AM 650*	0.50% w/w
Polymer Gafquat 755 N[#2]	5.00% w/w

Ethanol	10.00% w/w
Water (deionised)	77.50% w/w
Perfume /colour etc.	q.s.
Propellant (propane/butane)	5.00% w/w

* Copolymer of vinyl acetate and vinyl pyrrolidone 50% active (Rhone Poulenc)

#1 Copolymer of vinyl pyrrolidone and an amino-methacrylate-50% active (ISP)

#2 Ethoxylated fatty alcohol (Rhone Poulene)

Formula VI is related to Formula III, whereas the latter is an aerosol mousse optimised for hair conditioning. VI is intended to be competitive with lotion styling aids. It can be used to produce a roller curl or for blow drying, and with the roller curl the blow drying process involves applying the product to towel dry hair. Blow drying involves simultaneous drying and brushing combing. It has proved an ideal way of dealing with problem of hair conditions such as fineness, limpness and unmanageability. It will be noted that Formula VI contains an additional polymer comprising both vinyl pyrrolidone and vinyl acetate. The proportion of the former monomer is relatively high (30–35%) which means that the film of the PVP/VA copolymer is relatively rigid. It is the extra stiffness contributed by this copolymer which makes Formula VI more suitable to be used as a wave set than as purely styling product. Another feature of this product is that it is discharged from the container by a mixture of propane and butane hydrocarbons, today considered 'ozone friendly' (instead of the former CFC, Freon 12).

In many respects aerosol mousses are, when the container is shaken, oil in water emulsions in which the propellant is the internal oil phase. On release from the container the propellant droplets expand rapidly forming the characteristic foam.

Other non aerosol, blow-drying products have been developed. Formula VII is representative of earlier ideas on formulating them.

Formula VII

Blow drying aid

Cetyl trimethyl ammonium bromide (CTAB)	0.2% w/w
Ethanol	50.0% w/w
Perfume	0.3% w/w
Colour	0.3% w/w
Water (deionised)	49.2% w/w

A number of modifications have been successfully introduced. These include the substitution of CTAB by a whole range of different quaternary ammonium salts as wetting agents. Other modifications of the basic formula have been made for example, by increasing the level of CTAB or its analogues and the incorporation of small amounts of urea (carbamide). The urea is intended to swell the hair shaft and thereby increase the uptake by hair of the wetting agent. Formula VII and its derivatives have been proposed as adjuncts to heated curler methods for producing temporary waves in hair.

Marketed Products

- Anti-Dandruff Shampoo and 2–1 Anti-Dandruff Shampoo and Conditioner
- Body and Volume Root Lifting Treatment Spray (Garnier)
- Body and Volume Shampoo and Cream Conditioner (Garnier)
- Body and Volume Shampoo and Cream Conditioner (Garnier)
- Color Treated or Perfumed Shampoo and Cream Conditioner (Garnier)
- Curl and Shine Leave-in Conditioner (Garnier)
- Curl and Shine Shampoo and Cream Conditioner (Garnier)
- Dry or Damaged Shampoo and Cream Conditioner (Garnier)
- Fine Shampoo and Cream Conditioner (Garnier)
- Fortifying Deep Conditioner (Garnier)
- Hard Curl Gel (Garnier)
- Length and Strength Anti-split Ends Treatment (Garnier)
- Length and Strength Shampoo and Cream Conditioner (Garnier)
- Normal Shampoo and Conditioner and 2–1 Normal Shampoo/Conditioner (Garnier)

- Sleek and Shine Anti-frizz Serum (Garnier)
- Sleek and Shine Deep Conditioner (Garnier)
- Sleek and Shine Leave-in Conditioning Cream (Garnier)
- Sleek and Shine Shampoo and Cream Conditioner (Garnier)
- Soft Curl Cream (Garnier)
- Surf Hair (Garnier)
- H_2O Sea Plankton Restructuring Conditioner
- Schwarzkopf Professional Bonacure Moisture Conditioner
- Hydromilk (Wella System Professional)
- Tenderils Hair Wash
- Lakme Hair Col Our Fresh Conditioners.

13 Shampoo

> *Beauty.....is the shadow of God on the universe.*
>
> — Gabriela Mistral, Desolacion

Shampoo is a preparation meant for cleansing hair of dust, grime, crust and to impart lustre and gloss to hair. The word shampoo in English usage dates back to 1762, with the meaning "to massage". The word derived from Anglo-Indian shampoo, in turn from Hindi *châmpo* imperative of *châmpnâ* "to smear, knead the muscles, massage". It itself comes from Sanskrit/Hindi word *champâ*, the flowers of the plant *Michelia champaca* which have traditionally been used to make fragrant hair-oil. Previously soap cake was used for cleaning and washing of hair. Today, a plethora of shampoos are available for men and women.

IDEAL PROPERTIES OF A GOOD SHAMPOO

- It should be easily removable on rinsing of hair.
- It should effectively remove dust, dirt, soil, sebum and residues of hair setting lotions or oils from the scalp.
- It should have a pleasant fragrance.
- It should be non-toxic and non-irritant to skin and eyes.
- It should produce good amount of foam to satisfy psychological needs of customer.
- It should provide good finish after washing hair and make hair manageable, lustrous, silky and soft.
- It should perform its function in small amount.
- It should be able to form foam in hard and soft water.
- It should have pH 5.0 to 9.0 or slightly acidic pH, since a basic environment weakens the hair by breaking the disulfide bonds in hair keratin.
- It should have good biodegradability and cause no damage to hair
- Feels thick and/or creamy

- It should be clear and transparent if liquid and free from any agglomerated particles if paste, it should have free flowing property if in powder form.
- Many shampoos are pearlescent. This effect is achieved by addition of tiny flakes of suitable materials, e.g. glycol distearate (a wax).

CLASSIFICATION OF SHAMPOOS

According to the criteria employed for this taxonomy, two generally acceptable systems can be proposed. Where the criterion is based mainly on the appearance of the product the following will suit:

- Clear liquid
- Liquid creams
- Clear gel
- Paste (opaque)
- Powder shampoos
- Dry shampoos
- Pressurized pack
- Mousse shampoos.

A more modern classification is based upon functionality:

- Conditioning
- Therapeutic
- Antidandruff
- Mild (often referred to as 'Baby') shampoos
- Acid -balanced shampoos (closely related to mild shampoos).

Reference to these systems especially the classification related to functionality, will be made in the section of this chapter devoted to shampoo formulation.

Shampoos are generally prepared by two methods, i.e. using synthetic detergents or using soaps.

Synthetic detergents based shampoos are very much popular. Shampoos are available in liquid, emulsion, paste, gel or powder, these are suitably coloured and perfumed.

Colours used in shampoos must conform to the provisions of relevant Indian standard subject to provisions of schedule Q of the Drug and Cosmetics Act and Rules.

THE CLEANING OF HAIR BY SHAMPOOS

Basic components of hair soil have been recognised as:

1. Sebum itself, the oily secretion of the sebaceous glands
2. Pertinacious matter, originating from the cell debris of the *stratum corneum* layers of scalp skin, and also the protein content of sweat.
3. Soil from the atmosphere and from hair care products.

Table 13.1: Composition of sebum

Component	*% w/w*
Cholesterol	8.5
Fatty acids (free)	22.0
Triglycerides	35.0
Wax and wood wax esters	18.6
Squalene	11.3
Sundry hydrocarbons	5.0

There is an extensive literature on the subject of sebum which deals with the physiology of the secretion. In order to gain an insight into the *in situ* properties of sebum a number of studies were made. In the course of their investigations they reached the conclusion that the free fatty acids of sebum may well be linked to the protein surface of hair through calcium atoms.

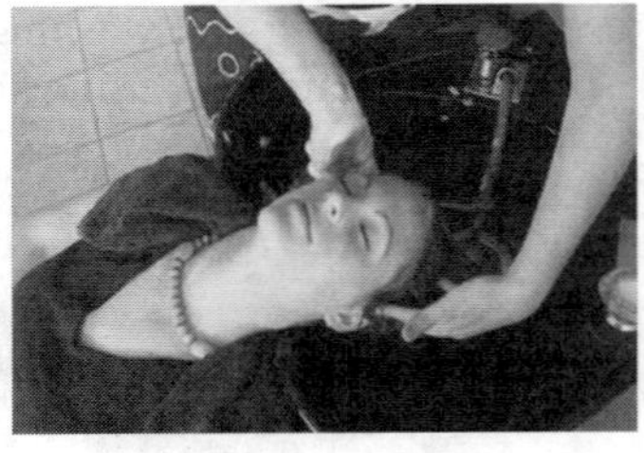

Among the methods which have been used to monitor the spreading of sebum was a sophisticated optical technique which measured changes in hair fibre spacing as the sebum spread in a parallel assembly of hair fibres. He also established, in other experiments, that the rate of sebum spreading, when the hair had been dried in a current of hot air from an electric hair dryer, was considerably greater than when the hair had been allowed to dry at room temperature.

THE MECHANICS OF SOIL REMOVAL

In essence, the removal of lipid soil from hair is controlled by the same processes as those that have been identified in detergency, in spite of the differences from textile laundering in respect of times and temperature. The principal agency is generally considered to be the roll-up mechanism, i.e. the displacement of surface soil by the detergent solution.

Myclenisis can be observed (through a microscope) when a layer of lipid is immersed in water. The lipid layer develops peninsular like processes which penetrate into the aqueous medium. These appear to function like pipes, transporting the lipid progressively into the bulk water phase. In the region the former surface lipid coexists within micelles of the detergent solution as co-micelles of detergent and lipid. The effect is best seen when the lipid is highly polar, e.g. in the phospholipids, lecithin and cephalin. The rate at which the lipid migrates into the aqueous phase is very temperature dependent.

SUMMARY OF CLEANSING

Although detergency plays an important role in the cleaning of hair with shampoo, other factors must be considered. For example, gaps exist in our knowledge of the physico-chemical nature of the ageing of sebum and how this is related to its theological properties. We would also wish to know the extent to which captured particles of soil from the atmosphere modify the fluidity of sebum.

It is also important to be able to apportion the individual contributions of surface energy and surface rigidity to the rate at which hair is regressed. Another element of fundamental information which is not currently available to us attaches to the relative importance of tactile and visual factors in determining the perception of the cleanliness of hair.

THE FOAMING OF SHAMPOOS

The signal to which the user responds when applying a shampoo is how quickly it builds up lather and how copious that lather is. Three well defined stages appear to be involved in the rapidity with which the foam is formed; the peak volume of the foam; and the consistency of the lather. High consistency foam is judged as being 'creamy'. It is not surprising, therefore, that the shampoo formulator needs to be able to measure the important foaming properties, even though the fundamental properties of a foam, e.g. interfacial tension and film modulus, do not loam a reliable guide to the performance of the shampoo in practice.

A standard volume of a shampoo solution is translated to a tap funnel. The solution in the funnel is run in a standard time into a large measuring cylinder which already contains a set volume of the solution or merely the dilution water. The result of the stream of solution from the funnel impacting upon the liquid in the cylinder is to generate foam, the volume of which can be read directly. The procedure can be modified; for example, the cylinder may contain a suitable quantity of sebum treated hair, or the gravity feed from the tap funnel can be replaced by a pump.

The Ross Miles method like some other methods for quantifying foam, usually ranks the foaming of shampoos in the same order as human judges do (users, panellists and hairdressers), but not invariably. Other methods of quantifying shampoo foaming capacity utilize proper stirring at air injection to generate foam but are in general, less reliable.

The quantification of the consistency of foam (creaminess) by *in-vitro* laboratory techniques is less well provided for than the measurement of foam volume. The principle of the method is that high consistency foam will take considerably longer to flow out via the stem (broad) of a powder funnel than foam which is thin and dubbed as non-creamy.

An *ideal* laboratory method for predicting the foaming power of shampoos would closely simulate the practical shampooing process. It would ensure that the foam was generated in a way similar to its formation on the head. Similarly, the composition of the system in which the experimental foam is produced would be as similar as possible to that of the hairdressing situation in terms of the materials present (hair, sebum, detergent, water).

The temperature and humidity profile would also be modelled upon that met with in hairdressing practice. Perhaps the most important simulation would be that of the mechanics and dynamics of foam generation. Lather production in practical shampooing is not by cascading water, mechanical stirring or gas injection. It is in fact achieved by a process of compressing and shearing hair when it is saturated with shampoo solution. The foam produced by compression and shear is then modified by the practice of separating by finger action a particular mass of hair fibres and shampoo solution into smaller 'swatches' before recombining them. The engineering problems of designing a machine to meet the requirements described above are formidable, but not insurmountable.

SHAMPOO INGREDIENT

A shampoo is basically a solution of a detergent modified by additives to render it easier to apply and to safeguard against deterioration of the hair condition after the shampoo has been rinsed away. The following list classifies the materials of shampoo formulation; the subdivision are not, however, mutually exclusive, e.g. viscosity modifiers can some times be used to stabilize or boost the loam and some specifying agents can also improve foam quality. Likewise amphoteric wetting agents can be used as the main detergent for specialist shampoos. They are also valued as hair conditioning agents.

Summary of shampoo additives, functions with examples:

- **Detergents (surfactants):** They clean the hair, remove dirt, soil and debris from hair and scalp. They perform well in hard water also in which soap do not give foam, e.g. sodium lauryl sulphate, alkyl benzene polyoxyethyl sulphonates, sodium lauryl sarcosinate, triethanolamine lauryl sulphate, potassium oleate, triethanolamine alkyl sulphate etc.
- **Foam boosters (Sudsers) and stabilizers:** They increase the quality, volume and stability of the foam. They also enhance the viscosity and leave slight conditioning effect, e.g. isopropanol amides of fatty acids, amine oxides, ethanol amides, etc.
- **Preservatives:** Preservatives are added to preserve shampoo from microbial growth, e.g. methyl and propyl parabens, alcohols, etc.
- **Opacifiers or clarifying agents:** These are added to turn shampoo opaque. For example, higher fatty alcohols, salts of fatty acids, mono- or di- glyceryl stearates, ethylene or propylene glycol stearates, etc.
- **Hydrotropes:** They are solubilizing agents added to solubilize poorly soluble ingredients and hence to avoid any hazyness in shampoos, e.g. surfactants, urea, alcohol, glycols, sodium benzoate, sodium o- or p- hydroxy benzoate, sodium salicylate, etc.
- **Viscosity modifiers, including hydrocolloids and electrolytes (thickening agents):** They are added to enhance the viscosity of the shampoos; their quantity is adjusted to provide desired consistency of the preparation, e.g. methyl and ethyl cellulose, sodium CMC, sodium alginate, polyvinyl alcohol, sodium or potassium chloride, etc.
- **Special additives for hair condition (conditioning agents or emollients):** They are added to improve the texture of the hair and to render them manageable, lustrous and silky, e.g. glycol esters, lanolin and its derivatives, fatty alcohols, etc.
- **Special additives for scalp health (antidandruff substances):** They are used to reduce dandruff from the hair, e.g. hexachlorophene, selenium sulphide, selenium disulphide, etc.
- **Sequestering agents or chelating agents:** They are added to prevent the formation of lime soap due to presence of hard water, e.g. sodium salts of EDTA, sodium polyphosphates, etc.
- Miscellaneous additives like colours, perfumes to improve their cosmetic value.

Detergents (Surfactants)

Development of synthetic detergents has revolutionized the shampoo market. These will be classified according to the way in which they ionize.

CLASS I ANIONICS SURFACTANTS

Anionic surfactants are principal most widely used surfactants employed in shampoos formulation. They have good foaming ability and have low cost.

Commonly used anionics are further subdivided into five major chemical classes and additional subgroups:

ACYLAMINO ACIDS AND SALTS

1. *Acylglutamates:* Mild, used in skin cleansing and shampoos.
2. *Acyl peptides:* Mild cleansing agent available as aqueous and alcoholic solutions for hair and skin products, they can tolerate hard water.
3. *Sarcosinates and taurates:* Basically similar to the isethionates are taurates, which are also sulfonates, and the carboxylate sarcosinates, the structural formula for which is as follows:

Acyl lactylates: These class shows good foaming, cleansing and thickening properties when acceleration of length of fatty acids occurs.

$$R{-}CO{-}CH_2{-}\underset{\displaystyle OH}{\underset{|}{CH}}{-}CH_2{-}OSO_3\ M$$

CARBOXYLIC ACID AND SALTS

1. Alkanoic acids and alkanoates
2. Ester carboxylic acids
3. Ether carboxylic acids

Phosphoric Acid Esters and Salts

Sulphonic Acids and Salts

1. Acyl isothionates
   Interesting properties apart from detergency and foaming potential. The isethionates, for example, are claimed to be very mild to skin and eyes:
   $$R{-}COOCH_2{-}CH_2{-}SO_3M$$
   Additionally, they are particularly tolerant of hard water.
2. Alkyl aryl sulfonates
3. Alkyl sulfonates
4. Sulfosuccinates: Less used than the above are the **mono and di-sulfosuccinates**: These are used in baby shampoos and has low incidence of irritation to eye and scalp. The synthesis of a dialkyl sulfosuccinate is given below:

Sulphuric Acid Esters

1. **Alkyl ether sulfates:** The alkyl radical R is Cl_2, Cl_4 unless otherwise specified. The counter ion symbol is M, which can be sodium, ammonium, mono-, di-or tri-ethanolamine. The commonest are alkyl sulphates and alkyl ether sulphates with the latter becoming more popular because of their superior solubility and temperature stability.
2. **Alkyl polyethylene glycol sulphates:** It has good foaming property but foam collapses in the presence of grease.
   Another group of sulphate detergents are the **alpha olefin sulphates**. These shampoo ingredients have considerable potential due to their high foaming properties and low cloud point. Similarly placed are alkylaryl sulfates, depicted below, which currently are much used in dish washing preparations.
   M is almost invariably sodium
   R attached to any C, R is typically C8 and C9

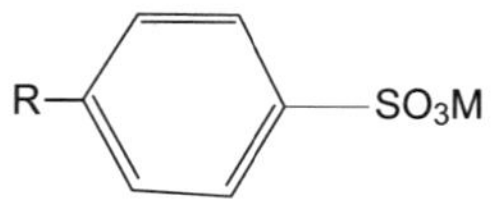

$$R{-}C_6H_4{-}SO_3M$$

Another acceptable group of compounds, although less used than alkyl sulfates and **alkyl ether sulfates**, are the monoglyceride sulfates.

2. **Alkyl sulphates:** Alkyl sulphates are most widely used surfactants. They are available as:

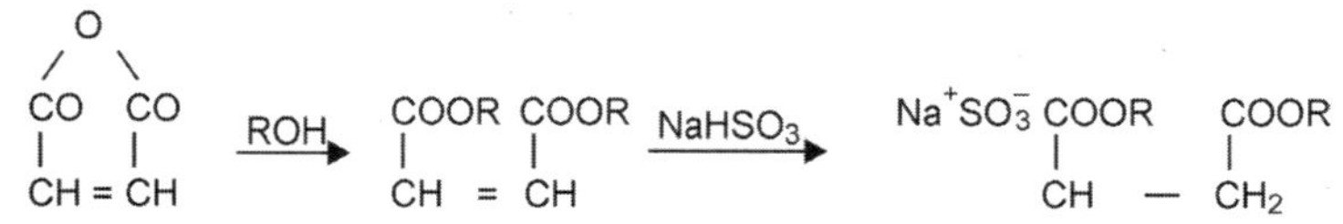

$$\underset{\text{Maleic anhydride}}{\text{(CO–O–CO) CH = CH}} \xrightarrow{ROH} \text{COOR–CH = CH–COOR} \xrightarrow{NaHSO_3} \underset{\text{Sodium dialkyl sulfosuccinote}}{Na^+SO_3^-\ \text{COOR–CH} - \text{CH}_2\text{–COOR}}$$

a. **Lauryl sulfates and myristyl sulphates:** greater volume of lather produce by lauryl sulphates and Myristyl sulphates provides richness to shampoo.
b. **Octyl and decyl sulphates:** These are foam suppressant.
c. **Sodium lauryl sulphate:** It is most widely used surfactant. It increases solubility at 35–40°C. It has sufficient viscosity and cloud point suitable for paste shampoo. Its HLB value is 40.

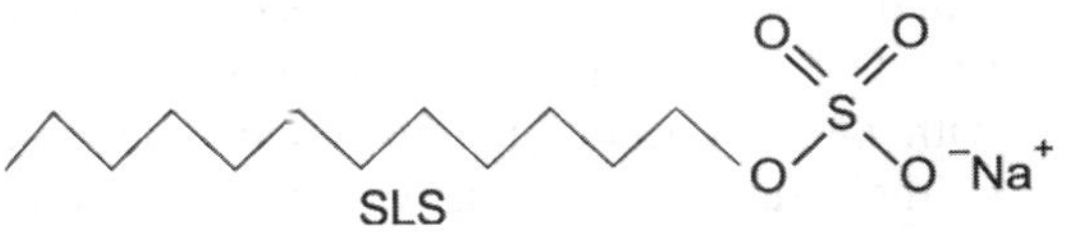

$$CH_3(CH_2)_{11}-O-S(=O)_2-O^-\,Na^+$$

SLS

CLASS 2 NON IONICS SURFACTANTS

Surfactants of this category are not often used as the shampoos main ingredient because they do not have good foaming ability. Soaps were used earlier as they are cheap but they are highly alkaline and they make hair dull, have sufficient cleaning power, and are foam booster and stabilizers. Further they also leave deposits of calcium and magnesium with hard water. The nearest to this definition are certain block co-polymers of ethylene and propylene oxides often referred to a pluronics. The structural formula is:

$$R\text{–}CO\text{–}CH_2\text{–}\underset{\displaystyle OH}{\underset{|}{CH}}\text{–}CH_2\text{–}OSO_3M$$

The number of ethylene oxide units can vary from 100 to 200 and the number of propylene oxide units can vary from 15 to 30. Like isothionates, pluronics are held to be mild to skin and eyes.

The main use of non ionic surfactants in shampoos is as modifiers of foam characteristics, especially foam quantity and consistency. The degree to which the superamides, mono and dialkanolamides, adjust foam properties needs to be quantified, although it is known that modest addition of these materials, e.g. 1.3% by weight increases the viscosity of the product making it easier and safer to apply during the shampooing process. Stearic ethanolamides are regarded as effective pearlizing agents and are also said to boost lathering. Structural formula for mono- (I) and di- (II) are given below.

$$R-CONH-CH_2-CH_2-OH-R-CON\begin{matrix} \diagup CH_2\,CH_2\,OH \\ \diagdown CH_2\,CH_2\,OH \end{matrix}$$

A radically different molecule which performs the same role as the alkanol amides is represented by the N-alkylpyrrolidone, trade name Surfadone; the CTFA name is caprylyl (or lauryl) pyrrolidone. This improves the solubility of Sodium lauryl sulphate. It softens hair after washing and improves volume and richness of foam.

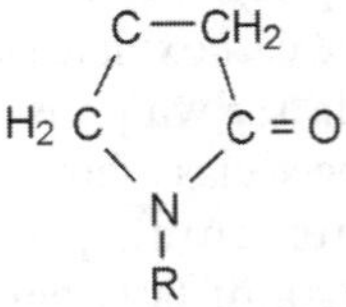

$$\begin{matrix} & C-CH_2 & \\ / & & \backslash \\ H_2C & & C=O \\ \backslash & & / \\ & N & \\ & | & \\ & R & \end{matrix}$$

R = Caprylyl or lauryl

CLASS 3 AMPHOTERICS SURFACTANTS

These are defined as having both acidic and basic properties (**Zwitterions**) and used in mild shampoos. Some examples of amphoteric components are:

1. Long chain **alkyl N substituted amino acids**: These are two types-**b-amino acid derivatives**: these produces foaming in alkaline pH and hair manageability in acidic pH. **Asparagines derivatives**: they exhibit good cleaning and conditioning properties.
2. Long chain **alkyl N-substituted betaines**: Zwitterions compounds have high foaming properties
3. Long chain **alkyl derivatives of imidazoline.**

The structural formula for the betaine derivative is:

$$R-\underset{\displaystyle CH_3}{\underset{|}{\overset{\displaystyle CH_3}{\overset{|}{N}}}}-(CH_2)_n-COOM$$

Hence it is capable of functioning as an anion in alkaline solution or as a cation in acid solution.

It has been demonstrated to be mild, substantive to hair and to produce a foam which is largely unaffected by pH changes. The imidazoline derivatives have similar characteristics; they are sold under the trade name of Miranol. These wetting agents can be constructively combined with practically all surfactants; they are also tolerant of wide ranges of electrolyte concentration. In respect of their mildness they afford an interesting comparison with isothionates, pluronics and indeed betaines as candidates for 'baby' (or frequent use) shampoos.

CLASS-4 CATIONICS SURFACTANTS

Cationic surfactants are only used as conditioner and not as principal surfactants because of their irritation potential. In the search for novel benefits for anionic based shampoos many new cationic materials have been evaluated. They produce foam well with reasonable cleaning power and leave hair lustrous and free of electrostatic charge. Disadvantages with cationic detergents are they cause injuries to corneal eye tissues and weigh down hair.

Cationic surfactants are subdivided into four major classes:

A. **Alkylamines:** These are waxy solid, salts is effective hair conditioners and have antistatic property.

$$(R—CH_2NH_2)$$

B. **Alkyl imidazolines:** These are used in cosmetics as emulsifiers or as substantive hair conditioning agent, used in aqueous media.

C. **Ethoxylated amines:** These are waxy solids melts at low temperature, used as emulsifiers and hair conditioning agents.

D. **Quaternaries:** These are used extensively as surfactants and antimicrobial agents, causing ocular and topical irritation.

1. **Alkyl benzyldimethylammonium salts:** used importantly in hair conditioning, skin degerming agents, deodorants and anti-microbial salts, e.g. **Benzalkonium chloride**.

$Cl^{\ominus}$ F $N^{\oplus}$

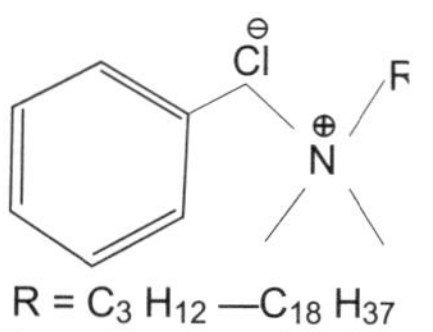

$R = C_3H_{12} — C_{18}H_{37}$

Benzalkonium chloride

2. **Alkyl betaines** are the N-alkyl derivatives of N-dimethylglycine, they foam even in hard water and do not become cloudy at low temperatures, mildness account for their wide use in baby shampoos and facial cleansing products.
3. **Heterocyclic ammonium salts:** These are derived from imidazoline and morpholine are used primarily as hair conditioning and anti static agents, they also exhibit good stability in cosmetics.
4. **Tetra alkyl ammonium salts:** They possess the general structure $[R_1\ R_2\ R_3\ R - N^+]X$, where R_1, R_2 and R_3 represent identical or separate alkyl groups and X^- represents an anion. Used in cream rinses hair and skin conditioners and as antistatic agents.

SHAMPOO ADDITIVES

These materials, known as product 'Builder' additives, do not in the main affect the hair washing performance of shampoos. Their function is to give the product in the bottle a desirable appearance and consistency. It was felt, with a certain amount of evidence in support that an opaque shampoo appealed to consumers who saw their general hair problem as dryness. The opposite view prevailed amongst people who saw their problem in hair dressing, thus, potential users with greasy hair required a product which did not appear fatty, i.e. a clear product. The other group whose hair was deficient in natural grease., felt more confident in using what was essentially the same clear product but with an greacifying material added to give a cream like appearance. To achieve this effect the additive must be insoluble in the shampoo and form a stable suspension. There is a wide range of such materials available to the formulator including;

- Adducts of higher fatty acids with ethylene oxide;
- Stearic monoethanolamide;

- 1,3-propylene glycol stearate;
- Magnesium stearate

Stearic monoethanolamide and 1, 3-propylene glycol stearate can give rise to a pearlescent effect arising from controlled crystallization.

To achieve the opposite effect, clarification, a small addition of a solvent such as ethanol may be added. The same result may be achieved, however, using another, less expensive hydrotrope, namely urea.

It is often necessary to increase the viscosity of a liquid shampoo. The opacifier and pearlescent agent will generally achieve this, but will not be suitable if the shampoo must remain transparent. In this case, viscosity is achieved by the addition of an electrolyte, normally sodium chloride.

Although it has not been substantiated that the alkanolamide superamides boost foam stability they do increase the viscosity of shampoo liquids without adversely affecting its optical properties. Hydrocolloids such as polyvinyl alcohol and methyl cellulose can also achieve increases in product consistency without affecting appearance. The incorporation of a cellulose derivative requires some skill; it must be predispersed in water before final incorporation in the shampoo liquid. It is advisable to follow the manufacturer's instructions faithfully.

Hydrotropes such as urea are often effective in increasing the dispersion of shampoo components, and in the process increasing their activity if they are functional materials.

FUNCTIONAL ADDITIVES

Antimicrobial substances added to a shampoo must be subjected to long term tests to assess their effectiveness as preservatives, their stability in the product and consumer safety in use. The functional but not the biologically active additives to be described in the following pages are those which promote the good condition of the hair. Hair in good condition is easy to comb in both wet and dry states. In the dry state it is free from the electrostatic phenomenon known as flyaway and is not dulled by the product treatment. It may even be enhanced in gloss. More success in achieving good hair condition is being obtained by using the newer synthetic materials than was possible with traditional materials. It must not be forgotten, however, that the acid balanced shampoos are essentially conditioning treatments which do not contain the newer materials listed below:

1. Novel amine oxides
2. Polyquaterniums (CH-A designations) of which several are currently available.
   a. Polyquaternium 5 (Retene 220 of Hercules Chem. Corp.)
   b. Polyquaternium 7 (Merquat 550 of Merck)
   c. Polyquaternium 10 (Polymer JR of Union Carbide)
   d. Polyquaternium 11 (Gafquat 734, 735 of GAF Corp.)
   e. Polyquaternium 24 (Quatrisoft LM200 of Union Carbide)
3. Silicone additives of which the following are representative;
   a. Silicone surfactants (designation dimethicone copoly of block or graft copolymers of dimethyl silicone and ethylene oxide)
   b. Reactive silicones containing an active amine group l; CHA designation; Amodimethicones. These have obvious advantages in terms of substantivity to hair.

Amine Oxides

Smith, Johannessen and Bauer have reported on the properties of some novel amine oxides. It will be remembered that the incorporation of amine oxides in shampoo formulae is generally held to enhance most of the aspects of good hair condition. Smith *et al.* investigated didecylmethylamine oxide (DI) and stearyldimethylamine oxide (ST). Preliminary experiments with space filling models revealed considerable chain branching in the case of the didecyl compound; the opposite was true of the stearyl compounds. A further comparison was made, this time in terms of their respective effect on shampoo performance using shampoos of the following formula:

Ingredients	% w/w
Ammonium lauryl sulfate	15.0%
Lauryl diethanolamide	3.0%
Amine oxide	2.0%
Water	80.0%

The results of the comparison is summarized in the Table 13.2. The amine oxide (DI) with the highly branched alkyl chains is superior in all respects except for the adverse effect on shampoo foaming.

Table 13.2: The influence of amine oxide structure upon hair conditioning performance

Useful properties	*DD*	*ST*
Combing	√	
Foaming	√	√
Flyaway control	√	
Solubility acid pH	√	
Where DD = *didecylmethylamine oxide;* ST = *Stearyldimethylamine oxide,* √ = *best performance*		

Water soluble polyquaternium mechanics of deposition from solution

Table 13.3: Effect of the toxicity of surfactant and polymer upon surface tension reduction

Surfactant	*Polymer*	
	JR 400	*Cellosize*
Sodium lauryl sulfate	√	×
Sodium alkyl aryl sulfate	√	Not tested
Sodium laurate	√	Not tested
C14 Betaine derivative	×	Not tested
Non-ionic	×	Not tested
Where √ = Change × = no change		

SURFACE TENSION VERSUS CONCENTRATION MEASUREMENTS

Surfactant concentration measurements were obtained in the presence and in the absence of 0.1 per cent of the cationic polymer JR400. This procedure was repeated for several surfactants. For some combinations of JR and surfactant there was a considerable lowering of the surface tension at the lowest surfactant concentrations. For the rest there was no real change from the plot for the surfactant alone. **Table 13.3** summarizes the results which are illustrated in **Fig 13.1.** It can be seen that a reduction in surface tension occurs only when an anionic surfactant is used in association with Polymer JR400. No change in surface tension is measured when an anionic detergent is tested with a non-ionic, neutral polymer substituted for polymer JR.

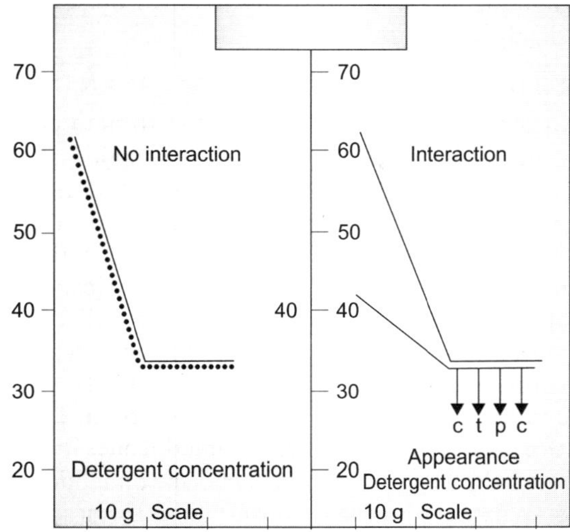

Fig. 13.1: Surface tension versus concentration plots can reveal association between surfactant (detergent) and polymer Appearance c clear L turbid; p, precipitate

Change inlets a strong interaction between the anionic surfactant and the cationic polymer. Visual examination reveals a pattern of precipitation near the critical micelle concentration. It is reasonable to assume that the maximum precipitation occurs at ratios of polymer to surfactant at which the polymer charge is balanced by that of the surfactant, compliant with the expression.

$$(PSt)\ n - 1 - S = (PS_1 + 1)\ n - 1$$

$$pn - 1 + nS = PS^n$$

n = positive charges

S = Surfactant

P = Polymer

Finally as the surfactant is increased the precipitate is redissolved. The technique is useful in screening additive detergent systems for their propensity to deposit a complex which is potentially beneficial in a shampoo formulation. Having identified where a useful level of precipitation occurs the next step is to determine how substantive (to hair) is the precipitated material. It has been established that certain cellulose based cationics gave a high level of deposition but that it was fairly easily washed away by a dilute detergent solution. This would have the virtue for a product of avoiding the troublesome build up of an active ingredient, i e. one residing removal by several shampooing

SILICONES AS CONDITIONING AGENTS

Silicone surfactants such as the dimethicone copolyols, when incorporated into shampoo formulations, greatly improve combing and antistatic properties at surprisingly low concentrations 0.1–0.5%. The structural formulas for these materials have already been presented. Alexander also described another novel series of silicones, the amino functional amodimethicones which, surprisingly, have a good compatibility with anionic vehicles. Their chemical structure has already been discussed. Amodimethicones impart much the same benefits to hair as the dimethicone co-polymers but have as an added feature of a very good substantivity to its surface. This is not totally unexpected as the functional amino group is capable of forming an amine salt linkage with the free carboxyl groups of the hair surface; rather in the manner that the carboxyl groups of certain hairspray polymers are neutralized by treatment with amino alcohols. Starch has described how the substantivity of the amodimethicones has been demonstrated using ESCA methods.

A class of shampoo component which is difficult to fit into the classification adopted for this chapter are those known under the trade name of Mirapol. *Their general structural formula is given below*

Mirapols have two properties which recommend them to the shampoo formulator; they are compatible with anionic detergents and they can enhance the conditioning properties of the shampoo into which they are incorporated. In some respect they resemble the polyquaterniums which were described earlier. One difference is that they are of considerably lower molecular weight (in the above formula $n = 6$) giving a molecular weight of approximately 2200 in oligometic material rather than a true polymel. Another difference from polyquaterniums is that each repeat unit has three quaternary nitrogens.

ANTIDANDRUFF AGENTS IN SHAMPOO BASES

Various agents are used as antidandruff agents like selenium sulphide, zinc pyridinium thiol N-oxide, zinc undecylnate, bithinol, resorcinol, etc.

Three agents associated with the treatment of scalp disorders have the chemical structures shown here. **Zinc pyrithione (Zinc pyridinium thiol N-oxide)** or ZPT was the first scientifically-based organic therapeutic agent to offer alleviation of the scalp disorder known as dandruff. Dandruff manifests itself as the detachment of flakes of scalp skin. Almost contemporary with ZPT was another antidandruff agent, pyroctone olamine (PO). Structurally it has little in common with ZPT except the presence of a pyridine ring.

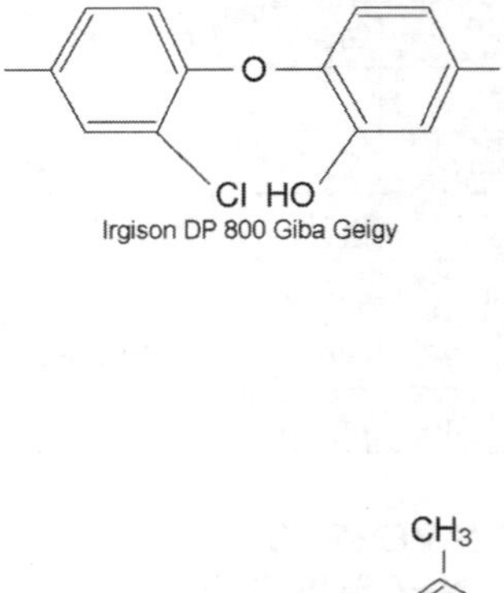

Cl HO

Irgison DP 800 Giba Geigy

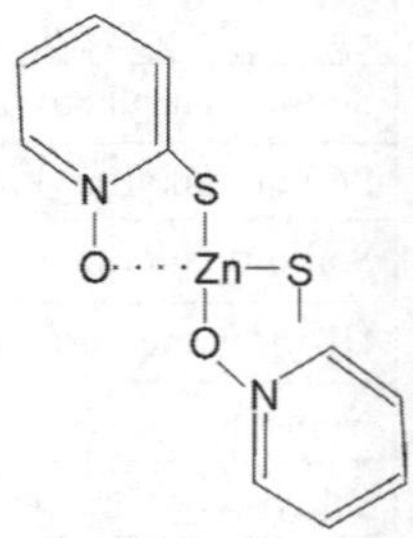

Zinc pyrithione (ZPT)

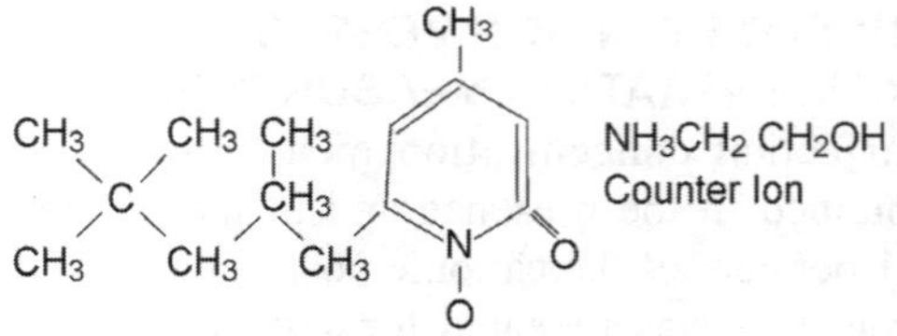

$NH_3CH_2CH_2OH$
Counter Ion

Pyroctone olamine

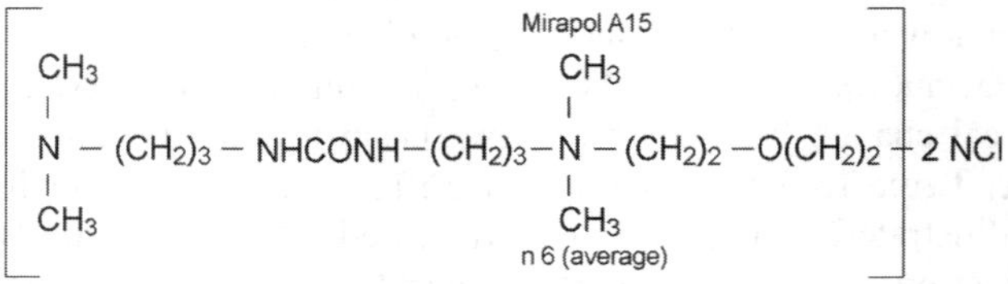

Mirapol A15

$$\left[N(CH_3)_2 - (CH_2)_3 - NHCONH - (CH_2)_3 - N(CH_3)_2 - (CH_2)_2 - O(CH_2)_2 \right]_n \; 2\,NCl$$

n 6 (average)

Table 13.4: Properties of healthy and seborrhoeic sebum		
Properties	*Sebum*	
	Healthy	*Seborrhoeic*
Quantity*	Less	Somewhat more
% Squalene	9%	12%
Iodine number	80	.100
Palmitic/Oleic acids**	1.0	0.7
Viscosity proportional to	1.5	0.7

* For the same extraction procedure.
** Ratio of saturated to unsaturated acids.

Shampoos are the most acceptable vehicle for an antidandruff treatment. Futterer found that reductions in dandruff level of the order of 68% could be achieved for ZPT treatments but, under the same conditions 82% when PO was the biologically active agent. The statistical significance of this difference in performance corresponds to $p < 0.05$. He also experimented with different concentrations of the biologically active materials and found that 0.5%. PO gave only a marginally different antidandruff performance to 0.75% ZPT. The lower concentration of PO needed for the desired result makes it preferable to the formulator.

The sebum samples were removed by a shampoo of the following constituents:

Formula I

Sodium lauryl sulfate	1.3%
Betaine derivative	2.0%
Lactic acid	0.24%
Water to	100%

PREPARATION AND MANUFACTURE OF SHAMPOOS

Relative to most other personnel products the preparations of shampoos is uncommunicated and straight forward. Nevertheless, extreme care at the development stage of the formulation is necessary to ensure the long term stability, compliance with microbial guidelines, regulations concerning consumer safety and consumer acceptability.

In addition the preparation of shampoos has some specific problems.

Shampoo Preparation

Care must be exercised with regard to the solubility of various components. Stability tests can be made which all determine whether deactivation of functional additives through interaction between components is taking place.

Example I

A typically straightforward case where functional ingredients are inert towards the other components.

1. The main detergent, foam booster (usually a 'super amide'), the hair functional additive and water are mixed together with gentle stirring to minimize frothing.
2. Citric acid is added carefully to the above mixture to adjust the pH to within the limits 6.5–7.0.
3. A consistency adjuster, say *N-alkyl* betaine, is added to the pH adjusted blend with more vigorous stirring until the desired viscosity is attained.

Example II

Where some of the components are difficult to solubilize:

1. Dissolve the main detergent in the water.
2. Add to the above whilst it is being moderately stirred the foam 'booster', difficulty dispersible materials such as opacifier and a functional ingredient which may also present problems of dispersion functional ingredient which may also present problems of dispersion.
3. Adjust the pH with citric acid to 6.5–7.0.
4. Adjust the viscosity with electrolyte additive (sodium chloride).

Example III

Where heat is needed to obtain solution

1. Mix by propeller stirring the functional ingredient and the foam 'booster'.
2. Using the same mixing regime add the main detergent to (say) half the formulation water.
3. Add the mixture obtained in step (2) to that of step (1).
4. Separated, use heat to disperse any difficultly soluble ingredient in the remainder of the water.
5. Add the product of step (1) to that of step (3).

6. Adjust the pH of product (5) to about 6.8 by means of citric acid addition.

Usually, with ingredients of the above solubility characteristics, no upwards adjustment of viscosity is needed.

Note that examples I and III represent clear shampoos and example II is an opaque product.

Representative Shampoo Formulations

Clear and opaque lotions, creams and gels

Early shampoos came in paste form of various combinations of clarity and opacity with low consistency and gel – or paste – like rheological properties. Formula A represents a clear free-flowing product with a number of interesting features.

Formula II

'Traditional' clear lotion shampoo

Monoeth, lauryl sulfate (30 %)	45.0% w/w
Sodium chloride	3.5 % w/w
Demineralized water	51.5 % w/w
Pefume, preservatives, colour	q.s.

This formulation (I) is one of very **few not** to make use of a foam 'booster' and consistency builder. The latter requirement is met by the incorporation of an electrolyte (sodium chloride).

Formula II is contemporary with I, but contrasting in almost every other way, it incorporates opacifiers which also contribute to its relatively greater consistency.

Formula III

'Traditional' cream shampoo

Sodium lauryl sulfate (30 %)	30.0 % w/w
Fatty acid allanolamide	2.0 % w/w
PEG glycol 100 distearate*	4.0 % w/w
Magnesium stearate	2.5 % w/w
Oleyl alcohol	0.75 % w/w
Deionized water	60.75 % w/w
Perfume, preservatives, colour	q.s.

* Opacifiers which also thicken the product

It will be noted that Formula B makes use of superamide foam 'Booster', a practice which is now almost universal.

Formula IV

Cream based shampoo

Sodium lauryl sulfate	38.0 % w/w
Cetyl alcohol	7.0 % w/w
Deionized water q.s.	100.0 % w/w
Perfume, preservatives, colour	q.s.

Formula IV, described as a traditional paste shampoo is, as its name implies an opaque product of high consistency.

Formula V

'Traditional' Paste (Opaque) Shampoo

Sodium lauryl sulfate (28%)	33.0 % w/w
Bentonite (clay)	8.0 % w/w

Formula VI

Early Conditioning Shampoo with a poly-quaternium

Cocoamidopropyl-3-dimethione betaine	5.0 % w/w
N-lauryl sarcosinate	3.5 % w/w
Ethoxy tridecyl alcohol (20 EO)	12.0 % w/w
Sodium lauryl sulfate (28%)	17.0 % w/w
Polyquaternium 10 (CTFA name)*	0.5 % w/w
Deionized water	62.0 % w/w
Perfume, preservatives, colour	q.s.

* Polymer JR 400 from Union Carbide Corporation.

More recent studies reveal that the polyquaternium need not be supported by the betaine derivative and the sarcosinate, etc. in order to produce a provable hair-conditioning benefit.

Formula VII

Gel based shampoo

Triethanolamine lauryl sulphate	25.0 % w/w
Coconut diethanol amide	10.0 % w/w
Methyl cellulose	2.0 % w/w
Perfume	q.s.
Distilled water ad	100.0 % w/w

Acid-balanced shampoos and their mechanism and advantages have been discussed earlier. Formula VII is a fairly typical example of this type of shampoo.

Formula VIII

Acid-balanced shampoo

Ammonium lauryl sulfate (32%)	10.0 % w/w
Betaine derivative	12.0 % w/w

Lauric diethanolamide	3.5 % w/w
Phosphate ester	2.5 % w/w
Protein hydrolysate	1.0 % w/w
Citric acid	to pH 5.5
Deionized water	41.0 % w/w
Perfume, preservative, colour	q.s.

e.g. Cocoamidopropyl betaine, PPG5 Ceteth 10 (CTFA).

Formula VII represent an effective shampoo but almost certainly contain superflows materials. It is felt that the protein hydrolysate could be omitted and probably also the phosphate ester.

Powder Shampoos

Formula VIII and IX are examples of a powder shampoo to which the user adds water and mixes until the solution is clear.

Formula IX

Powder shampoo

Sodium lauryl sulfate (100%)	30.0 % w/w
N-lauryl sarcosinate	8.0 % w/w
Sodium bisulfate	12.0 % w/w
Sodium sulfate	50.0 % w/w

Powder shampoo used as 3–4g per shampooing.

The example given above is not necessarily typical of the class. A similar product recently available in India, in the form of 5 g sachets, contained mainly sodium lauryl sulphate with about 10% by weight each of N-lauryl sarcosinate and ethylene diamine tetracetic acid (EDTA).

Formula X

Powder shampoo

Soap powder	50.0 % w/w
Sodium bicarbonate	20.0 % w/w
Henna powder	5.0 % w/w
Borax	25.0 % w/w
Perfume	q.s.

Dry-use Shampoo

This product is to be used in a totally dry way. It is applied to the hair and brushed through, cleansing by absorbing the excess sebum and taking the atmosphere soil with it. No water is used at any stage.

Formula XI

Dry-use powder shampoo

Corn starch	10.0 % w/w
Talc	15.0 % w/w
Tripolite	15.0 % w/w

Talc is a mineral of magnesium silicate. This ingredient can carry among other microorganisms, spores of *Clostridium* spp. bacteria which are pathogenic to main. Good quality suppliers sterilize this earth product and guarantee that it is sterile when delivered. It should be delivered in clean dry bags and kept dry in store.

Tripolite is another material mined from the earth and therefore must be guaranteed by the supplier. It consists of naturally-occurring silica, a diatomaceous earth (Fuller's earth, Kieselguhr) and is purified talc, free from iron oxide. It is a good lipid absorbent and its purpose in this formula is to absorb the sebum. The talc is a good lubricant and will help to lubricate the passing of the brush or comb through the powder-treated hair.

The role of the corn starch is more obscure; it should have sebum absorptive properties, and it may also help to trap some of the kieselguhr dust. This would be an advantage in what is an essentially dusty operation.

Aerosol (mousse) Shampoo

Formula XII represents at increasingly popular sub-division of the shampoo market, namely the aerosol mousse form.

Formula XII

Aerosol Shampoo

Sodium laureth 2 sulphate 50%	50% w/w
Cocamide DEA	3.0 % w/w
Co polymer 932	5.0 % w/w
Deionised water	37.0 % w/w
Hydrocarbon propellant	5.0 % w/w

EVALUATION OF SHAMPOOS

Shampoos are evaluated for

- **Foaming ability:** Foaming is essential for consumer acceptance although it is not a measure of cleaning action. With the help of Rose-Milles' foam column method foam height and foam stability is measured.

- **Cleansing properties:** Also known a detergency action, since shampoos are used for the cleaning purpose so it is the real measure of property of shampoo. It is done on wool-yarn and grease. Place 5.0 g of wool–yarn covered in grease and put it in 200.0 ml of water (at 35°C) containing 1.0g of shampoo in a flask. Shake the flask for four minutes at 50 revolutions per minute. Remove the sample from solution, dry it and weigh. Now calculate the amount of grease and dirt, remove under experimental conditions.
- **Skin irritation and sensitization test:** Various methods have been described to predict the potential of substances to induce sensitization and irritation. Most of the methods involve use of animals. Some methods also involve use of human being. These predictive tests give fair idea about the harmful effects a substance can have. However, it should be understood that human diversity is so great that a few persons may always be sensitive to a particular substance. For detection of potential primary irritation, Draize test or its slight modification is used, in this test albino rabbits are clipped and the substance to be tested is applied to:
  - Intact skin
  - Abraded skin or
  - Lightly scarified skin.

All of them are covered with a patch for 24 hours, the sites of application are examined at intervals and changes are assessed and recorded. The skin of rabbit is more susceptible than man; however, this method of testing can lead to false positive or false negative. It is advisable to compare results of test substance with the results of known harmless substance. Bureau of Indian Standards in IS:4011–1982 recommends that if there is no reaction in any of the animals the same test should be performed on 10 human volunteers applying the substance on the skin of the forearm. Sensitivity testing can be done by the following methods.

1. **Patch test:** On humans sensitivity testing of cosmetics may be performed either as a diagnostic or as prophylactic test. By diagnostic test it is intended to discover whether the cosmetic used has caused dermatitis, if cosmetic is known, the ingredient which has caused it. It is known as diagnostic patch, prophetic test is done to assess whether a new cosmetic should be placed in market or not. This test is known as prophetic patch test, general procedure for patch test is:

Place about 0.1–0.3 gm of cosmetic to be tested on a piece of cotton fabric or flannel (2–3 sqcm in size) and apply this to the skin or arms, thighs or back. This patch is covered with a patch of cellophane (about 5 sqcm) and sealed with adhesive plaster (about 40 sqcm). Apply several patches at one time. These patches are allowed to remain on the skin for 24–72 hours.

Sites of patches should be examined after 30 minutes of removal of patch. Usually the skin under adhesive tape gets inflamed. The skin under cellophane tape remains clear, the skin under test patch may not have reaction or may have reaction like erythema, erythema with papules, papulo-vascular reaction or ulceration or necrosis. Patch test reactions, usually, are graded as under:

Description of observation	Symbol
No reaction	–
Erythema only	+
Erythema with papules	++
Papulo-vascular reaction	+++
Ulceration or necrosis	++++

a. *Open patch test:* In case of cosmetic containing higher percentage of potential irritants like hair dyes, shampoos, hair tonics, patches should be used as open patches. Open patch test is performed on sensitive part of the skin, e.g. bend of elbow, popliteal space and skin behind the ears. The site of patch is inspected after 24 hours. If there is no reaction, the test is repeated once more on the same site. If still no reaction, then the test is repeated for third time. If still no reaction is observed, the person may be taken as not hypersensitive.

b. ***Prophetic patches test:*** For full scale prophetic patch test, 200 normal subjects are used. The cosmetics to be used is placed on the skin of the subject for one to five days depending upon the nature of the cosmetics and judgment of investigator. The patch sites are examined and observations are made. Subjects are observed for three more days for development of any late reactions. After 7–10 days, patches are again applied to the same area of those subjects who did not show reaction to the test.

2. **Repeated insult test:** The repeated insult test consist in applying the finished cosmetics or if the test substance is an ingredient of the cosmetic, it is applied in the same concentration as is found if the finished formulation incorporated in a bland base. Ordinarily 0.5 gm or 0.5ml of the test sample is applied by usual test procedure. The test material is maintained *in-situ* for 24 hours. Allowing 15 to 20 minutes after the removal of the patches, readings are made and the reactions recorded. The test subject is allowed 24 hours of the rest and the second test patch is applied to different test site, otherwise the procedure is identical to that of the first application. Each individual is thus subjected to a service of 10 such consecutive exposures, exclusive of Sundays and non-work days. Following the 10 individual exposures the subject is given 10 to 14 days rest after each time a "retest application" is made similar to one of the original 10 applications. A comparison of the reaction following "retest" with the average reading of the 10 original applications permits an appraisal of the sensitizing propensities of the substances.
3. **Photo patch test:** Certain substances are not harmful by themselves but they become harmful when expose to sunlight. Substances that absorb light wavelength between 300–800 nm have potential of phototoxicity. Incase, a substance is considered phototoxic; photo patch test may be performed. To perform this, the substance to be tested is applied in duplicate patches in the same manner as for standard patch test. After 24 hours, one of the patches in the pair is exposed to sunlight for 30 minutes. Alternately the patch can also be exposed to ultraviolet light. The light exposed patch is covered again. One additional site in the adjoining area of the skin is exposed to sunlight or ultraviolet as has been done in case of one patch of the pair. The time of exposure is also same as for the exposed patch. This site acts as control. After further 24 hours the patches are opened and examined. If the patch is not exposed to light and the skin area exposed to light do not show reaction but the patch site which has been exposed to light shows reaction, the test indicates that substance is phototoxic. If no reactions are observed on patch site and control skin site the substance may be taken as non-phototoxic.
4. **Test for sensitizing potential:** Standard patch tests with same chemical or cosmetic are repeated in the same 200 volunteers after an interval of 10–14 days. The number of persons who will show positive reaction will represent the sensitizing potential of the substance or cosmetic.
5. **Use test:** In use test the cosmetic to be test is actually used and its adverse effects, if any, are observed. IS: 4011 suggests that 15 volunteers should be asked to use the cosmetic and they should make 15 applications of it. If there is no adverse reactions, cosmetic can be released for trial.
   - *Color:* Shampoo color selection has evolved to the increasing use of light, bright, eye-catching shades. Introduction of improved color stabilizing systems, and merchandising techniques that have made product and package mutually supporting. Any coloring of shampoos must be done with certified colors.
   - *Consistency:* Thick products may be advantageous in reducing spoilage or loss during application but may be difficult to disperse through the hair. Thinner products may hold an advantage where rapid dispersion and cleaning are desirable and where loss due to inadvertent handling is unlikely.
   - *Package:* Shampoos should be contained in packages of good barrier properties, passage of water vapor, essential oil, and air through the container poses a threat to product stability.
   - *Eye irritation test:* This is also a type of consumer acceptance test, consumers prefer only those shampoos which are less irritant

and cause no stinging sensation in eyes. Eye can be accidentally exposed to shampoos, other hair preparation and bath preparation. Draize and Kelly designed test on eye mucosa of albino rabbit. Such test can be conducted as follows:

Instill 0 1 ml of test substance in conjunctiva sac of one eye of nine albino rabbits (the other eye works as control) divide rabbits in three groups comprising of three rabbits each.

Group 1: Leave the eyes of this group unwashed.

Group 2: Wash the treated eyes with 20 ml of lukewarm water after 2 seconds of instillation of test substance.

Group-3: wash the treated eyes after 4 seconds of instillation of test substance with 20 ml of lukewarm water.

Read the ocular reactions with hand slit lamp for seven days or till any residual injury persist.

Any preparation, which leaves corneal or iris lesions for more than 7 days, is considered severe eye irritant. Eye irritant effect on combination of synthetic detergents is greater than produced singly by them. All synthetic preparation used on hair or during bath should be assessed for their eye irritating effect.

- *Stability study*

  Foam and foam stability: The Ross-Miles foam column test. In this test, 200ml. of a surfactant solution is dropped into a glass column containing 50ml.of the same solution. The height of the foam generated is measured immediately and again after a specified time interval, and is considered proportional to the volume.
- *Surfactant content and analysis:* The surfactant content of shampoos frequently runs from 15 to 25%, with the gel and crème concentrates at the higher level, whereas the liquid formulas tend to the lower side.
- *Rising:* Technique is to employ skilled beauticians to make comparisons of the performance of several shampoos. Here rinsing can be more easily related on a comparative basis.
- *Conditioning action:* The degree of conditioning given to hair is ultimately judged by the shampoo user who is making the evaluation on the basis of past experience, present expectations, and a continuing change in the individual scalp and hair situation.
- *Luster and softness:* Measuring luster of hair with an amplified photo volt photometer and a polarizing filter.
- *Lubricity:* Determine it for combing ease and subjective "handle" with frictional measurement. Waggoner and Scott have devised instrumentation for measuring dry hair responsiveness with an electronic comb.
- *Body, texture set retention:* These preparations are intended to modify the hair fiber properties in a favorable manner. The proper amount of 'body' depends on an individual subjective evaluation. The hair length and style have an important bearing on the rating.
- *Effect of water hardness:* Testing shampoos over a range of water hardness is necessary, particularly since soaps are susceptible to the impact of calcium and magnesium ions.
- *Surface tension and wetting:* The DuNouy ring tension meter is a classical means of measuring this value. Normally quite low concentrations of surfactants are used, in the range of 0.1 to 0.25%.

Some Marketing Available Brands

- Avon Naturals Shampoo
- Ayur Herbal Shampoo
- Cetrilak Baby Shampoo
- Clinic Plus
- Dabur Vatika
- Emami
- Garnier Fructis—Conditioner Shampoo
- Garnier Ultra Doux
- Head & Shoulders—Anti Dandruff Shampoo
- Johnson & Johnson Baby Shampoo
- Keune—after Colour Shampoo
- Lakmé
- Lambency Antidandruff Shampoo
- L'oréal Elvive Frizz Control Shampoo and Conditioners

- L'oréal Liss Control Serum
- Mediker—Antilice Shampoo
- Nizoral Anti Dandruff Shampoo
- Nyle—Active Herbal (Deep Moisturising) Shampoo: Contains Aloe Vera, Tulsi, Amla, Henna, Reetha and Shikakai.
- Pantene Pro-V Shampoo
- Ponds
- Satinique
- Selesin Antidandruff Shampoo
- Sunsilk 9 to 9
- Vasmol Shampoo
- Tenderils Baby Shampoo.

14

Hair Sprays

> *I've never seen a smiling face that was not beautiful.*
>
> — Anonymous

INTRODUCTION

Hair spray is a common personal care aqueous grooming product that is used to keep hair stiff or in a certain style. Hair sprays are weaker than hair gels or wax-type products as those films are very flexible and soft. The product is usually sprayed using a pump or aerosol spray nozzle. Hair spray was first developed and manufactured in 1948 by Chase Products Company based in Broad view, IL.

In terms of the size of their share of the hair care market, hair sprays, rank second only to shampoos. Most of them are sold in conventional pressurized packs. Unlike the aerosol mousses they produce a particulate spray and contain a small proportion of true aerosol droplets, but in common with mousses they have been until recently dependent upon chlorofluorocarbon (CFC) type propellants. Although the term aerosol product is used interchangeably with pressurized pack product, the aerosol mousses are not in fact, aerosols. A true aerosol is a colloidal system like a mist or fog in which the dispersion medium is a gas. A genuine aerosol particle is regarded by many authorities as having a diameter of 1 μm or less. A hair spray contains a small proportion of true aerosol droplets, the insecticide spray which is regarded, technically, as a genuine space spray significantly more. The medical spray for the alleviation of asthma has a majority of its droplets in or close to the aerosol limit.

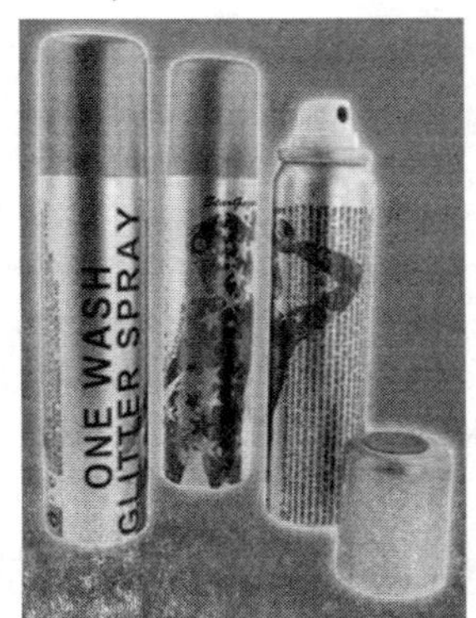

Profound changes in the performance of aerosol hairsprays can be made whilst keeping their chemical composition constant. These changes can be brought about by alteration of the pressurized package mechanical.

Figure on next page shows log on probability particle size distribution curves for three pressurized pack products an asthma relief inhaler, an insecticide and a hair spray. The particle sizes were determined by air (slow) impactation methods. Mass mean diameter may be calculated from the curves features, often known as the hardware. They relate almost exclusively to the

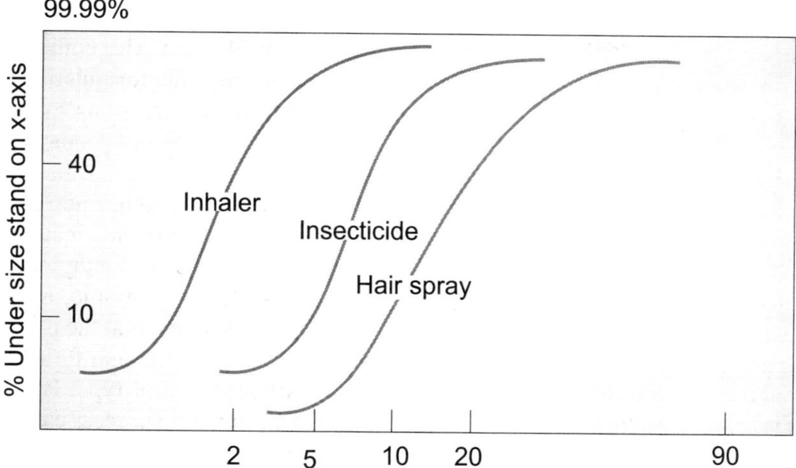

very complex specifications of the valve and spray head. A finer spray for example may be obtained by a variety of modifications including reducing the spray orifice, enlarging the expansion chamber, increasing the diameter of the swirl chamber, etc. Other simple modifications can adjust the rate of spraying and the coverage of the target at the spraying distance, which is approximately 20 cm in hair spray use.

The manner in which hair spray droplets interact with hair fibres has been investigated in depth. When the droplets arrive on the hair most of the propellant has been lost through evaporation. The droplets on the hair are therefore fairly concentrated polymer solutions. Depending upon their surface energy, viscosity and the volatility of the solvent the droplets interact with the mass of hair fibres in two primary ways. If they are not spread efficiently they will bind adjacent hair fibres by **'beads'** of polymer. Conversely, if they are less concentrated and less viscous, efficient spreading occurs and the hair mass is bound by seams of dried polymer. The former effect is known as **'spot welding'**, the latter **'seam welding'**. Sometimes formulators seek to improve the holding power of a hair spray by increasing the concentration of polymer. This can have exactly the opposite effect as the extra polymer can make the droplet too viscous to spread sufficiently to form seam welds.

Until recently hair sprays have been intended for use on dry hair. An ideal one is visualized as having a relatively fine spray which is gentle enough not to disturb the configuration of the hair. Additionally it is expected to cover the hair fairly quickly. Thus the apex angle of the spray should be relatively large, i.e. about 40. The dried spray should be easily removed by shampooing and brushing otherwise the dulling and coarsening effect known as build-up will occur. The dried spray on the hair should not be intrinsically tacky nor should it become tacky owing to absorption of water vapour.

The functional ingredient in a hair spray is a suitable polymer, usually an acrylic, which is soluble in alcohol and water to form liquid elastic film that keeps the hair stiff and firm without snapping. Typical polymers include it or polyvinylpyrrolidone (PVP) or other types of acrylates copolymer resin combinations such as the Luviskols from BASF and the Resyns from National Starch & Chemical Company.

Excessive use of or lack of washing after using hair spray may lead to dull or damaged hair.

Hair sprays can be scented or unscented. The formula also contains emollients and proteins to keep the hair shiny and soft-feeling as the alcohol tends to dry the hair out by stripping it of its natural oils. Dry hair can be a problem for people with already damaged hair from excessive coloring or excessive washing without the use of a conditioner. Small levels of quaternium ammonium compounds or cationic surfactants are sometimes added in hair sprays as they act as conditioning agents for the hair.

The historical level of alcohol as volatile organic content (VOC) in a hair spray was around 80%. Nowadays, due to strict environmental laws—especially in US, low VOC hair sprays are on the market that contain about 55% alcohol. They still function well but formulators use co-solubilizers to get the desired amount of polymer into the formulations without it kicking out of solution. This is critical, as the amount of hold the product has is dependent on the level of polymer used in the formula.

HAIR SPRAY FORMULATIONS

Formula I

Hair spray (M)

Amphomer*	2.0% w/w
AMP	0.33% w/w
Silicone surfactant	0.11% w/w
Methylene chloride	20.00% w/w
Ethanol (anhydrous)	57.57% w/w
90/10% butane/propane	20.00% w/w

* Cross linking polymer, claimed to have good holding power (National Starch Corporation).

2 amino-2-methyl- 1- propanol (neutralizing agent) by Angus Chemical Co.

(Dimethicone copolymer (Dow corning)

Formula II

Regroomable hair spray

Gantrez ES 425*	5.00% w/w
Tri-isopropanolamine (TIPA)	0.50% w/w
Mulgofen AM 310#	0.50% w/w
Ethanol (anhydrous)	68.98% w/w
Prepare/butane	25.00% w/w

* Copolymer of methylvinylether and maleic anhydride (50% solution) ISP

\# Ethoxylated fatty alcohol (Rhone Poulenc)

Formula III and IV illustrates a number of factors important in the formulation of contemporary hair sprays. Both of the formulae make use of a hair control polymer which can be neutralized with an amino alcohol, AMP or TIPA to render the spray residue more clearly removable in the shampooing process. Also both are an adaptation of traditional hair spray formulation which depended upon chlorofluorocarbons as propellants. The lower concentrations of propane butane mixtures necessitate an increase in the concentration of ethanol, heavily taxed in some countries. Formula III offsets this cost of raw materials by replacing some of the ethanol by methylene chloride, which is permissible, for example, in Germany.

It should be noted that the product contains not the familiar polysiloxane fluid, but a surfactant of the silicone plycol type. It is possible that it renders the combing process easier in the dry state. The Amphomer polymer is reputed to produce good holding properties at low concentrations, acceptable even as low as 2%.

Formula IV pays more attention to the more modern concept of a hair spray which states that hairsprays should be suitable for application to damp hair. This explains (a) the high level of neutralization and (b) the inclusion of ethoxylated fatty alcohol in the formula. As the polymer is more comparable with the residual water in the hair it is possible at this stage to create a style or wave. This particular formula then can be justifiably described as a regroomable hair spray.

Formula III

Aerosol hair spray

Gantrez/ES 225* (50%)	4.0% w/w
AMP (see formula VIII)	0.10% w/w
Ethoxylated lanolin	0.10% w/w
Ethanol (anhydrous)	75.30% w/w
Perfume	0.50% w/w
Propane/butane (90:10)	20.00% w/w

* Gantrez ES 225 is the monoethyl ester analogue of ES 335 and ES 425.

Formula IV

Aerosol spray–ozone friendly

Resyn 28-2930*	2.25% w/w
AMP (neutraliser) #	0.18% w/w
Silicone	0.12% w/w
Methylene chloride	26.00% w/w
Anhydrous ethanol	66.75% w/w
Perfume	0.20% w/w
Carbon dioxide	4.50% w/w

* Resyn 28–2930 is a terpolymer of vinyl acetate, crotonic acid and a third monomer which is of a hydrophobic nature. Manufactured by National Starch Corporation.

AMP 2- amino-2-methyl-1-propanol (neutralizer) Angus

Formula V

Pump driven hairspray

Gantrez FS 225 (50%)	3.5% w/w
DC Silicone *	1.0% w/w
AMP (see formula III) #	0.2% w/w
Ethanol (anhydrous)	95.3% w/w

* Dispersion of a silicone surfactant in a volatile silicone

AMP 2-amino-2-methyl-1-propanel (neutralizing agent)

The Resyn 28–2930 of Formula IV neutralized to approximately the 10% levels has also been found to be compatible with mixtures of CFC type and hydrocarbon propellants. In Formula V the incorporation of the silicone surfactant should have the effect of producing a some what fine spray. Pump dispersal hair sprays will be inevitably rather coarse at this stage of their development.

Available Market Products

- Lakme Hair Next Instant Bounc Volume Spray
- L'oréal Volumizing Mousse
- Garnier Hair Spray.

15

Hairdressings

Beauty is not in the face; beauty is a light in the heart.

— Khalil Gibran

INTRODUCT ON

Hairdressings are unique amongst prod.icts as they are usually purchased by women. There are a great many types some very traditional some making use of modern pa:kaging and ingredients. The list below covers the main ones:

1. Brilliantine
   a. Jelly
   b. Pomade
   c. Liquid
2. Emulsions which can be of the oil in water (o/w) or the water in oil (w/o) kinds.
3. Modern lotion, (unrelated to the brilliantine)
4. Modern gel hair-dressings.
5. Aerosol ha.r dressings.

BRILLIANTINE

Brilliantines are employed to impart luster to the hair and also keeping them in proper place. Brilliantines have to be applied to the dry hair after it has been waved. Jelly and pomade or viscous type of brilliantines is generally preferred by men and liquid type brilliantines are preferred by women on their long hair.

Pomade brilliantine typically contain about 10% of uncrystallizable paraffin wax, the remainder being petroleum jelly (about 90%) and an acceptable amount of an appropriate perfume.

Jelly brilliantine

Formula I

Spermaceti	15.0%
Myristic acid	5.0%
Oleic acid	25.0%
White mineral oil	54.0%
Perfume	1.0%

Spermaceti, myristic acid and oleic acid melt in one beaker. Heat the mineral oil separately, then mix together thoroughly at the same temperature. Mix perfume at about 45 °C.

Liquid brilliantine are composed of about 75 per cent pharmaceutical grade light liquid paraffin BP (mineral oil) with about 24 per cent of isopropyl myristate; the rest being largely lanolin and perfume to suit.

Formula II

Olive oil	45.0%
Sweet almond oil	42.0%
Castor oil	5.0%
Parahydroxybenzoic acid ester	0.05%
Perfume	1.0%
Alcohol	6.95%

Dissolve perfume and parahydroxybenzoic acid ester in alcohol and then mix sweet almond oil, in this mixture add olive oil and castor oil. Allow it to stand for three days and then filter.

EMULSION HAIRDRESSINGS

Emulsion hairdressings, the w/o types of which formula I is representative are traditional formulae but have some unique properties.

Formula III

Emulsion hairdressing

Beeswax	2.00% w/w
Light liquid paraffin BP (Mineral oil)	30.8% w/w
Paraffin wax	0.2% w/w
Petroleum jelly	6.00% w/w
Stearic acid	0.50% w/w
Perfume	0.50% w/w
Limewater	60.00% w/w

Emulsions conforming to the above basic formula are reasonably stable yet when rubbed briskly they break giving rise to freed droplets of water. The liberated water appears to act as a grooming aid and also lessens the greasiness of the product. These features appear to have a strong consumer appeal. Investigations into the mechanism of breaking indicate that the higher fatty acids of the beeswax (C22) and the paraffin wax play an important role. Microcrystalline paraffin waxes may be substituted for the standard wax.

LOTION HAIRDRESSING

Hairdressings of this type found appeal in the sixties and began to supplant the emulsion type.

Formula IV

Lotion hairdressing

Resin 28	13.10% w/w
AMP (see Formula VIII)*	1.5% w/w
UCON Oil#	0.1% w/w
Ethanol	40.0% w/w
Water (deionised)	28.4% w/w
Perfume	q.s.

* AMP 2-amino–2-methyl-1-propanol (neutralizer).

UCON oil based on block copolymers of ethylene and propylene oxides (Union Carbide Corporation).

The substitution of the UCON oil for the more familiar mineral paraffin oils is claimed to remove the greasy feel which was inherent in the earlier gel-type hairdressings.

AEROSOL HAIRDRESSINGS

These are basically similar to women's hairsprays but with a lower polymer concentration to avoid holding properties which are too rigid. The presence of polyethylene glycol (PEG) laurate and PEG 400 afford a measure of regroomability. Formula XV is representative of the mainstream of aerosol hairsprays, perhaps biased to obtaining softer and less harsh hold than would be given by a conventional hairspray. It uses the traditional CFC propellant mixture which should be replaced by modern acceptable 'ozone friendly' propellants according to the Montreal agreement made by cosmetics manufactures. Experimental work would have to be done to ensure that the ingredients were soluble and that no precipitation occurred on ageing.

Formula V

Aerosol hairdressing

Gantrez FS 225 (50%)*

AMP	2.00% w/w
PEG laurate	0.40% w/w
PEG 400	0.20% w/w
Ethanol (anhydrous)	77.30% w/w
Propellant F 11/12 (50/50)	20.00% w/w

* Gantry ES 225.....ISP

GEL HAIRDRESSINGS

Typical gel hairdressing conforms closely to Formula IV.

Formula VI

Gel hairdressing

Luviskol VA 37 E*	
Carbopol 940[#1]	0.50% w/w
Triethanolamine (TEA) [#2]	0.25 % w/w

Perfume 0.20 % w/w
Ethanol (anhydrous) 26.00 % w/w
Water (deionised) 72.05 % w/w

* PVP/VA copolymer (BASF)

#1 Vinyl polymer with unesterified carboxyls - Goodrich)

#2 TEA, neutralizing agent

This class of hairdressing product has much in common with the styling gels for the women's hair-care market. Whereas the majority of the hair-care products we have discussed are prepared simply by making multicomponent solutions, gels, like conclusions, must follow a more complicated procedure. A part of the difficulty is in the use of gelling and thickening agents. Natrosol HR 250 is an example of the latter. Unless directions are followed closely the product is likely to finish looking lumpy and uneven. The following instructions if followed carefully should avoid the pitfalls.

Step 1 Mix the ethanol and water.

Step 2 Take half of this solution and dissolve the Luviskol in it.

Step 3 Add to Step 2 with gentle stirring the TEA and perfume.

Step 4 Disperse the Carbopol 940 in the other half of the ethanol/water solution with high speed stirring.

Step 5 Gradually reduce speed and allow mixture to clear.

Step 6 Filter the mixture from step 5 and add slowly to Step 3.

Step 7 Stir slowly until the product is homogeneous.

Avaliable Market Products

- L'Oréal Hair Dressing
- Garnier Hair Dressing.

16

Permanent Waving Preparations

> *Wisdom is the abstract of the past, but beauty is the promise of the future.*
> — Oliver Wendell Holmes

HAIR CARE PRODUCTS WITH CHEMICAL MECHANISMS

Permanent Waving Preparations

Hair waving preparations are used to curl or wave the hair. A proper treatment is given to hair for curling them in beauty parlour. Mechanism of hair curling is a great strain along the length of one side of hair shaft than the other so it turns into curls.

For curling hair, they have to soften and stretched so as to take the proper wave. Hair is stretched on rolling them on heat rod so as to cause unequal strain on one side and then permitting the hair to cool the hair thus curled.

Various preparations are used for waving like:

- Permanent wave fluids
- Curling concentrated powders
- Hair wave liquids, etc.

Materials used for preparations for hair waving are:

- Mucilage of gums like gum karaya, gum tragacanth with borax or alkali carbonates.

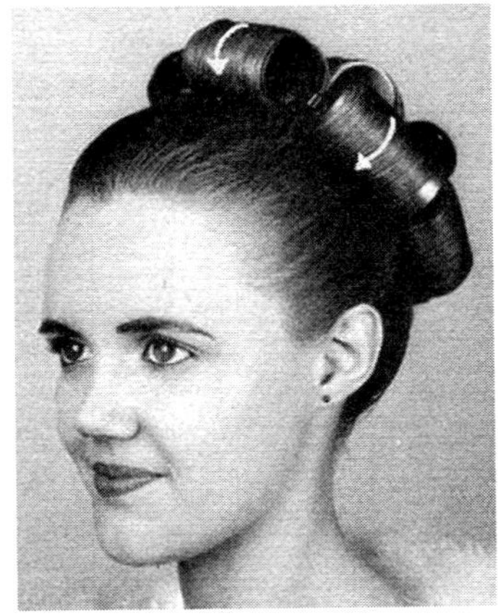

- Some preparation contains only alkalis like mixture of sodium carbonate and sodium bicarbonate.
- Other bases like ammonia, triethanol amine, trisodium phosphate, tertasodium pyrophosphate.

Supporting materials in hair waving preparations are:

- Glycerine
- Alcohol
- Lanolin and its derivatives
- Surfactants (soaps, lecithin)
- Sulphonated oils
- Glycols
- Oils and fatty acids, etc.

Permanent and Temporary Wave Set of Hair

A temporary set is achieved by use of water waving techniques followed by treatment with wave sets, hairsprays, etc. These treatments work by physical as opposed to chemical mechanisms. From their nature the set will span days rather than weeks or months.

Permanent sets are obtained through the use of chemical treatments, such as thioglycollate based hair-waving and -straightening products. Processes which make use of other keratin reducing agents also fall into this category as do caustic lye products. The latter are used almost exclusively in straightening products for Negroid hair. Basically these products giving rise to a permanent set introduce new chemical bonds into the hair structure, especially new disulfide bonds. This topic has already been discussed in the section of this chapter devoted to the chemical reactivity of hair.

Interestingly the wool industry would regard the above chemical treatment as producing a temporary set. The explanation is that textile technologists realize that the set obtained by inter chain disulfide bonding can be eroded by subjection to mechanical strains, particularly repeated strains. In structural chemistry, strains of sufficient size and persistence can rearrange the disulfide bonds responsible for the set. This effect is known as the sulfhydryl disulfide interchange. When native hair keratin is sufficiently stressed lission of the disulfide bonds occurs by means of the above interchange reaction I virtually alpha keratin becomes irreversibly denatured (effectively destroyed). A technically permanent set can be obtained by replacing disulfide bonds with other covalent cross linkages which cannot participate in the interchange. Two examples of mechanically stable cross linkages are (a) the lanthionine keratin structure by a controlled treatment in alkaline conditions as set out below:

$CH-CH_2-S-S-CH_2-CH \xrightarrow{OH} CH=CH_2 + S-S-CH_2-CH$

(Cysteine) (Dehydro alanine)

$CH-CH_2-SH$ (Cysteine) $CH-(CH_2)_4-NH_2$ (Lysine)

$CH-CH_2-S-CH_2-CH$ (lanthionine) $CH-CH_2-NHR$ (lanthionine)

$CHCH_2NH(CH_2)_4-CH$ (lysino alanine)

Formula I

Gum karaya	2.0%
Deionized water	90.0%
Alcohol	7.75%
Methyl paraben	0.25%
Perfume	q.s.

Mix the gum karaya in alcohol and stir the mixture into water. Add the perfume and preservative dissolved in remaining alcohol. Filter if necessary.

There are two stages in the process, reduction or breaking of the disulfide bonds using Formula XVII, followed by oxidation (neutralization) to reform them into the new configuration.

Formula II

Permanent wave relaxing cream, part I

Thioglycollic acid*	8.0% w/w
Ammonia (88)	1.0% w/w
Tween 40#1	3.0% w/w
Lanolin	1.5% w/w
Span 80#2	1.0 % w/w
Mineral oil	0.5 % w/w
Water (deionised)	85.0 % w/w
Preservative	q.s.
Perfume	q.s.

* Thioglycollic acid $CH_2SH{\cdot}COOH$ is the keratin reducing agent.

#1 Tween 40 is an emulsifying agent-ethoxylated sorbitan monopalmitate (ICI Speciality Chemicals)

#2 Span 80 is also an emulsifying agent-sorbitan mono-oleate. (ICI Speciality Chemicals).

A typical neutralizer (i.e. Oxidant) formulation to be used in conjunction with Formula II -part II is shown below:

Formula III

Neutralizer/oxidant, part II

Cetyl trimethyl ammonium bromide (CTAB)	0.3 % w/w
Hydrogen peroxide (20 Volume)	27.0 % w/w
Water (deionised) at pH 4.0 to	100.0 % w/w

Note on part I

The ammonium thiglycollate which is formed in the reaction between the first-two ingredients of Part I, is an active reducer of keratin at pH values near 10. Permanent-waving products of this kind are generally used in an emulsion vehicle. There are two main advantages:

a. The viscous vehicle moderates the reaction rate of the thioglycollate with hair and
b. In the emulsion form application of the product on to the hair is safer and more controllable. These properties are particularly useful to the home user, but well appreciated also by the professional salon operators.

The emulsions are of the oil in water type (o/w) would normally require only one emulsifier which would be of a high HLB value. HLB measures the affinity of the emulsifying agent for water. This formulation uses two emulsifiers whose combined HLB values are additive and give the desired stability.

Thus the effective combined HLB is given by the expression:

$$\text{HLB (combined)} = \text{HM} = \text{wHL} + (1 - \text{w})\ \text{HH} \quad (3)$$

Where,

HL is the HLB of Span 80 = 4.3

HH is the HLB of Tween 40 = 15.6

Formula ratio percentage of each is = 1:3

w is the weight fraction of Span 80 = 0.25

(1 – w) is the weight fraction of Tween 40 = 0.75.

Evaluating equation (3)

$$\text{HM} = 0.25 \times 4.3 \times 0.75 \times 15.6 = 1.0775 + 11.7 = 12.8$$

The HLB value of approximately 13 means that the combined effect of the two agents is equivalent to one of about 13. Tween 21, HLB 13.1 would in theory be just as effective. However it has been discovered, in practice, that mixed emulsifiers have definite advantages in ease of preparation and the subsequent stability of the emulsion.

EVALUATIONS

The performance of hair waving preparations depends on alkalinity and chemical integrity of the reducing agent.

Avaliable Market Products

- Fuel for Man
- L'Oréal Professional Shine Curl
- Schwarzkof Shine Curl
- Wella Biotouch
- Brylcream
- Himalaya Hair Detangler
- FX Special Effect Curl Miracle.

17 Hair Straighteners

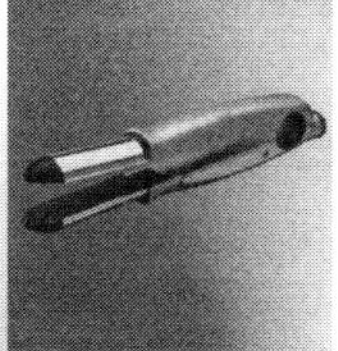

Beauty - in projection and perceiving - 99.9% attitude.
— Grey Livingston

Hair straighteners are the preparations which are used to take curl or kink out of the hair and at the same time also provide sheen and perfume. Hair straighteners are applied with the help of hard brush which aids in removing the curls of the hair. These products are mainly, but not exclusively for the Afro-Caribbean market.

Hair straighteners generally contain materials like gums, castor oil, rosin, beeswax, and paraffins.

Press Oils

The simplest form of hair straightener product is the press oil of Formula I. This product bears a strong resemblance in composition to the pomade brilliantine; it would be surprising if the pomades had not at sometime been substituted for genuine press oils.

Formula I

Press oil

Petroleum jelly	90.0 % w/w
Paraffin wax	10.0 % w/w
Perfume, etc.	q.s.

The hair is anointed with the press oil rather in the manner of applying pomade. It is advantageous, but not totally necessary, that the hair should have been shampooed and dried prior to its application. A metal comb which has been heated to a surface temperature of between 200 and 300 °C is then used to comb and stretch the oiled hair. In the short term the treatment is very effective but the hair does not remain decrimped for very long. The press oil method cannot be applied too frequently because it causes a good deal of hair breakage. The mechanism of the straightening may be likened to the process of steam pressing of wool textiles, where it is assumed that sulfonamide bonds (R' — SONH — R" are formed within the hair structure.

Caustic Lye Hair Straighteners

To obtain a more permanent effect chemical straightening techniques are used, e.g. with caustic lye and thioglycollate-based products. Two examples of the former are described below:

Formula II

Caustic lye hair straightener

Stearic acid	15.0 % w/w
Oil	
Oleic acid	5.0 % w/w

phase A

Glycerol	8.0 % w/w
Aqueous	
Sodium hydroxide	4.5 % w/w
Phase B	
Water (deionised)	67.5 % w/w
Perfume	q.s.

Procedure

1. Heat the fatty acids of the oil phase (A) to 70 °C in a jacketed pan with slow stirring and maintain the temperature at 70 °C
2. Dissolve the sodium hydroxide in water and then add the glycerol with minimal stirring.
3. Heat part B to 65 °C and maintain this temperature.
4. Add A to B with vigorous stirring taking care not to allow the temperature to drop below 50°C.
5. Cool to 45 °C with slow stirring.
6. Add the perfume while slowly stirring until cooled to 35 °C.
7. Homogenise if necessary.

In Formula II the emulsification is brought about by the soap formed between the sodium hydroxide in B reacting with the fatty acids in A. To increase the stability of the product Formula III uses, in addition, a non-ionic emulsifier (Tween 40 HI.B–15.6) from Formula III, the hair relaxing treatment.

Formula III

Formula II with added emulsifiers

Stearic acid	15.0 % w/w
Oil phase A	
Oleic acid	5.0 % w/w
Tween 40	3.0 % w/w
Glycerol	8.0 % w/w
Sodium hydroxide	4.5 % w/w
Aqueous phase B	
Water (deionised)	64.5 % w/w
Perfume	q.s.

The caustic lye formula II, is a traditional hair-straightening product. The caustic lye is presented in an emulsion form because caustic lye is even less safe for hair, skin and eyes, than thioglycollic acid. It is however very effective and reasonably permanent in the hands of highly skilled hairdressers. The hairdresser's expertise must be supplemented by very good salon equipment, especially a reliable supply of water at the required temperature.

In the interests of the best performance and safety the procedure for straightening hair with a caustic lye preparation is quite complex.

1. Apply a basic shampoo followed by combing in a petroleum jelly preparation.
2. Apply the caustic lye preparation working quickly so that the processing time is as short as possible.
3. Treat with 'neutralizing' solution.
4. Treat the hair with pomade.
5. Wash with a regular (not necessarily a specialist) shampoo.

Thioglycollate Hair Straighteners

Formula IV

Thioglycollate cream straightener

Glycerol monostearate	15.0% w/w
Oil phase A	
Stearic acid	8.0% w/w
Ceresin	1.5% w/w
Paraffin wax	1.0% w/w
Thioglycollic acid	6.5% w/w
Aqueous phase B	
Ammonia (88)	20.0% w/w
Detergent	1.0% w/w
Water (deionised)	67.0% w/w

The formula for the neutralizer follows:

Formula V

Thioglycollate cream neutralizer

Lanolin	2.0% w/w
Sorbitan trioleate (Span 85)*	2.0% w/w
Sodium bromate	14.0% w/w
Propylene glycol	2.0% w/w
Sorbitan monopalmitate (Span 40)*	1.0% w/w
Detergent	3.0% w/w
Water (deionised)	76.0% w/w

* Span 40 (HLB = 6.7), Span 85 (HLB = 18)

Procedure

1. Blend together the lanolin and Span 85 in a jacketed pan and heat to approximately 55° C.

2. Mix the rest of the ingredients in the water and raise the temperature to 50° C.
3. Slowly add the lanolin and Span 85 of (1) to the aqueous solution of item (2) with constant stirring.
4. Cool and adjust the pH to 4.5–5.0 by gradually adding citric acid crystals while stirring.
5. Homogenize if necessary.

Perusal of Formulae IV and V reveal how similar are the hair straightness to the more familiar permanent wave preparations. In the straightening process the hair will be treated as above but wound on large diameter curling rollers. The oxidant of the neutralizer in V is sodium bromate instead of the hydrogen peroxide and the product is in emulsion form for extra safety.

Avaliable Market Products

- L'Oréal Professional
- Schwarzkopf
- Himalaya Hair Detangler
- Fuel for Man Hair Strengthening Gel
- FX Special Effect Curl Miracle.

18

Hair Removal Agents

> *As we grow old, the beauty steals inward.*
>
> — Ralph Waldo Emerson

Many men and women choose to remove unwanted body hair for cosmetic, social, cultural, or medical reasons. Medical indications for hair removal include hirsutism, which is excess terminal hair in the distribution of hair growth influenced by androgens (i.e. face, chest, back, abdomen), or hypertrichosis, which is congenital or drug-induced increase in hair growth in areas that are not androgen dependent. Other medical indications include pseudofolliculitis, hair growth from a grafted donor site, and sex-charge operations performed in men.

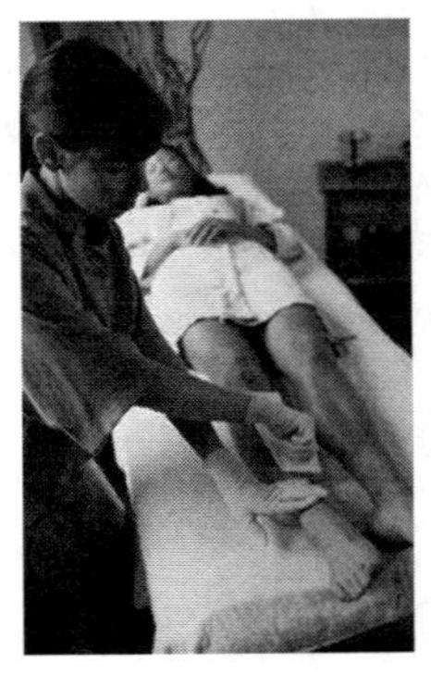

Many methods are available for temporary or permanent hair removal, each with its own relative efficacy and adverse effects. Different methods for the removal of body hair include the following:

- **Temporary hair removal:** Depilation (by depilatories), bleaching, epilation, plucking, shaving, waxing, threading, abrasives, sugaring.
- **Temporary hair reduction:** Eflornithine hydrochloride (13.9%), laser-assisted hair removal, intense pulsed light sources for hair removal.
- **Permanent hair removal:** Electrolysis, galvanic electrolysis, thermolysis.

Temporary and permanent methods of hair removal or reduction are important components in the treatment of patients with unwanted hair. No single method is perfect for all patients. Factors such as the underlying medical conditions causing excessive hair growth, size and location of treatment area, desire for temporary versus permanent hair removal, and expertise of the technician providing treatment should be considered when choosing a method for hair removal.

TEMPORARY HAIR REMOVAL

Depilation

Depilatories are the preparations which are use to remove excess or unwanted hair from body parts like face, armpit, axilla, groin and thigh. Depilatories used in the forms of:

1. *Crème hair:* Shea butter, olive oil
2. *Wet tissue:* For skin care
3. *Retarder:* For delaying the re-growth of hair.

Chemical depilatories remove part of the hair shaft and are easy and painless to use.

The standard chemical depilatory agents, available in:

- Gels,
- Creams,
- Lotions,
- Aerosols, lor
- Roll-on forms

Chemical agents are barium sulfide, sodium and potassium sulfide, the tin salts, calcium thioglycerol, and the salts of thioglycolic acid (sodium or calcium thioglycolate).

The carrier materials like talc, terra alba, titanium dioxide, barium sulphate, corn starch and zinc oxide, etc. pulverized soap is added as binding agent. Perfumes used for depilatories are aromatic alcohol, rose oils, anise, saffron, ketones, ionones used alone or in combination because depilatories have a very unpleasant odour.

Thioglycolate depilatories work dissolving hair by hydrolyzing and disrupting disulfide bonds of hair keratin, causing the hair to break in half and allowing the hair to separate from the skin. Depilatories are good for use on the legs, bikini line, face, and underarms, and they perform best when hair is at a reasonable length. Before using a depilatory, carefully read the manufacturer's instructions. Test a small site before use to assess for irritation or allergic reactions. Do not use these agents on eyebrows, near mucous membranes, or on broken skin.

Adverse effects include skin irritation, burns, folliculitis, ingrown hairs, and allergic contact dermatitis to either thioglycolate or fragrances. The preparations must be so formulated as to complete their reaction is not more than five minutes. After the action has been completed, depilatories must be properly washed with soap and applied cold cream over affected area for moistening effect.

Formula I

Barium sulfide	30.0%
Titanium dioxide	30.0%
Corn starch	38.0%
Perfume	2.0%

Blend and shift the barium sulfide, titanium dioxide and corn starch in ascending order of their weight and add perfume in it. Constant mixing must takes place at least for half an hour.

Bleaching

Bleaching is not a method of hair removal, but many women use bleaching as an inexpensive method of disguising the presence of unwanted hair by removing the hair's natural pigment. Common sites for bleaching include the upper lip, beard area, and arms. The active ingredients in over-the-counter bleaching agents are hydrogen peroxide and sulfates as activating agents, a combination that bleaches, softens, and oxidizes hair. A variety of commercial bleaches are available, and the manufacturer's instructions are easy to follow. As with chemical depilatories, perform a small patch test to assess for allergic reaction.

The disadvantages of bleaching include skin irritation, temporary skin discoloration, pruritus, and the prominence of bleached hair against tanned or naturally dark skin. Reports exist of generalized urticaria, asthma, syncope, and shock in reaction to the persulfate activator added to boost the effect of hydrogen peroxide bleach.

Epilation

Epilation involves the removal of the entire hair shaft and is the most effective method for temporarily removing hair. Epilation includes waxing, plucking, threading, sugaring, and using abrasives or mechanical devices (e.g. Epilady). For epilation to be effective, treated hairs should be long enough for the device to grasp. The long-term effects of epilation on the hair follicle are not known, and whether this practice may result in long-term reduction of hair regrowth is unclear. Because epilation wounds the hair follicle, repetitive epilation may result in permanent matrix damage, resulting in finer or thinner hairs.

Shaving

Shaving is the method used most frequently to temporarily remove unwanted hair. Shaving is fast, easy, painless, effective, and inexpensive. The results are temporary, lasting 1–3 days, and shaving requires a constant commitment to maintaining a hair-free appearance.

Shaving is performed with a razor on wet skin using shaving cream or other lubricants, with the razor oriented against the direction of hair growth. For sensitive areas, shaving with the direction of hair growth may reduce cuts. Contrary to a widespread misconception, shaving does not result in increased hair growth. The primary disadvantages and/or adverse effects of shaving include skin irritation, cuts in the skin, ingrown hair pseudofolliculitis, the need to shave daily, and stubble.

Plucking

Plucking is best performed using tweezers and is a beneficial and economic method for removing the occasional coarse hair or a small group of hairs, such as those found on the eyebrows, chin, or nipples. The results of plucking last longer than shaving because hair is pulled from the hair shaft, as in waxing. This method is time consuming, tedious, and painful. The reaction of the hair follicle to plucking can be unpredictable, possibly resulting in folliculitis, hyperpigmentation, scarring, ingrown hairs, and distorted follicles. Adverse effects from plucking include pain, hyperpigmentation, scarring, folliculitis, and ingrown hair pseudofolliculitis.

Waxing

Waxing is similar to plucking and involves applying warm or cold wax onto hair-bearing skin and quickly stripping off the hardened wax and embedded hairs against the direction of hair growth. Waxing is the most expensive yet most effective method of epilation because hair is removed completely from the hair shaft in large quantities. Often, hair can take 2–3 weeks to regrow. The effects on the hair follicle of long-term waxing are unknown. However, theoretically, this modality may reduce regrowth because repeated waxing may destroy follicles. Although many kits are offered for use at home, faster and more successful results are obtained by an experienced salon-based operator.

Although no formal studies have been conducted, the recommendation is that patients using systemic retinoids (i.e. isotretinoin [Accutane], acitretin [Soriatane]) refrain from waxing until treatment has been discontinued for a minimum of 6 months to 1 year to avoid tearing of the skin and scarring. Patients using topical retinoids (i.e. tretinoin [Retin-A, Avita], adapalene [Differin]) should be careful to avoid injuring the skin. Waxing should not be performed on moles or skin that is irritated, sunburned, or broken. Pay special attention to the temperature of the wax to avoid burning the skin. Adverse effects from waxing include pain, hyperpigmentation, scarring, folliculitis, and ingrown hair pseudofolliculitis.

Threading

Threading is an ancient manual technique, popular in many Arabic countries, that involves the use of a long twisted loop of thread rotated rapidly across the skin. By maneuvering the twisted string, hairs are trapped within the tight entwined coils and are pulled or broken off. Adverse effects from threading include pain, hyperpigmentation, scarring, folliculitis, and ingrown hair pseudofolliculitis.

Abrasives

Abrasives such as pumice stones and devices or gloves made of fine sandpaper work by physically rubbing the hair away from the skin surface. This method can be irritating to the skin and is not commonly used today for hair removal.

Sugaring

Sugaring is similar to waxing. The sugar mixture is prepared by heating sugar, lemon juice, and water to form syrup. The syrup is formed into a ball, flattened onto the skin, and then quickly stripped away. Similar to waxing, the hair is removed entirely from the hair shaft, and sugaring is an alternative to waxing for people sensitive to wax. Adverse effects from sugaring include pain, hyperpigmentation, scarring, folliculitis, and ingrown hair pseudofolliculitis.

TEMPORARY HAIR REDUCTION

Eflornithine

Eflornithine, a novel method for temporary hair reduction in women, is a topical cream available by prescription only and recently approved by the US Food and Drug Administration for the reduction of unwanted facial hair in women.

Eflornithine is not a hair remover or depilatory, but is a topical cream that decreases the rate of hair growth. It works by inhibiting the enzyme *ornithine decarboxylase*, an enzyme in human skin that stimulates hair growth. When this enzyme is blocked by the medication, metabolic activity in the hair follicle decreases and hairs grow in more slowly. It has been studied only on the face and the adjacently involved areas under the chin; therefore, it should be used only in those areas. Because eflornithine does not remove hair, it must be used in combination with the patients' normal hair removal methods (e.g. shaving, waxing, and plucking). It is rubbed onto the affected areas on the face twice daily.

Laser-assisted Hair Removal

Laser-assisted hair removal is a relatively new method available for long-term hair reduction. The different lasers available for hair removal are:

- The ruby laser (694 nm),
- Alexandrite laser (755 nm),
- Diode laser (800 nm), and
- Nd:YAG laser (1064 nm).

These lasers target melanin and subsequently produce selective photothermolysis of the hair follicles. The longer wavelengths are safer for darker skin types. Refer to Laser-Assisted Removal for more information.

Intense Pulsed Light Sources for Hair Removal

Intense pulsed light sources use the same principle of selective photothermolysis used with lasers to target melanin in hair follicles; however, a noncoherent filtered flash lamp that emits wavelengths ranging from 500–1200 nm is used in this process rather than a laser. Different cutoff filters are used to select the appropriate wavelength for each patient. Studies have shown that the hair removal efficiency rate (i.e. percent of the number of hairs present compared with baseline counts) is best after 1–3 treatments. Adverse effects are minimal, and the hair removal efficiency rate is greater than 50% when evaluated more than 12 months following the last treatment.

PERMANENT HAIR REMOVAL

Electrolysis

Electrolysis, also termed electrology, involves the insertion of a small, fine needle into the hair follicle, followed by the firing of a pulse of electric current that damages and eventually destroys the hair follicle. Multiple treatment sessions are required to achieve a clinically significant result. The 2 types of electrolysis are:

- Galvanic electrolysis (direct current electrolysis) and
- Thermolysis (alternating current electrolysis).

Galvanic Electrolysis

In galvanic electrolysis, a direct electric current is passed down a needle inserted into the hair follicle, where it acts on tissue saline to produce sodium hydroxide (lye), a caustic agent that destroys the hair bulb and dermal papilla (chemical reaction $2\ NaCl + 2\ H_2O = 2\ NaOH + H_2 + Cl_2$). During the procedure, the patient holds a metal rod covered with conductive cream or gel or a metal plate attached to a moistened pad. The current (milliamperes) is set by the technician based on the patient's pain threshold, and the duration of the pulse is controlled by how long the technician presses down on the hand or foot pedal. Galvanic electrolysis is slow and may require a minute or more for each hair, including repeated insertions into the follicle.

Thermolysis

Most modern electrolysis machines use thermolysis or the blend method, a combination of galvanic electrolysis and thermolysis. Thermolysis uses a high-frequency alternating current that is passed down the needle into the follicle. The high-frequency alternating current produces heat in the hair follicle via molecular vibration, resulting in destruction of the hair bulb by thermal, not chemical, means.

Significant evidence indicates that the region of the erector pili muscle insertion is the site of the stem cells responsible for hair regeneration. More research is needed to determine the effects of electrolysis on this region.

Proper electrolysis requires accurate needle insertion technique and appropriate intensities and duration of current. In addition, only anagen-phase hairs should be treated because telogen-phase hairs are believed to be more resistant to damage. Anagen-phase hairs can be distinguished easily from telogen-phase hairs by shaving the area to be treated and, in a few days, treating only those hairs visible on the skin surface (anagen-phase hairs).

Electronic tweezer devices have been developed for home use; however, because hair is not an electric conductor, current cannot be transmitted via hair to the hair bulb.

Adverse effects of electrolysis: Adverse effects include those which are dependent on the duration and intensity of the current and technician experience.

- Scarring (i.e. keloid formation) and
- Post-inflammatory hyperpigmentation or hypopigmentation,
- Pain, a primary adverse effect of electrolysis, can be diminished with the use of new topical anesthetic creams 1 hour prior to the procedure.
- Local bacterial and viral infections. The spread of hepatitis or HIV has not been reported with electrolysis. Electrolysis is not safe for patients with pacemakers and should not be used on these patients.

HAIR GEL

Hair gel is a hairstyling product that is used to stiffen hair into a particular hairstyle. It was invented in 1936 in North America. The results it produces are usually similar to but stronger than those of hair spray and weaker than those of hair glue or hair wax.

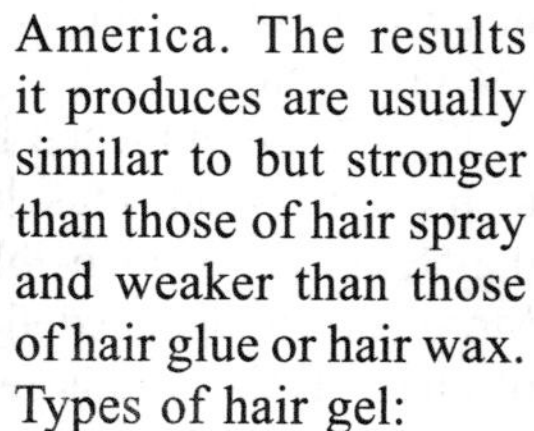

Types of hair gel:

i. **Higher numbered gels:** It maintains a greater "hold" on hair, while
ii. **Lower numbers gel:** Do not make the hair as stiff and in some products give the hair a wet look.

Some forms of hair gel include temporary coloring for the hair, including variants in unnatural colors associated with various subcultures, and is popular within the Gothics and Raver subcultures.

ANTIDANDRUFF PREPARATIONS

Most preparations for the treatment of dandruff largely depend on antimicrobial agents for their therapeutic effect. One leading antidandruff lotion is a combination of benzethonium chloride and N-trichloroethyl mercapto-4-cyclohexene-1,2-dicarboximide. Biphenamine hydrochloride provides an excellent bacteriostatic antifungal property and local anesthetic action.

Zinc Pyrithione

Zinc-2-pyridine-thiol-1-oxide is one of the most widely used antimicrobial agents, while lauryl isoquinolinium bromide and bislauryltrimethylammonium polythionate are 2 other popular agents. In addition to their antiseborrheic properties, 2,2′-thiobis-4-chlorophenol and diiodohydroxyquin also exhibit antiseptic qualities.

A anti-dandruff medicated shampoo is required to have following characteristics:

- It should be non-irritant to sebaceous gland.
- The active ingredients used in shampoo should not sensitize the scalp.
- It should clean both the hair and the scalp.
- It should leave the hair manageable and lustrous.
- It should contain anti-microbial to prevent the growth of microbes.
- It should reduce the degree of itching and scaling.

Table 18.1: Active ingredients used in the treatment of dandruff		
S.No.	*Name of active ingredients*	*Usual concentration%*
1.	Zinc pyrithion	upto 2.0
2.	Thymol	0.05–0.2
3.	Resorcinol	0.05–1.0
4.	Selenium sulphide	1.0–2.5
5.	1, 2-Bithionol	0.5–1.0
6.	Coal tar (prepared)	2.0–2.5
7.	Quaternary ammonium compounds (QAC)	15–20 when used as shampoo base

Avaliable Market Products

- Jolen Hair Removal Kit-three varieties are available:
  1. Crème Hair Removal-with shea butter and olive oil.
  2. Wet Tissue-for skin care.
  3. Retarder-for delaying the regrowth of hair
- Care-o-hair (by Flavita).
- Hairnext-Easy Nourish Spray (by Lakme).
- Techni-Art Liss Control (by L'Oréal).
- Roghan Badam Shirin (Hamdard).
- Nutrísse (Garnier).
- Dercos Aminexil SP94 (Vichy's Lab.)-hair loss preventing formula.
- Fructis Fortifying Shampoo (Garnier).
- Head & Shoulders Natural Shine.
- VANIQA cream (for temporary hair reduction).
- ELA-Max-local anaesthetic cream for thermolysis.

19 Hair Colouring Products

> *We ascribe beauty to that which is simple, which has no superfluous parts; which exactly answers its end; which stands related to all things; which is the mean of many extremes.*
>
> — Ralph Waldo Emerson

HAIR COLOURING PRODUCTS (HAIR DYES)

Hair turns grey due to loss of pigment from hair shaft and progressive loss of tyrosine activity from hair bulbs. It is a normal part of aging. White or grey hair first appears at the temples at an average age of about 35–40 years. After the age of 50 years about 50% of hair turns grey. The age at which this occurs varies from person to person, but in general nearly everyone 75 years or older has grey hair, and in general men tend to become grey at younger ages than women.

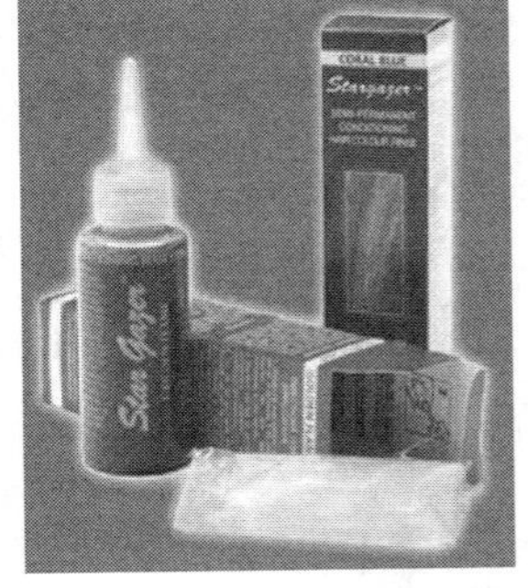

There are several other reasons also responsible for hair turning into grey like emotional disturbance or stress, febrile illness (telogen effluvium). In telogen effluvium sudden shedding of dark hair from scalp and leaving grey hair. There is no obvious medical known treatment available for reversing of graying of the hair. Therefore hair dyes have been developed to colour grey hair.

Hair dying or colouring products generally fall into four categories:

- Temporary,
- Semi permanent,
- Deposit only/demi, and
- Permanent.

All these hair colour products, except for temporary colour, require a patch test before application to determine if the client is allergic to the product.

"Hair lightening," referred to as **"bleaching"** or **"decolourising"**, is a chemical process involving the diffusion of the natural colour pigment or artificial colour from the hair. This process is central to both permanent hair colour and hair lighteners.

All permanent hair colour products and lighteners contain either a developer, or oxidising agent, and an alkalizing ingredient as part of their ammonia or an ammonia substitute. The purpose of this is to:

- Raise the cuticle of the hair fibre so the tint can penetrate,
- Facilitate the formation of tints within the hair fibre,
- Bring about the lightening action of peroxide.

- When the tint containing the alkalizing ingredient is combined with the developer (usually hydrogen peroxide), the peroxide becomes alkaline and diffuses through the hair fibre, entering the cortex, where the melanin is located. The lightening occurs when the alkaline peroxide breaks up the melanin and replaces it with new colour.

Temporary Hair Colour

The pigment molecules in temporary hair colour are large and, therefore, do not penetrate the cuticle layer, allowing only a coating action that may be removed by shampooing. An example of use of temporary hair colour is for Halloween costumes.

Acid dyes are used to coat on the surface of hair, since acid dyes have a low affinity to hair, thus can be removed after a shampoo.

Temporary hair colour is available in various product forms including rinses, shampoos, gels, sprays, and others. This type of hair colour is typically used to give brighter, more vibrant shades or colours such as orange or red that may be difficult to achieve with semi-permanent and permanent hair colour. This phenomenon is due to the fact that temporary hair colourants do not penetrate the hair shaft itself. Instead, these dyes remain adsorbed (closely adherent) to the follicle and can be easily removed with a single shampooing. However, even temporary hair colouring agents can persist if the user's hair is excessively dry or damaged, conditions that allow for migration of the dye from the exterior to the interior of the follicle. While temporary hair colour products hold a lesser market than semi-permanent and permanent agents, they have value in that they can be easily and quickly removed without bleaching or application of a different dye (Herbal hair colouring agent will be discussed in chapter 27 Herbal Cosmetics).

Semi Permanent Hair Colour

Formulated to deposit colour on the hair shaft without lightening it. This formula has smaller molecules than those of temporary tinting formulas, and is therefore able to penetrate the hair shaft. Has no developer, may be used with heat for penetration. It also lasts longer than temporary hair colour, keeping intact up to 8–14 shampoos.

Demipermanent Hair Colour

Uses a mild, creamy developer of a lower volume (3 to 7 volumes or 1 to 3% H_2O_2) than permanent colour that lasts for 2 to 3 months. Some demi-products contain MEA's [an ammonia substitute] which helps with penetration and can lift natural colour, but not seriously. Penetrates the hair shaft slightly, leaves hair shiny, and covers/blends some grey hair.

The American Board of Certified Haircolourists and most major manufacturers of hair colour now say one should colour the new growth area with a permanent colour to cover gray and touch up or refresh the ends and length of the hair with a compatible shade of demipermanent colour to protect the condition of the hair.

Most hair colour manufacturers offer a demipermanent hair colour tube and a permanent hair colour tube within their product line. However, lately, some hair colour manufacturers like Compagnia Del Colore from Italy have possibly come up with a better and more cost-efficient solution for hair colourists. By using an activator or 7 Volume Hydrogen Peroxide (2.1% H_2O_2) you can now use the same permanent hair colour tube and convert it into a semipermanent hair colour tube.

If one is using a demi-permanent hair colour, especially one that is not of a natural tone, they should be cautioned. While the colour may cover some gray hair, on blonde it is known to give the appearance of gray as the colour fades.

Permanent Hair Colour

This is mixed with developer and remains in the hair shaft until new growth of hair occurs. It's used to match, lighten, and cover grey hair. Permanent hair colour generally contains ammonia, oxidative tints, and peroxide. The allergic reaction that comes from hair dye is generally one of sensitization to p-phenylenediamine (PPD). The reaction will most likely occur each time one dyes one's hair and will probably get worse each time. The sensitization from the ingredients in hair colour can extend to sensitization of other products of same or similar composition, including but not limited to the dye used in textiles, sunscreen, rubber, and/or certain medications.

Henna is a deposit-only hair colour whose active component, lawsone, binds to keratin and is therefore permanent. Henna may be removed with mineral oil; however, it is considered "permanent" because it does not wash out with shampoos or rinses. It is often mixed with other plant dyes, such as indigo, turmeric, and henna, to change the colour. Allergy to henna is much rarer than allergy to permanent hair colours. It is also considered a conditioning treatment.

Using a plant-based colour, specifically henna, can cause problems later when trying to do a permanent wave (perm) and other permanent hair colour. Discoloration can occur on hair that has been previously tinted with henna; hennaed hair typically cannot be curled. Breakage could also be an issue.

SPECIAL EFFECTS

Special effects include highlighting and vivid, unusual hair colours such as green or fuchsia. Highlighting can range from temporary to permanent, using the techniques listed above and a special application process. The techniques required to apply highlighting can be difficult for an individual to perform upon him/herself. One can create looks that range from subtle highlights acquired during a day at the beach, to more dramatic looks, such as bold, chunky highlights.

The more exotic, bright dyes typically contain only tint, and have no developer. These are typically sold in punk-themed stores (such as comic book and music stores), but are rarely available at commercial hair dressers. Colours range from blood red to sea foam green. Many shades are black light reactive. Individuals with darker hair (medium brown to black) are advised to use a bleaching kit prior to tint application for the full effect of the colour. Some people with fair hair may benefit from prior bleaching as well, as the yellow undertones of blonde hair can make blue dye look green. These dyes are less permanent, and tend to "bleed" onto other fabric even when dry. Users should anticipate staining of light-coloured pillows for a week or so after application.

Problems Related to Colouring Hair

When colouring one's hair it is always advisable to visit a professional hair colourist as there are many mistakes a person could make, as well as some serious consequences. The following are some of the problems that may occur as a result of applying hair colour:

Different colour outcome compared to what was expected and some of the problems people have had from using hair dyes are:

- Breakage of hair strands.
- Loss of hair.
- Dry scalp.
- Burning
- Redness
- Itchy or raw skin
- Swelling in the face
- Trouble breathing.

Safety Concern of Hair Dyes

The decision to change hair color may be a hard one. Some studies have linked hair dyes with a higher risk of certain cancers, while other studies have not found this link. Most hair dyes also don't have to go through safety testing that other cosmetic color additives do before hitting store shelves. Women are often on their own trying to figure out whether hair dyes are safe. When hair dyes first came out, the main ingredient in coal-tar hair dye caused allergic reactions in some people. Most hair dyes are now made from petroleum sources. But FDA still considers them to be coal-tar dyes. This is because they have some of the same compounds found in these older dyes.

Cosmetic makers have stopped using things known to cause cancer in animals. For example, 4-methoxy-m-phenylenediamine (4MMPD) or 4-methoxy-m-phenylenediamine sulfate (4MMPD sulfate) is no longer used. But chemicals made almost the same way have replaced some of the cancer-causing compounds. Some experts feel that these newer ingredients aren't very different from the things they're replacing.

Experts suggest that you may reduce your risk of cancer by using less hair dye over time. You may also reduce your risk by not dyeing your hair until it starts to gray.

Lead acetate is used as a color additive in "progressive" hair dye products. These products are put on over a period of time to produce a gradual coloring effect. For safe use we must follow the directions carefully. This warning statement must appear on the product labels of lead acetate hair dyes:

"**Caution:** Contains lead acetate, for external use only. Keep this product out of children's reach. Do not use on cut or abraded scalp. If skin irritation develops, discontinue use. Do not use to color mustaches, eyelashes, eyebrows, or hair on parts of the body other than the scalp. Do not let into eyes. Follow instructions carefully and wash hands thoroughly after use."

Safety of hair dyes during pregnancy: It's likely that when applying hair dye, only a small amount is absorbed into system. So very little chemicals, if any, would be able to get to baby. In the few animal and human studies that have been done, no changes were seen in the developing baby.

Hair Dye Safety Procedures

- Do a patch test before using dye on your hair. Here's how: Rub a tiny bit of the dye on the inside of your elbow or behind your ear. Leave it there for two days. If you get a rash, do not use the dye on your hair. You should do the test each time you dye your hair. (Salons should also do the patch test before dyeing your hair.)
- Do not leave the dye on longer than the directions say you should.
- Follow the directions in the package. Pay attention to all "Caution" and "Warning" statements.
- Keep hair dyes out of the reach of children.
- Never dye your eyebrows or eyelashes. This can hurt your eyes. You might even go blind. FDA does not allow using hair dyes on eyelashes and eyebrows.
- Never mix different hair dye products. This can hurt your hair and scalp.
- Rinse your scalp well with water after dyeing.
- Wear gloves when you apply the hair dye.

Avaliable Market Products

- Lakme Hair Next Colour Fresh Conditioners
- Croma Reflect (Kerastase)
- Garnier Hair Colour
- L'oréal Plum Mystique
- Wella Bio Touch Colour Conditioning Spray
- Garnier Multilight Street Wear Hights
- Schwarzkopf Igora Action Paint.

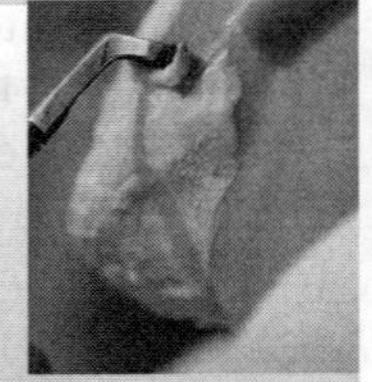

20 Shaving Preparations

It is amazing how complete is the delusion that beauty is goodness.

— Leo Tolstoy

SHAVING PREPARATIONS

Shaving soaps or aids as such were first described over 125 years ago and as would be supposed, were prepared by the saponification of cooking fats with potash and soda lye. Since these early attempts, man has been trying to make this process of shaving a more pleasant, convenient, and comfortable necessary ritual.

The primary aim of shaving preparation is hair softening. Shaving media serve the purpose of removing hair from men's face or also from legs, under arms, groin, chest and other conspicuous places of the women.

Shaving Preparations

A. Shaving soap (cakes, sticks, etc.)
B. Shaving cream (aerosol, brushless, and lather)
C. Beard softeners
D. Aftershave lotion
E. Preshave lotions (all types)
F. Men's talc
G. Other shaving preparations

A shaving preparation irrespective of its physical form must have following properties:

1. It should be non-irritating to the skin.
2. It should retain moisture during the period it remains on the face.
3. It should soften the beard so that hair is cut easily.
4. It should provide sufficient lubricity so that razor glides along the face.
5. It should have sufficient viscosity to hold individual hair erect so it cut close to surface.
6. It should be stable over a wide range of temperature.
7. It should be non-corrosive to the container in which it is packed.
8. It should be able to wash down the drain without creating the clogging problem.

SHAVING SOAP

The potassium soap alone is usually softener but a certain proportion of sodium soap is present in shaving soaps. Shaving soaps are presented in bar, cake, stick mug, bowl or cup form.

They all have to be applied with a shaving brush to produce a sufficient quantity of lather.

Formula I

Tallow	15 to 30 parts
Coconut oil	10 to 15 parts
Stearic Acid	10 to 50 parts
Sodium hydroxide	5 to 10 parts
Potassium hydroxide	10 to 15 parts

Additive

Super fatting agents: monodecyl, dodecyl ether of diglycerol of amine.

Talc: to increase persistence of lather helps razor to glide on surface smoothly.

Formulation

Shaving cup/sticks contains stearic acid, coconut fatty acids which are saponified with potassium hydroxide and sodium hydroxide in the ratio of 5:1.

Shaving bar/cake contains toilet soap and fatty acids which are saponified with potassium hydroxide and sodium hydroxide.

Pilomotor: it causes contraction of hair follicle muscle which causes hair fiber erect.

LATHER SHAVING CREAM

Shaving cream softens and moistens the skin and the hair, thus making shaving more comfortable and contributing to smoother skin. Shaving cream is a substance applied to the skin to facilitate removal of hair. The advantages of using shaving cream, rather than soap, oil, or just water, are many. Shaving with a modern bar of soap approximates shaving with cream but doesn't provide all of the benefits: soap is only one element of many in a modern shaving preparation.

Modern shaving cream began with Burma Shave, by Frank Rowsome, Jr. Prior to that time, lather was produced from a bar, and was basically another form of soap.

Manufacturing soap itself is an ancient craft—the word comes from the Old English word sape. In the fourteenth and fifteenth centuries soap was made at Savona, Italy. The modern French, Spanish and German words for soap (savon, jabon, and seife, respectively) are cognates of the name of that town.

The early American settlers manufactured soap at home, using a method which called for mixing and heating animal fat with lye in a pot set over a fire, usually outdoors. This "open kettle" method of soap making was popular for years. Later adapted for large scale production, its use continued through the first half of the twentieth century. Now 64 million men use shaving cream.

SHAVING CREAM BY TYPE

- Regular-a large, undifferentiated group.
- Medicated-older, downscale.
- Moisturizer/conditioner-younger than average.
- Sensitive skin-another young user group.

Two distinct age groups use tough beard shaving cream:

- Aloe/lanolin users-young, mildly upscale
- Menthol users-oldest group.

By the eighteenth century, soap makers realized that they could enhance their product by improving the quality of the fat and the purity of the lye they used. **Castile soap**, made in Spain and still available today, soon achieved eminence as face soap because of its smoothness and quality. Castile soap originally used olive oil rather than animal fat, and the modern version uses other fats and oils in addition to olive oil.

The first soap maker to render (purify by melting) fats at his own operation was **William Colgate**, who had learned his trade in the early 1800s in New York City. The company that today bears his name is a Major producer of soap and other cosmetic preparations.

In the nineteenth century storekeepers purchased soap from manufacturers in large blocks, from which their customers in turn cut smaller chunks. Jesse Oakley of Newburgh, New York, became the first manufacturer to sell wrapped soap in a cake form that was a good size for home use.

Soap was used for shaving through the early 1800s. In 1840, a concentrated soap that foamed was sold in tablets by Vroom and Fowler, whose Walnut Oil Military Shaving Soap was probably the first soap made especially for shaving.

In addition to raising concerns about the quality of soap, World War II contributed to the invention of the spray can. The first aerosol shaving cream appeared in 1950, it captured almost one fifth of the market for shaving preparations within a short time. Today, aerosol preparations dominate the shaving cream market.

Raw Materials

The main aim of any shaving preparation is:

- To wet and soften the hair to be shaved,
- Cushion the effect of the razor, and
- Provide a residual film to soothe the skin. This film should be of the proper pH value: neither excessively alkaline nor overly acidic, it should correspond to the skin's pH level.

Many manufacturers would have us believe that the recipes for shaving cream are carefully guarded secrets. However, the secrecy revolves mostly around the quantities in which standard ingredients are used, and the choice of substitutes for the few ingredients that are variable. By law, ingredients are listed right on the container, except for perfumes.

A standard formula II

Contains approximately:

Stearic acid	8.2%,
Triethanolamine	3.7%,
Lanolin	5%,
Glycerin	2%,
Polyoxyethylene sorbitan monostearate	6%,
Water	79.6%.

Essentially, a lather-type shaving cream contains ingredients similar to those of the bar shaving soap.

Formula III

Stearic acid	20 to 40%
Coconut oil	6 to 10%
Glycerin	5 to 15%
Potassium hydroxide	2 to 6%
Sodium hydroxide	1 to 3%
Vegetable or Mineral Oil	1 to 5%
Water	q.s.

Shaving sticks are usually produced in a very dry, firm form which is rubbed onto the moistened skin, then worked into lather with a brush.

Formula IV

Stearic acid	10.5 parts
Coconut oil	3.0 parts
Potassium Hydroxide	5.5 parts
Sodium Hydroxide	0.4 parts
Glycerin	1.0 parts

Two major ingredients in this formula are common in many of today's preparations. Stearic acid is one of the main ingredients in soap making, and triethanolamine is a surfactant, or surface-acting agent, which does the job of soap, albeit much better. While one end of a surfactant molecule attracts dirt and grease, the other end attracts water. Lanolin and polyoxyethylene sorbitan monostearate are both emulsifiers which hold water to the skin, while glycerin, a solvent and an emollient, renders skin softer and suppler.

Common substitutes for the third, fourth, and fifth ingredients listed above include laureth 23 and lauryl sulfate (both sudsing and foaming agents), waxes, cocamides (which cleanse and aid foaming), and lanolin derivatives (emulsifiers). Most ingredients are powdered or flaked, although lanolin, lanolin derivatives, and cocamides are liquids.

The differences between one brand of shaving cream and another amount to adjustments in the proportions of ingredients and in the processing method (longer or shorter heating times, storage of the finished product, and so on), and choice of ingredients such as emulsifiers or perfumes. Also important is the choice of aerosol propellant. Some mixtures contain more than one propellant; most common are butane, isobutane, and propane. Though the wide range of choices for ingredients is well known, the exact combinations of ingredients represent the highest level of "magic" in modern chemistry.

BRUSHLESS SHAVING CREAMS

Brushless shaving creams are very much popular because of their convenience of use and comfort in shaving. They require no brush for application, applied to face more frequently and required very less space. They leave the face with the thin layer of the oil or grease after shaving.

The most common materials used for brushless shaving creams are: stearic acid, caustic potash, glycerin, boric acid, wetting agents, vegetable oil, mineral oils, sulphonated oils, borax, triethanolamine, propylene glycol, alcohol, stearyl alcohol, cetyl alcohol, lanolin, spermaceti, bees wax, etc. and suitable perfume is also added.

Formula V

Triple pressed stearic acid	25.0%
Glycerin	3.0%
Lanolin (anhydrous)	3.0%
Borax	0.5%
Alcohol	2.0%
Triethanolamine	1.0%
Deionized water	65.0%
Perfume	0.5%

Heat triethanolamine, borax, glycerin and water in a baker at 70 °C and melt stearic acid, lanolin in another beaker on a water bath. Add both solutions with constant stirring until smooth and white emulsion I formed. Add perfume dissolved in alcohol.

The Manufacturing Process

The modern manufacture of shaving cream is a carefully controlled process. Although carried out on a large scale, its manufacture resembles a laboratory procedure involving only small quantities of ingredients. There are two main phases to the manufacturing process.

In the first phase, the fatty or oily portions of the formula-stearic acid, lanolin, and polyoxyethylene sorbitan monostearate-are heated in a jacketed kettle to a temperature of approximately 179° to 188°F (80° to 85°C). The jacketed kettle, which can hold as little as 300 gallons or as much as 10,000 gallons, resembles a double boiler: one container, placed inside another, is heated when steam is circulated through the outer container. Inside the interior kettle are blades that revolve to mix the oils as they are heated.

After the first group of ingredients has turned smooth over a period of roughly 40 minutes, the steam is released from the outer container of the kettle, and the mixture is allowed to cool.

The second phase of manufacture begins when the mixture has cooled to about 152 °F (65 °C). Most of the remaining ingredients-water, glycerin, and triethanolamine-are added now, and mixing continues for approximately 40 minutes.

When the mixture reaches a temperature of 125° to 134 °C (50° to 55 °C), perfumes or other scents can be added. Because perfumes consist primarily of highly volatile oils, they would evaporate if added when the blend was still warm. The formulas for perfumes, which can contain more than 200 different ingredients, come closer to being trade secrets than information about shaving cream itself (though textbook and handbook formulas for perfume are not hard to come by). In recognition of this, manufacturers do not have to disclose information about fragrances.

The mixture, still being stirred, is allowed to cool further, until it reaches a temperature of 89 °F (30 °C). Now a thickening white mass of highly viscous liquid, it is forced through a silk or stainless steel screen to eliminate any lumps that may have formed in the mixing process, and to catch the rare impurity or foreign object such as a small wood splinter.

If this particular mixture is designated for tube packaging, it is now placed in a tube and fitted with a cap. After the bottom of the tube has been crimped, the product is ready for shipment and stocking on a store shelf.

When the desired product is an aerosol spray, the shaving cream is poured into an open can. Next a valve and a cover are fitted onto the can and forced downward to form a seal. Propellant is then forced into the can through the valve. Most shaving preparations contain between four and five percent propellant; a larger amount would dry the shaving cream as it came out of the can, rendering it unusable. A small amount of material is intentionally released (purged) to relieve excess pressure, and the can is tested in water to make sure that the valve is holding tightly. The can is now ready to be shipped.

AFTERSHAVE LOTION OR COLOGNE

After shave lotion are used by men to provide emollient effect, astringent effect, softening of beards and antiseptic effects after shaving and also in case of razor cuts. They also provide cooling and refreshing feeling after application, which is obtained by high content of alcohol and addition of other ingredients like capsicum and cantharides. The most used perfumes in after shave lotions are lavender, lilac, bay rum, etc. 59 Million American men use aftershave or cologne differences among light, medium, and heavy users.

Formula VI

Boric acid	2.0%
Menthol	0.05%
Glycerin	3.0%
Ethanol	10.0%
Witch hazel	84.0%
Perfume	0.5%

Method of preparation: Dissolve boric acid in witch hazel and add glycerin to it. Menthol and perfume are dissolved in ethanol in a separate beaker. Add alcoholic solution to aqueous solution slowly with constant stirring.

PRE-SHAVING LOTIONS

Pre-shaving lotions are designed to make beard soft before applying any kind of razor (electric, safety, or other) to make it smooth and close shaving without irritation. Beard softener not only softens hair but also help to allay irritation. Sodium cholate is among the most common ingredients of beard softener preparations. These preparations are available in lotion and cream forms.

Formula VII

Sodium cholate	0.50%
Sodium lauryl sulphate	3.00%
Glycerin	10.0%
Deionized water	86.0%
Perfume	0.5%

Dissolve mixture of sodium cholate and sodium lauryl sulphate in glycerin and equal volume of water. Heat this mixture at 60–70 °C. After cooling of solution add remainder of water and perfume in it. Keep it aside for a week and then filter it.

EVALUATION OF SHAVING CREAMS

Today's soaps, shaving creams, and lotions are all manufactured under strict quality control, and regulated by various federal agencies including the Food and Drug Administration (FDA). Some states have their own regulatory agencies, though state agencies are more likely to focus on environmental concerns than product safety. Batches of shaving cream are examined and analyzed both at the manufacturing site and in the laboratory. Individual containers of shaving preparations are coded so that a manufacturer knows exactly which batch any given can or tube came from, and can identify its distribution history.

Hair-softening Studies

These were primarily measurements of the swelling of hair by water and its effect on the softening or cutting strength. Hollander and Gasselman determined that with the use of 120 °F water for presoftening of the beard, a minimum of 2½ to 3 min. would be needed to attain satisfactory softening prior to shaving. This preparation time increased with a decrease in temperature.

Many shavers have found that the passage of the razor edge across the face was invariably accompanied by the production of superficial cutaneous abrasion of the skin. These abrasions more often than not were invisible but resulted in subsequent irritation was made by careful examination of the scrapings and residues obtained from the faces of men who had shaved. The skin scrapings were found to be composed of hair with large amounts of varied epithelial components, both nucleated and non-nucleated.

A manufacturer of shaving cream needs to be certain that each batch meets quality standards. Among the things tested for are:

- pH value (the acidity or alkalinity of the product),
- The height of the foam when sprayed, and
- Its absorption rate (spray the foam on a piece of paper-how long does it takes till the bottom of the paper shows moisture).
- Water quality must also be checked carefully. Most manufacturers make sure the water they use is pure by exposing the water to ultraviolet light or using distilled water. Having a microbiologist on site to test the water and the final product is common in the industry.

Avaliable Market Products

- Gillette
- Bic
- Palmolive
- Denim
- Axe
- Schick
- Old spice
- Vi John
- Willium.

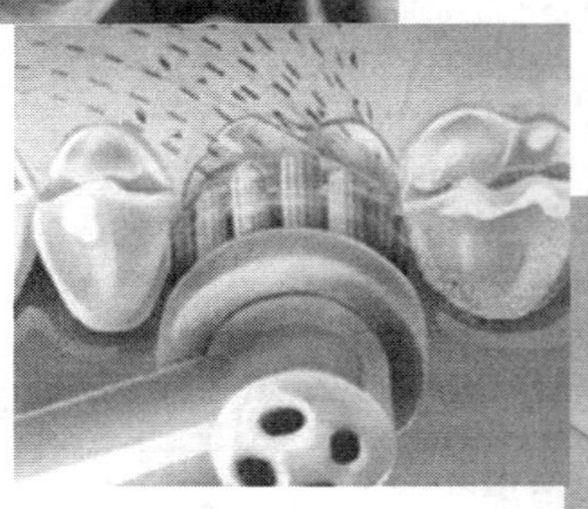

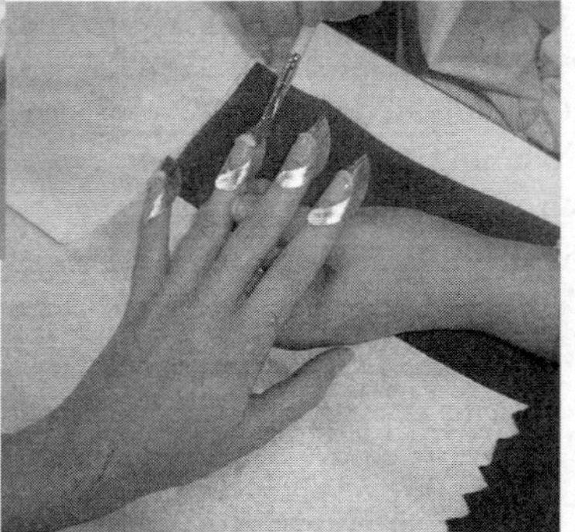

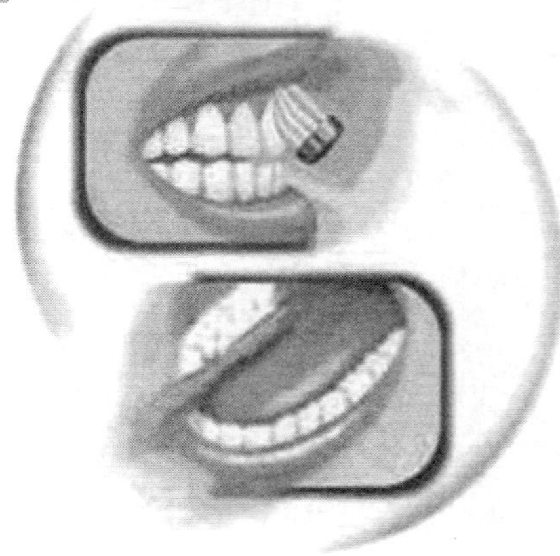

Cosmetics for Eye, Nail and Teeth

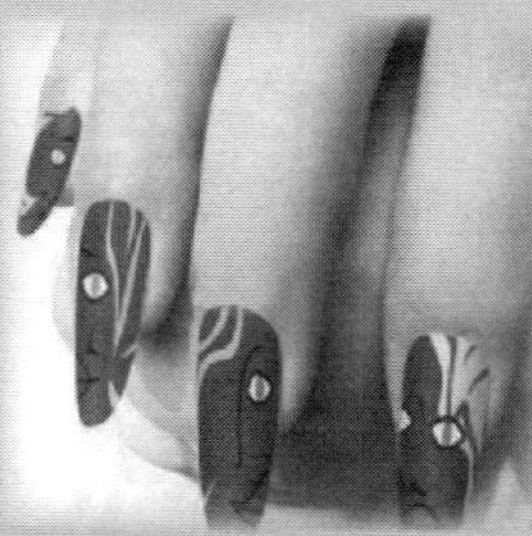

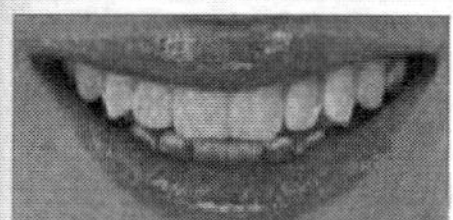

Before

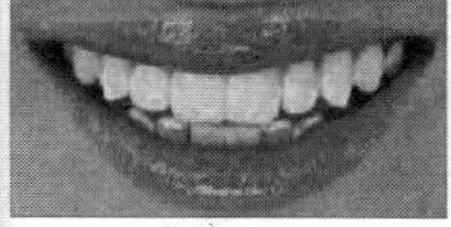

After

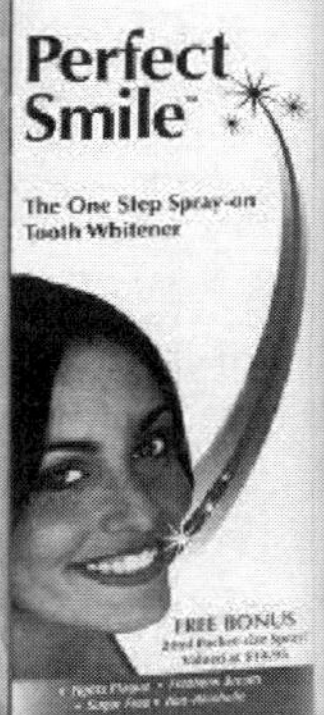

21

Cosmetics for the Eye

> *Beauty lies in the eyes of beholder.*— Shakespeare
> *Beauty comes as much from the mind as from the eye.*— Grey Livingston

INTRODUCTION

Eyes are the most attractive, very much sensitive and delicate part of the body. Eyes are the doors of perception and window to our soul. The life and motion of a beautiful face is in the eyes. The eyes mirror a myriad of emotions—they interact from a beautiful face. They are the epitome of expression. Eye makeup application technique is one of the most important parts in the overall cosmetic application. If one know how to apply eye makeup which is suitable; one could highly cheer up appearance and attractiveness. Now a days the latest eye makeup application techniques and beauty products are available.

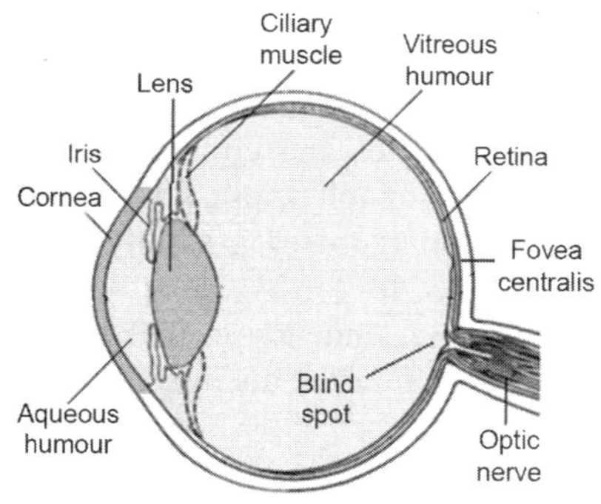

HUMAN EYELASHES

An eyelash or simply lash is one of the hairs that grow at the edge of the eyelid. Eyelashes protect the eye from debris and perform some of the same function as whiskers do on a cat or a mouse in the sense that they are sensitive to being touched, thus providing a warning that an object (such as an insect or dust mite) is near the eye (which is then closed reflexively).

The eyelashes of the embryo develop between the 7th and 8th week. An eyelash takes about four months to grow. Their color may differ from that of the hair and they take longer to grow white.

The follicles of eyelashes are associated with a number of glands known as the glands of Zeiss and the glands of Moll.

Long eyelashes are considered a sign of femininity in most if not all cultures. Kohl has been worn as far back as the Bronze Age to protect and enhance lashes.

COMPLETE EYE MAKEUP INCLUDES

1. Eyeliner,
2. Mascara and

3. Eye shadow to emphasize the eyelashes. The twentieth century saw the beginning of convincing-looking false eyelashes, popular in the 1960s.

EYE LINER

Eye liner is a makeup used to define the eyes, to change their perceived shape or to create a certain mood: a razor-sharp matte line or smoky kohl contour may change the entire look of the face.

Depending on its texture, eyeliner can be softly smudged or clearly defined. There are four main formulas available on the market, each one for producing a different effect:

i. **Harder powder-based eye pencil** draws a clean and precise line and is easy to apply. It is available in dark matte shades.

ii. **Liquid eye liner** gives the most intense and precise line that perfectly defines the eyes and stays sharp over time. It is usually available in dark matte or iridescent shades and come in small bottle with a brush or felt applicator. It is difficult to apply and requires a steady hand. It is recommended for top eye lid only, as not only does it have a tendency to irritate the eye if incorrectly applied, it also gives a very harsh line.

iii. **Kohl eyeliner** has the softest powder texture and is available in dark matte shades. Kohl is used to softly define the eye contour with a dark (usually black) colour. This product may come in several forms: pencil, pressed powder, or loose powder. It is usually applied to the inner rim of the eye (the pink area behind the eyelashes) unlike other eyeliners as it does not irritate the sensitive tissue in that area.

iv. **Softer wax-based eye pencils** contain waxes that ease application. They are good for smudging and come in a variety of colours. Bright blue, green, aqua, violet, or bronze liners are appropriate when a bright colour should be quickly and precisely applied in a line. Light beige or white liners are used to highlight eye lids and corners. This type of eye liner may come in a pencil, a cone or a compact with a brush applicator.

MASCARA

Mascara is a cosmetic used to improve the beauty of the eye also darken, emphasize, lengthen, thicken and define eyelashes. Mascara strengthens and lengthens Eyelashes. Mascara has been applied since Biblical times and today is the most commonly used eye cosmetic. The original mascara worn by women of many ancient civilizations was kohl, based on antimony trisulfide.

The first mascara product was invented by Eugene Rimmel in the 19th century. The word "rimmel" still means "mascara" in several languages, including French and Italian.

The word mascara derives from the Spanish word ***máscara*** or the Italian ***maschera***, which means "**mask**". Modern mascara was created in 1913 by a chemist named T. L. Williams for his sister, Mabel. This early mascara was made from coal dust mixed with Vaseline petroleum jelly. The product was a success with Mabel, and Williams began to sell his new product through the mail. His company Maybelline, a combination of his sister's name and Vaseline, eventually became a leading cosmetics company.

Mascara was available only in cake form, and was composed of colorants and carnauba wax. Users wet a brush and rubbed it over the cake, then applied it to the eyes. The modern tube and wand applicator did not appear until 1957, when it was introduced by and founded by Helena Rubinstein.

Early mascara from the modern era usually took the form of a pressed cake. It was applied to the lashes with a wetted brush. The ingredients typically were 50% soap and 50% black pigment. The pigment was sifted and combined with soap chips, run through a mill several times, and then pressed into cakes. A variation on this was cream mascara, a lotion-like substance that was packaged in a tube. To apply it, the user would squeeze a small amount of mascara out of the tube onto a small brush. This was a messy process that was much improved with the invention in the 1960s of the mascara applicator. This patented device was a grooved application rod that picked up a consistent amount of mascara when pulled from the bottle. The grooved rod was soon replaced with a brush. This new ease of application may have contributed to the increased popularity of mascara in the late 1960s. Mascara comes in three forms:

Liquid: They are suspension of colouring matter usually in light mucilaginous vehicle. Various agents used for mucilage like gelatin, gum acacia, tragacanth, karaya gum, etc. mucilage increases the ability to cling to the eyelashes.

Cake: This is the most convenient and popular form of mascara, it is prepared by milling toilet soap, sometimes with beeswax and petrolatum along with proper colour and

Cream: This is an emulsion prepared by glyceryl monostearates and water combined with suitable colour.

It also comes in many formulas, tints and colors. Mascara is available with tube and wand applicators.

Ingredients in mascara include:

- Water,
- Wax thickeners,
- Film-formers and
- Preservatives.

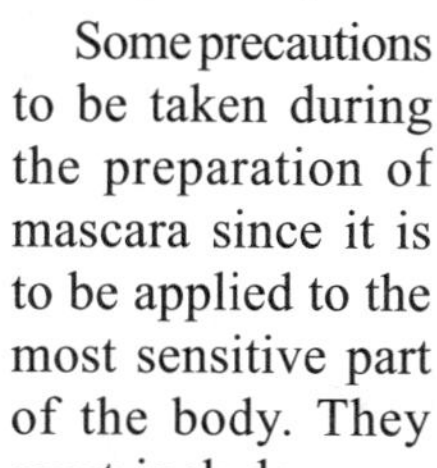

Mascara brushes can be straight or curved, to curl eye lashes, with fine or thick bristles. Some mascara wands contain rayon or nylon fibers to lengthen eyelashes.

Some precautions to be taken during the preparation of mascara since it is to be applied to the most sensitive part of the body. They must include:

1. Materials must be non-toxic and non-irritating to the eyes.
2. The colour must be harmless and intense in shade.
3. They should not run and cause eyelashes to stick together after application.

FORMULATION

Mascara must be formulated carefully to allow easy and even application without smudging, irritancy, or toxicity. The US FD&C Act prohibits the use of coal tar colors on the eyelashes. Therefore, mascara colorants must be selected from vegetable colors or inorganic pigments and lakes. Colors employed include iron oxide to produce black, ultramarine blue to create navy, and umber or burnt sienna or synthetic brown oxide to create brown.

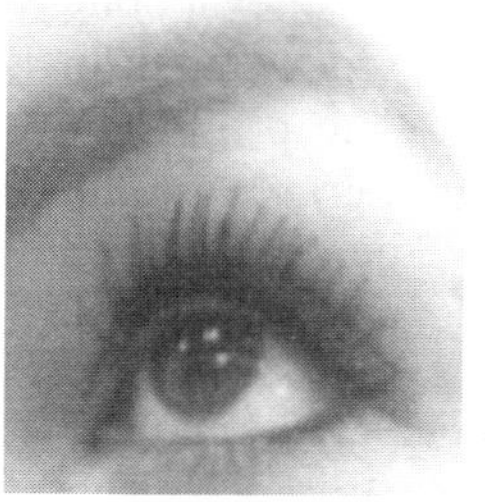

Formula I

Ingredient	%
Gelatin	2.0%
Stearic acid	15.0%
Triethanolamine	5.0%
Glyceryl monostearate	10.0%
Petrolatum	20.0%
Beeswax	25.0%
Prepared lamp black	8.0%
Purified water	15.0%

Soak the gelatin in hot purified water. Melt the beeswax, petrolatum and glyceryl monostearate, continuous stirring in prepared lamp black thoroughly. Add triethanolamine to the soaked gelatin and then add the mixture to the melted mass. Stir continuously to obtain uniform creamy mascara.

Liquid mascara is the most popular modern formulation. Liquids can be divided into:

1. Water-based,
2. Solvent-based, and
3. Water/solvent hybrid varieties.

These products are unique in that they are applied from an automatic mascara tube consisting of a round brush that is inserted through a small aperture to remove a metered amount of product.

1. Water-based mascaras: formulated from:

- Waxes (e.g. beeswax, carnauba wax, synthetic waxes),

- Pigments (e.g. iron oxides, chrome oxides, ultramarine blue, carmine, titanium dioxide) and
- Resins dissolved in water.
- Preservatives, usually parabens.

They are classified as o/w emulsions. The water evaporates readily, creating a fast-drying product that thickens and darkens the lashes. The product is water-soluble, allowing for easy removal, but unfortunately smudges with perspiration and tearing. Some water-based mascara is labeled water-resistant if they contain an increased amount of wax or a polymer to improve adherence of pigment to the lashes. Water-based mascaras are contaminated easily with bacteria, which readily grow in water, and must include preservatives, usually parabens. Thus, these products may cause an allergic reaction in individuals who are sensitive to parabens; however, water-based mascaras are generally the least sensitizing of the mascara types. Some patients may experience contact irritancy from the emulsifiers required to maintain the pigment in solution.

2. **Solvent-based mascaras:** Formulated with
   - Petroleum distillates
   - Pigments (e.g. carmine , chrome oxides, iron oxides, titanium dioxide, ultramarine blue) and
   - Waxes (e.g. candelilla wax, carnauba wax, ozokerite, and hydrogenated castor oil) are added, making them waterproof.
   - Preservatives

   As a result, the product performs well with perspiration and tearing, but removal is difficult and requires an oil-based lotion or cream. Deposits may form on the lashes if the product is incompletely removed. Care must be taken to avoid smudging the product immediately after application because solvent-based mascaras have a prolonged drying time.

   Preservatives are still added to solvent-based mascaras, but microbial contamination is not a great problem because the petroleum-based solvent is antibacterial. Some products also contain talc or kaolin to improve lash thickening and nylon or rayon fibers to lengthen lashes. Solvent-based mascaras can irritate the eye.

3. **Solvent-and water-based systems:** Some mascara combines both to form either a w/o or o/w emulsion. The idea is to create an optimal product that thickens with a short drying time, such as water-based mascaras, but provides waterproof lash separation, such as solvent-based mascaras. The water in the formulation requires incorporation of a good preservative system.

RAW MATERIALS

There are many different formulas for mascara. All contain pigments. In the United States, federal regulations prohibit the use of any pigments derived from coal or tar in eye cosmetics, so mascaras use natural colors and inorganic pigments. Carbon black is the black pigment in most mascara recipes, and iron oxides provide brown colors. Other colors such as ultramarine blue are used in some formulas. One common type of mascara consists of an emulsion of oils, waxes, and water. In formulas for this type of mascara, beeswax is often used, as is carnauba wax and paraffin. Oils may be mineral oil, lanolin, linseed oil, castor oil, oil of turpentine, eucalyptus oil, and even sesame oil. Some formulas contain alcohol. Stearic acid is a common ingredient of lotion-based formulas, as are stiffeners such as ceresin and gums such as gum tragacanth and methyl cellulose. Some mascara includes fine rayon fibers, which make the product more viscous.

The Manufacturing Process

There are two main types of mascara currently manufactured. One type is called anhydrous, meaning it contains no water. The second type is made with a lotion base, and it is manufactured by the emulsion method.

Anhydrous Method

In this method, ingredients are mixed in tanks or kettles, which make a small batch of 10–30 gal (38–114 1). The ingredients are first carefully measured and weighed. Then a worker empties them into the mixing tank. Heat is applied to melt the waxes, and the mixture is agitated, usually by means of a propeller blade. The agitation continues until the mixture reaches a semi-solid state.

Emulsion Method

In this method, water and thickeners are combined to make a lotion or cream base. Waxes and emulsifiers are heated and melted separately, and pigments are added. Then the waxes and lotion base are combined in a very high speed mixer or homogenizer. Unlike the tank or kettle above, the homogenizer is enclosed and mixes the ingredients at very high speed without incorporating any air or causing evaporation. The oils and waxes are broken down into very small beads by the rapid action of the homogenizer and held in suspension in the water. The homogenizer may hold as little as 5 gal (19 1), or as much as 100 gal (380 1). The high-speed mixing action continues until the mixture reaches room temperature.

The Following Steps are Common to Both Types of Mascara

Filling

After the mascara solution has cooled or reached the proper state, workers transfer it to a tote bin. Next, they roll the tote bin to the filling area and empty the solution into a hopper on a filling machine. The filling machine pumps a measured amount (typically about 0.175 oz [5 g]) of the solution into glass or plastic mascara bottles. The bottles are usually capped by hand. Samples are removed for inspection, and the rest are readied for distribution.

Using Mascara

Mascara may be used on all eyelashes, from inner to outer corners. The mascara wand is dipped into a clean tube of mascara, applied close to the base of the lashes and worked out to the tips. They have made new mascara wands that are of plastic and can be cleaned. Mascara can be applied to the top eyelashes for a 'heavy-lidded' look, or to the bottom lashes to widen the eyes. It is usually applied to curled lashes and may be preceded by a lash primer. The moisture in some mascaras and primers can cause lashes to uncurl during application, which is easily solved by using drier, waterproof mascara. Waterproof mascara is popular for this purpose, as well as its tendency to 'clump' less; however, it is more difficult to remove.

Mascara that contains nylon fibers can give lashes a fuller and longer appearance because it clings to the lashes like mini extensions. Pro-vitamin B5 in mascara acts as a conditioner for lashes, giving them a softer and more natural feel.

Various colour shades are used in mascara like lamp black, fluorescent green, synthetic ochre, brown shade.

EYE SHADOW

Eye shadow is a cosmetic which is applied primarily on the upper eyelids and under the eyebrows or around eyes to enhance their brilliance, to help make wearer's eyes stand out or look more attractive. Many people use eye shadow to simply improve their appearance or to impart background shadow to the eyes, but it is also commonly used in theatre and other plays, to create a memorable look, with bright and even ridiculous colours.

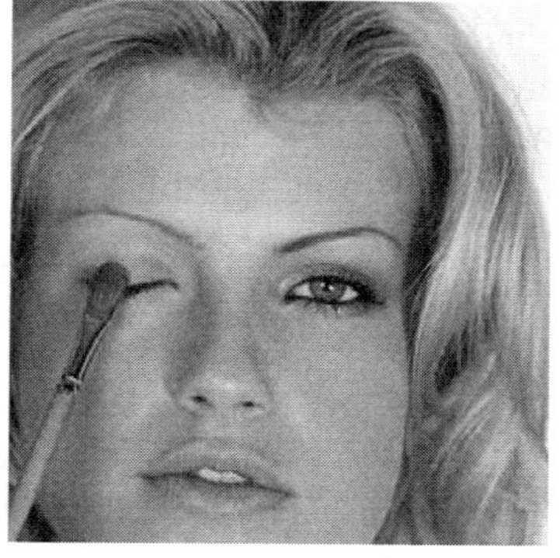

Civilizations across the world use eye shadow predominantly on females, but occasionally on males. Eyelid cosmetic use dates to antiquity, recorded as early as 4000 BC. Green powder made from malachite was applied heavily to both the upper and the lower eyelids. Eyelid glitter composed of ground beetle shells was also popular. Modern eye shadow cosmetics became popular between 1959 and 1962.

In Western society it is seen exclusively as a feminine cosmetic, even when used by men. In Gothic and some Punk subcultures, black or similarly dark-colored eye shadow and other types of eye makeup are popular amongst both genders.

The most common colours used in the eye shadows are grey, white, gold, silver, blue, green, etc.

Eye shadows are applied with the help of fine camel brush.

Formulation

Eye shadow comes in:

- Pressed powder,
- Liquid, anhydrous creams, emulsions
- Sticks, and Pencil or mousse form.

Eye shadow is available in many extensive colours, but no coal tar derivatives can be used in the eye area.

Pressed powder eye shadows are the most popular formulation and are applied to the eyelid by lightly stroking a soft sponge-tipped applicator across the skin. These eye shadows are predominantly talc with pigments and zinc or magnesium stearate used as a binder. Kaolin or chalk may be added to improve oil absorption and to increase wear ability.

Variation in eye shadow surface texture can range from matte to a pearled shine to a metallic shine. Titanium dioxide is used in pastel matte-finish eye shadows to improve coverage. However, it is not found in pearled shine–finish eye shadows because it tends to mask the desired pearled effect. Bismuth oxychloride, mica, and fish scale essence are the standard materials used to produce a pearly shine. A metallic shine is obtained by adding copper, brass, aluminum, or silver powders.

Formula II

Ultramarine	12.0%
Zinc oxide	31.0%
White petrolatum	45.0%
White beeswax	5.0%
Spermaceti	4.0%
Lanolin	4.0%
Perfume	q.s.

Add ultramarine (blue shadow) with zinc oxide and white petrolatum. Mill through ointment mill. Melt the beeswax, spermaceti and lanolin. Add the colour base and perfume. Mill the final mixture to obtain uniform product.

Adverse Effects Linked to Eye Makeup

The eyelid skin is the thinnest on the body and is frequently affected by both irritant and allergic contact dermatitis. The North American Contact Dermatitis Group has determined that 12% of cosmetic reactions occur on the eyelid, but only 4% of them could be linked to eye makeup use.

Modern liquid mascaras provide an applicator that is inserted into the tube between uses, providing numerous opportunities to inoculate bacteria into the mascara. The most feared adverse effect of mascaras is that of infection, particularly *Pseudomonas aeruginosa* corneal infections, which can permanently destroy visual acuity. *Staphylococcus epidermidis* and *Staphylococcus aureus* organisms also may proliferate in contaminated mascaras. Infections are more common if the eyeball is traumatized with the infected mascara.

Even though mascaras contain preservatives, discard all mascara tubes after 3 months and do not allow multiple persons to use the same mascara tube. Individuals with recurrent bacterial infections due to colonization should select solvent-based mascaras.

Fungal organisms also can contaminate mascaras and result in eye infection. This is rare and usually only found in patients who are immunocompromised or wear contact lenses.

The pigment contained within mascaras can result in conjunctival pigmentation, if the mascara is washed into the conjunctival sac by lachrymal fluid. This colored particulate matter can be observed on the upper margin of the tarsal conjunctiva. Histologically, the pigment is seen within macrophages and extracellularly with varying degrees of lymphocytic infiltrate. Electron microscopy suggests that ferritin, carbon, and iron oxides are present within the tissues. Unfortunately, no treatment is available for the condition, which fortunately is usually asymptomatic.

Allergic contact dermatitis has been reported to rosin (colophony) and dihydroabietyl alcohol (abitol) contained in some mascaras. Mascaras can be open or closed patch tested as is, but they should be allowed to thoroughly dry prior to closed patch testing to avoid an irritant reaction from the volatile vehicle.

Summary

Colored cosmetics are an important part of the dermatologic armamentarium. They can camouflage contour and pigment abnormalities, provide moisturization, and create a sense of personal well-being, but they can also induce disease. Familiarity with these products results in better patient care.

EYE PENCILS

Eyebrow pencils are high wax containing hard crayons for darkening of eyebrows and to impart pseudo-growth of the eyebrows when none is evident. Generally they contain black colour. They can be easily sharpened to apply exactly on the eyebrows.

Formula III

Yellow beeswax	10.0%
Paraffin wax	30.0%
Theobroma oil	30.0%
Petrolatum	20.0%
Carbon black	10.0%

All the ingredients are melted and mixed in a mill and then molded in pencil moulds.

EVALUATIONS OF MASCARA

Checks for quality and purity are taken at various stages in the manufacture of mascara. The chemicals are checked in the tank before the mixing begins to make sure the correct ingredients and proper amounts are in place. After the batch is mixed, it is rechecked. After the batch is bottled, representative samples from the beginning, middle, and end of the batch are taken out. These are examined for chemical composition. At this point they are also tested for microbiological impurities.

Droop test: This test is used for eye liner pencils to check the hardness and consistency of leads. Specially designed wooden slots are used to hold the uncased leads which are placed in an oven for two hours at 40ºC. The degree of bend or droop is measured and according to that grade is assigned.

Patch testing: Many substances, such as nail polish, can be transferred to the eye area by the hands.

Open or closed patch testing: If eye cosmetics are the source of the dermatitis, the distinction between irritant and allergic contact dermatitis must be made. Irritant contact dermatitis is more common than allergic contact dermatitis. Open or closed patch testing can be performed as with eye shadows.

Use testing: Use testing is performed by placing the eye shadow at the corner of the eye for five consecutive nights followed by evaluation of the skin for allergic or irritant contact dermatitis.

Some mascara on the market today boast all-natural ingredients, and their recipes vary little from products that might have been made at home 100 years ago. One development that may affect mascara manufacturing in the future, however, is the development of new pigments. Researchers in the plastics industry have developed bold, vivid pigments that have recently been introduced to lipsticks. Plastic-derived pigments may be of interest to mascara manufacturers as well.

If you use eye cosmetics, FDA urges you to follow these safety tips:

- If any eye cosmetic causes irritation, stop using it immediately. If irritation persists, see a doctor.
- Avoid color additives that are not approved for use in the area of the eye, such as "permanent" eyelash tints and kohl. Be especially careful to keep kohl away from children, since reports have linked it to lead poisoning.
- Avoid using eye cosmetics if you have an eye infection or the skin around the eye is inflamed. Wait until the area is healed. Discard any eye cosmetics you were using when you got the infection.
- Be aware that there are bacteria on your hands that, if placed in the eye, could cause infections. Wash your hands before applying eye cosmetics.
- Discard dried-up mascara. Don't add saliva or water to moisten it. The bacteria from your mouth may grow in the mascara and cause infection. Adding water may introduce bacteria and will dilute the preservative that is intended to protect against microbial growth.

- Do not allow cosmetics to become covered with dust or contaminated with dirt or soil. Keep containers clean.
- Don't share your cosmetics. Another person's bacteria may be hazardous to you.
- Don't store cosmetics at temperatures above 85 degrees F. Cosmetics held for long periods in hot cars, for example, are more susceptible to deterioration of the preservative.
- Don't use any cosmetics near your eyes unless they are intended specifically for that use. For instance, don't use a lip liner as an eye liner. You may be exposing your eyes to contamination from your mouth, or to color additives that are not approved for use in the area of the eye.
- Don't use old containers of eye cosmetics. Manufacturers usually recommend discarding mascara two to four months after purchase.
- Make sure that any instrument you place in the eye area is clean.
- When applying or removing eye cosmetics, be careful not to scratch the eyeball or other sensitive area. Never apply or remove eye cosmetics in a moving vehicle.

Avaliable Market Products

- Black Magic Mascara.
- Cake Mascara.
- Clarins Fix Mascara.
- Color Mascara.
- Christian Dior Blue Mascara, Colour Eye Shadow and Eye Liner.
- Helena Rubinstein Spectacular Mascara.
- Luxe (eyelash extending mascara by Lancaster).
- Maybelline Great Lash Mascara.
- Sisley Phyto Mascara.
- Tova Mascara.
- Clarius Eye Contour Gel.
- L'Oréal Revita Lift Eye Cream (Used for Crow's Feet means removing wrinkles of eye).
- Revlon Vitamin C Eye Contour Radiance Cream.
- Vichy Reti Eyes.
- Maybelline Unstoppable Waterproof Eye Liner.
- Nina Ricci Eye Pencils.
- Avon True Colours.
- Himalya Kazal Herbal Eye Definer.

22 Cosmetics for the Nails

> *There are no better cosmetics than a severe temperance and purity, modesty and humility, a gracious temper and calmness of spirit; and there is no true beauty without the signatures of these graces in the very countenance.*— John Ray

INTRODUCTION

Nails are body language at our fingertips. Natural pink sheen in nails is the indication of perfect health. Nails also known as **'window of life'** because nails change colour reddish-pink to bluish after death. In anatomy, a nail is a horn-like piece at the end of an animal finger or toe. Nails are protective covering of the fingers. Unlike many other cosmetics that have a history of hundreds or even thousands of years, nail polish (or lacquer, or enamel) is almost completely an invention of twentieth century technology. Nail coverings were not unknown in ancient times-the upper classes of ancient Egypt probably used henna to dye both hair and fingernails-but essentially, its composition, manufacture and handling reflect developments in modern chemical technology.

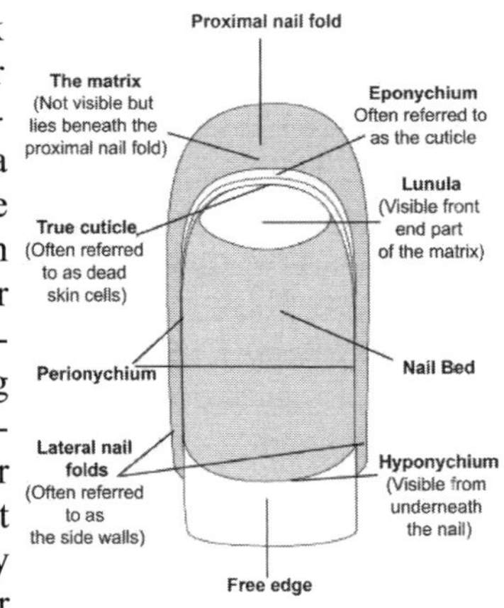

Nail products for both home and salon use are regulated by the Food and Drug Administration. Under the Federal Food, Drug, and Cosmetic Act (FD&C Act), these products are cosmetics [FD&C Act, section 201(i)].

Nails grow out of the cuticle or horny layer of the skin. Thus, it is necessary to take care of this growth which can create an odd situation if not been taken care of. The care of nails is known as **manicure**. The care of the toe-nails is termed as **pedicure**. Modern nail polish is sold in liquid form in small bottles and is applied with a tiny brush. Within a few minutes after application, the substance hardens and forms a shiny coating on the fingernail that is both water- and chip-resistant. Generally, a coating of nail polish may last several days before it begins to chip and fall off. Nail polish can also be removed manually by applying nail polish "remover," a substance designed to break down and dissolve the polish. Recently nail tatoos are also very common among women. It is very common fashion among girls that they have longer nails on one hand.

PARTS OF THE FINGER NAIL

Fingernails and toenails, which are made of a tough protein called keratin and are a form of modified hair are composed of:

- The free edge is the part of the nail extends past the finger, beyond the nail plate. There are no nerve endings within, thus it does not hurt to cut it.
- The nail matrix or the root of the nail—this is the growing part of the nail still under the skin at the nail's proximal end.
- Eponychium or cuticle is the fold of skin at the proximal end of the nail.
- Paronychium is the fold of skin on the sides of the nail.
- Hyponychium is the attachment between the skin of the finger or toe and the distal end of the nail.
- Nail plate is what we think of when we say nail, the hard and translucent portion, composed of keratin.
- Nail bed is the adherent connective tissue that underlies the nail.
- Lunula is the crescent shaped whitish area of the nail bed.
- Nail fold: a fold of hard skin overlapping the base and sides of a fingernail or toenail.

GROWTH OF NAILS

Nails grow at an average rate of 0.1 mm/day (1 cm every 100 days). Fingernails require 4 to 6 months to regrow completely. Toe-nails require 12 to 18 months. Actual growth rate is dependent upon age, season, exercise level, and hereditary factors.

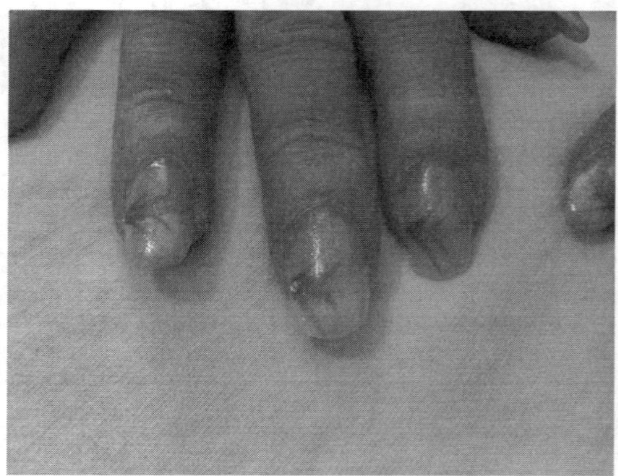

This growth record can show the history of recent health and physiological imbalances, and has been used as a diagnostic tool since ancient times. Major illness will cause a deep horizontal groove to form in the nails. Miscoloration, thinning, thickening, brittleness, splitting, grooves, Mee's lines, small white spots, receded lunula, clubbing (convex), flatness, spooning (concave) can indicate illness in other areas of the body, nutrient deficiencies, drug reaction or poisoning, or merely local injury. Nails can also become thickened (onychogryposis), loosened (onycholysis), infected

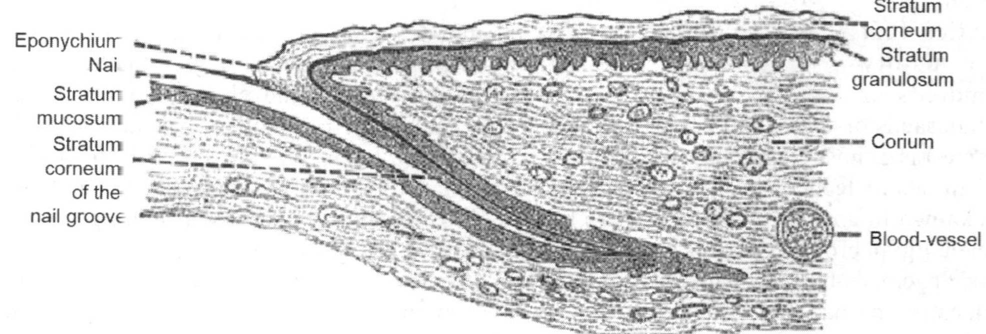

with fungus (onychomycosis) or degenerative (onychodystrophy).

Health and Care of the Nails

Nails can dry out, just like skin. They can also be infected: toe infections, for instance, can come from dirty socks, certain types of aggressive exercise as well as walking unprotected in an unclean environment.

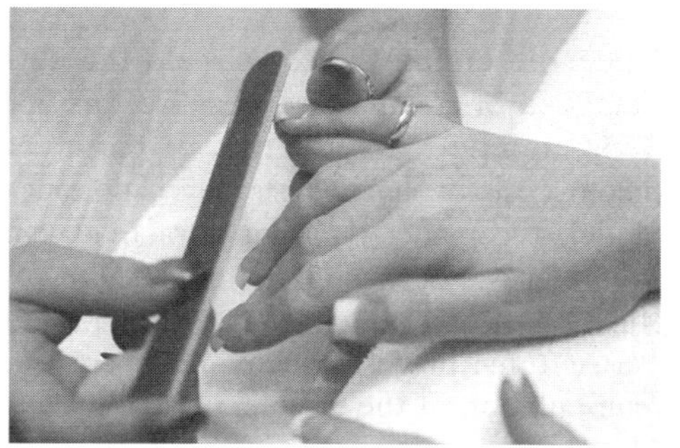

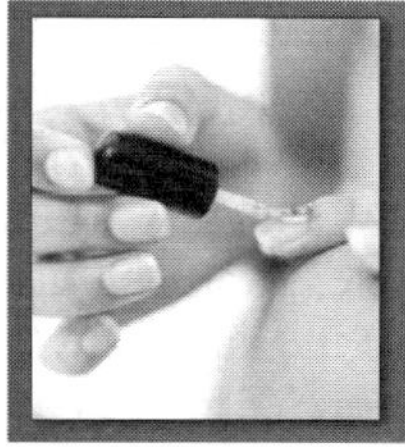

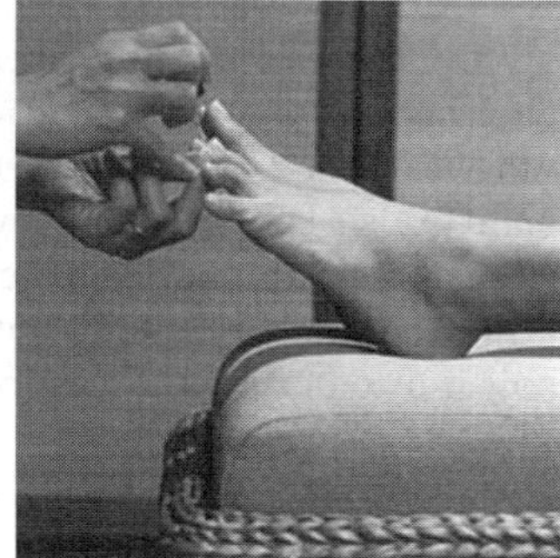

Manicures and pedicures are health and cosmetic procedures to groom, trim, and paint the nails and manage callous. They require various tools such as cuticle scissors, nail scissors, nail clippers, and nail files.

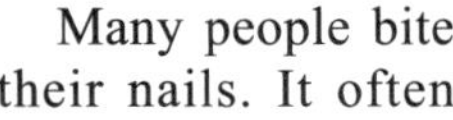

Many people bite their nails. It often indicates internal tension, stress, boredom, hunger, or it may simply be a habit.

However, biting the nails can result in the transportation of germs that are buried under the surface of the nail into the mouth. In fact, over 100 bacterial skin infections were traced to footbaths in nail salons. Thus, one can see that many pathogens have the ability to live beneath a nail, and because of this biting the nails can potentially cause health issues.

In western culture, long nails are a symbol of femininity, while short nails are a symbol of masculinity. Nail decoration is usually limited to females.

In some Asian cultures men may also grow long fingernails, or only the nail on the little finger, to show that they do not do much manual labor, but instead work in an office setting. However, this practice is gradually becoming unpopular and unrefined because a long fingernail on the little finger is variously associated with either nose picking or cocaine usage.

Some guitar players, notably classical and finger style players will purposely grow long nails on the hand they use to pluck the strings. Their longer nails serve as small, easily-maneuverable guitar picks (use of acrylic or "gel" nail enhancements is also growing in popularity, because the natural nail sometimes wears off faster by playing than it can grow back). For some serious musicians, daily nail care can become a mark of pride and dedication.

Someone whose occupation is to cut any type of nail, give artificial nails (or "nail arts") and care nails is generally called a nailist. The place where a nailist works is generally called a nail shop or a nail salon.

To paint the nails, nail lacquer (also known as nail polish) is manually applied and allowed to dry. Typical colours for women are red, pink, clear, and "natural" (an off-white intended to match the colour of an undecorated nail).

Health care and pre-hospital care providers (paramedics) often use the fingernail beds as a cursory indicator of distal tissue perfusion of individuals who may be in shock.

Procedure: Gently depress and release the fingernail bed with your finger. This act will briefly turn your nailbed white and normally return to a pink colour within 1–2 seconds. Delayed return to pink colour could be an indicator of certain shock states such as hypovolemia.

THE NAIL MANICURE

Nail cosmetics represent a form of self-adornment allowing personal expression dictated by current fashion trends. Their popularity is greatest among females; however, the market for male manicures is rapidly growing. Forms of adornment for the fingernails and the toenails include nail polish or enamel, artificial nails, nail elongators, and nail treatment preparations like polish remover, powder polish, paste polish, nail cream, nail bleach, cuticle remover, cuticle softener and nail tint. The goal of this article is to discuss the basic formulation and the use of these cosmetics in a medical framework.

In both males and females, professional grooming of the fingernails and the toenails is known as a manicure and a pedicure, respectively. The grooming includes cutting the nails according to current fashion standards, while improving their cosmetic appearance. The procedure for a manicure and a pedicure is essentially the same. The nails are first soaked in a soapy solution to remove any debris and to soften the nail plate prior to cutting. Softening the nail is important to prevent cracking, splitting, and horizontal layering (onychoschizia), which may occur when trying to cut a brittle nail plate.

Current fashion dictates that the nails should be trimmed to a delicate arc at the middle of the fingertip and filed to remove any corners at the medial and lateral parts of the nail. Although this shape is esthetically pleasing and serves to create the illusion of long, slender fingers, it predisposes to nail plate fracture, hang nails, and ingrown nails. Ideally, the nail should be trimmed with as slight a curve as possible and the corners of the nail left untouched. This technique is particularly important when trimming the toenails because they are frequently ingrown because of pressure from ill-fitting shoes or from trauma encountered during exercise. Recurrent ingrown toenails are best prevented by leaving the nail corners longer than the center of the toenail to create a concave shape.

Ideally, the nail plate should not be cut, but it should be frequently filed to prevent cracking through shearing forces generated by scissors or clippers. However, if cutting is necessary, the nails should be trimmed with sharp scissors or a nail clipper after softening. The cutting implement should be held perfectly perpendicular to the nail surface to avoid layering the nail plate, which predisposes to onychoschizia. Any remaining sharp edges should be filed with a diamond dust file.

Under no circumstances should the cuticle be removed or traumatized because this action may precipitate the formation of paronychia, onychomycosis, or onychodystrophy. Unfortunately, the cuticle is considered to be unattractive by most manicure artists because it complicates the even application of nail polish. Most of the problems that arise from a professional manicure are related to manipulation of the cuticle.

The last step in the manicure is grooming of the surface of the nail plate. Sometimes, this step is as simple as buffing the nail plate to a shine with creams containing finely ground pumice, talc, kaolin, or precipitated chalk as abrasives, with wax added to increase nail shine. A white pencil, also known as nail white, is sometimes stroked beneath the free edge of the nail plate to brighten the nail. Females may prefer to use nail polish or other nail decoration.

NAIL LACQUERS (NAIL POLISH)

A nail lacquer or enamel in order to be successful must contain a film former whose characteristics are ease of brushing application, fast drying and hardening and whose formulations should enhance adhesion to the nail without losing its resistance to chipping and abrasion. These characteristics can be complied by the proper formulation of the necessary constituents of nail enamel. Nail polishes generally consists of a base, like a resin, a plasticizer, triacetyl cellulose of nitrocellulose, a solvent, a dye and a suitable perfume. These are the following:

1. Primary film former
2. Secondary film forming resin.
3. Plasticizers.

4. Solvents.
5. Colorants
6. Speciality fillers.

Formula I

Nitrocellulose RS 1/2 sec dry	15.00%
Sulphonamide resin (Santolite)	7.5%
Dibutyl phthalate	3.75%
Butyl acetate	29.35%
Ethyl alcohol (from nitrocellulose)	6.4%
Butyl alcohol	1.10%
Toluene	2.9%

Formula II

Nitrocellulose	12.0%
Butyl stearate	20.0%
Ethyl acetate	50.0%
Diethyl phthalate	12.0%
Camphor	5.0%
Colourants	1.0%

Dissolve the nitrocellulose and camphor in butyl and ethyl stearates, add other ingredients along with colour and flavour and mix well.

FORMULATION

Prior to 1920, nails were manicured and then rubbed with abrasive powder to achieve a shine. Color was added through the use of stains. In 1930, Charles Revlon developed the first pigmented, opaque nail polish, which launched **Revlon**, still a major manufacturer of nail cosmetics today. Nail polish basically consists of pigments suspended in a volatile solvent to which film formers have been added. The ingredients are as follows:

- Primary film former (nitrocellulose, methacrylate polymers, vinyl polymers).
- Secondary film-forming resin (formaldehyde, p-toluene sulfonamide, polyamide, acrylate, alkyl, vinyl resins).
- Plasticizers (dibutyl phthalate, dioctyl phthalate, tricresyl phosphate, camphor).
- Solvents and diluents (acetates, ketones, toluene, xylene, alcohols).
- Colorants (organic D&C pigments, inorganic pigments).
- Specialty fillers (guanine, fish scale, titanium dioxide–coated mica flakes, or bismuth oxychloride for iridescence).

Table 22.1: Plasticizers for nail lacquers
• Benzyl benzoate
• Butyl acetyl ricenoleate
• Butyl glycollate
• Butyl stearate
• Camphor
• Castor oil
• Dibutyl phthalate
• Dioctyl phthalate
• Diamyl phthalate
• Dibutoxy ethyl phthalate
• Dibutyl tartrate
• Tributyl phosphate
• Tripheyl phosphate
• Triethyl citrate
• Tricresyl phosphate

RAW MATERIALS

There is no single formula for nail polish. There are, however, a number of ingredient types that are used. These basic components include: film forming agents, resins and plasticizers, solvents, and coloring agents. The exact formulation of a nail polish, apart from being a corporate secret, greatly depends upon choices made by chemists and chemical engineers in the research and development phase of manufacturing. Additionally, as chemicals and other ingredients become accepted or discredited for some uses, adjustments are made. For example, formaldehyde was once frequently used in polish production, but now it is rarely used.

The primary ingredient in nail polish is nitrocellulose (cellulose nitrate) cotton, a flammable and explosive ingredient also used in making dynamite. Nitrocellulose is a liquid mixed with tiny, near-microscopic cotton fibers. In the manufacturing process, the cotton fibers are ground even smaller and do not need to be removed. The nitrocellulose can be purchased in various viscosities to match the desired viscosity of the final product.

Table 22.2: Solvents for nail lacquers

Solvent	*Boiling point range in °C*
Low boiling point solvents	
• Diethyl ether	34–35
• Petroleum ether	40–60 and 60–80
• Methyl acetate	56–69
• Acetone	55–56
• Cyclohexane	81
• Ethyl acetate	74–79
• Methyl ethyl ketone	79
• Carbon tetrachloride	77
• Ethanol	78
• Isopropanol	80
Medium boiling point solvents	
• Amyl acetate	137–142
• Xylol	138
• n-Butyl alcohol	115–118
• Diethylene glycol monoethyl ether	135
• n-butyl acetate	124–128
High boiling point solvents	
• Ethyl lactate	150–160
• Cellosolve acetate	145–165
• Butyl cellosolve	166–173
• Carbitol	185–205

Nitrocellulose acts as a film forming agent. For nail polish to work properly, a hard film must form on the exposed surface of the nail, but it cannot form so quickly that it prevents the material underneath from drying. (Consider commercial puddings or gelatin products that dry or film on an exposed surface and protect the moist product underneath.) By itself or used with other functional ingredients, the nitrocellulose film is brittle and adheres poorly to nails.

Manufacturers add synthetic resins and plasticizers (and occasionally similar, natural products) to their mixes to improve flexibility, resistance to soap and water, and other qualities; older recipes sometimes even used nylon for this purpose. Because of the number of desired qualities involved, however, there is no single resin or combination of resins that meets every specification. Among the resins and plasticizers in use today are castor oil, amyl and butyl stearate, and mixes of glycerol, fatty acids, and acetic acids.

The colorings and other components of nail polish must be contained within one or more solvents that hold the colorings and other materials until the polish is applied. After application, the solvent must be able to evaporate. In many cases, the solvent also acts a plasticizer. Butyl stearate and acetate compounds are perhaps the most common.

Finally, the polish must have a color. Early polishes used soluble dyes, but today's product contains pigments of one type or another. Choice of pigment and its ability to mix well with the solvent and other ingredients is essential to producing a good quality product.

Nail polish is a "suspension" product, in which particles of color can only be held by the solvent for a relatively short period of time, rarely more than two or three years. Shaking a bottle of nail polish before use helps to restore settled particles to the suspension; a very old bottle of nail polish may have so much settled pigment that it can never be restored to the solvent. The problem of settling is perhaps the most difficult to be addressed in the manufacturing process.

In addition to usual coloring pigments, other, color tones can be added depending upon the color, tone, and hue of the desired product. Micas (tiny reflective minerals), also used in lipsticks, are a common additive, as is "pearl" or "fish scale" essence. "Pearl" or "guanine" is literally made from small fish scales and skin, suitably cleaned, and mixed with solvents such as castor oil and butyl acetate. The guanine can also be mixed with gold, silver, and bronze tones.

Pigment choices are restricted by the federal Food and Drug Administration (FDA), which maintains lists of pigments considered acceptable and others that are dangerous and cannot be used. Manufacturing plants are inspected regularly, and manufacturers must be able to prove they are using only FDA approved pigments. Since the FDA lists of acceptable and unacceptable pigments change with new findings and reexaminations of colors, manufacturers occasionally have to reformulate a polish formula.

THE MANUFACTURING PROCESS

Early methods of making nail polish used a variety of methods that today look charmingly amateurish. One common technique was to mix cleaned scraps of movie film and other cellulose with alcohol and castor oil and leave the mixture to soak overnight in a covered container. The mixture was then strained, colored, and perfumed. Though recognizable as nail polish, the product was far from what we have available today.

The modern manufacturing process is a very sophisticated operation utilizing highly skilled workers, advanced machinery, and even robotics. Today's consumers expect a nail polish to apply smoothly, evenly, and easily; to set relatively quickly; and to be resistant to chipping and peeling. In addition, the polish should be dermatologically innocuous.

MIXING THE PIGMENT WITH NITROCELLULOSE AND PLASTICIZER

The pigments are mixed with nitrocellulose and plasticizer using a "two-roll" differential speed mill. This mill grinds the pigment between a pair of rollers that are able to work with increasing speed as the pigment is ground down. The goal is to produce fine dispersion of the color. A variation of this mill is the Banbury Mixer (used also in the production of rubber for rubber bands).

When properly and fully milled, the mixture is removed from the mill in sheet form and then broken up into small chips for mixing with the solvent. The mixing is performed in stainless steel kettles that can hold anywhere from 5 to 2,000 gallons. Stainless steel must be used because the nitrocellulose is extremely reactive in the presence of iron. The kettles are jacketed so that the mixture can be cooled by circulating cold water or another liquid around the outside of the kettle. The temperature of the kettle, and the rate of cooling, is controlled by both computers and technicians.

This step is performed in a special room or area designed to control the hazards of fire and explosion. Most modern factories perform this step in an area with walls that will close in if an alarm sounds and, in the event of explosion, with ceilings that will safely blow off without endangering the rest of the structure.

Adding other Ingredients

Materials are mixed in computerized, closed kettles. At the end of the process, the mix is cooled slightly before the addition of such other materials as perfumes and moisturizers.

The mixture is then pumped into smaller, 55 gallon drums, and then trucked to a production line. The finished nail polish is pumped into explosion proof pumps, and then into smaller bottles suitable for the retail market.

Suspending Agents

Nitrocellulose is the most commonly used primary film-forming agent in nail lacquer. It produces a shiny, tough, nontoxic film that adheres well to the nail plate. The film is somewhat oxygen permeable, allowing gas exchange between the atmosphere and the nail plate; this gas exchange is important for ensuring nail plate health. Resins and plasticizers are then added to increase the flexibility of the film, minimizing chipping and peeling.

The most popular resin used to enhance the nitrocellulose film is toluene-sulfonamide formaldehyde; however, it is the source of allergic contact dermatitis in some nail enamels. Hypoallergenic nail enamels use polyester resin or cellulose acetate butyrate, but sensitivity is still possible. Plasticizers, such as dibutyl phthalate, are also used to keep the product soft and pliable. All of these ingredients are dissolved in a solvent, such as N-butyl acetate or ethyl acetate, with toluene and isopropyl alcohol added as diluents.

Variety in nail polish color can be achieved through the addition of coloring agents, such as organic colors, selected from a US Food and Drug Administration–approved list of certified colors. Inorganic colors and pigments may also be used, but they must conform to low heavy metal content standards. These colors can be suspended within the lacquer with suspending agents, such as stearalkonium hectorite, to produce a range of colors, including white, pink, purple, brown, orange, blue, or green. If the pigments are dissolved rather than suspended in the polish, nail staining is more likely.

Other specialty additives can also create variety. Guanine, fish scale, bismuth oxychloride, or titanium dioxide–coated mica flakes can be added to enhance light reflection and to give a frosted appearance. Chopped aluminum, silver, and gold can be added for a metallic shine. Nylon or rayon fibers can be added for nail-strengthening purposes. Other agents can also be added for nail treatment purposes.

Application

After a manicure, the patient may elect to have nail enamel applied. A professional nail enamel application requires 3 layers of polish:

1. A base coat,
2. A pigmented nail enamel, and
3. A topcoat.

The base coat ensures good adhesion to the nail plate and prevents the polish from chipping. It contains no pigment, less primary film former, and more secondary film-forming resins, and it is of a lower viscosity because a thinner film is desirable. The second layer is the actual pigmented nail enamel. The topcoat, or third layer, provides gloss and resistance to chipping. It contains increased amounts of primary film former, more plasticizer, and less secondary film-forming resins. Some topcoats may contain a chemical sunscreen, which is designed to prevent the nail polish color from fading, not to protect the nail bed from ultraviolet light damage.

Dermatologic Considerations

Dermatologic problems associated with nail enamels include nail plate discoloration and allergic contact dermatitis. The nail staining, as mentioned previously, is seen with dissolved rather than suspended pigments, and it is most common in deep red nail polishes that contain D&C Reds No. 6, 7, 34, or 5 Lake. The nail plate is stained yellow after 7 days of continuous wear. The staining fades without treatment approximately 14 days after the enamel has been removed. Scraping of the nail plate with a scalpel blade can be used to confirm that only the nail surface has been stained; this finding is an important distinction in nail pigmentation abnormalities.

Allergic contact dermatitis to nail polish may present as proximal nail fold erythema and edema, fingertip tenderness and swelling, and/or eyelid dermatitis. The North American Contact Dermatitis Group determined that 4% of positive patch test results were due to toluene-sulfonamide-formaldehyde resin. The allergic reaction is most commonly due to wet nail enamel. Allergic reactions can be severe, necessitating lost work time or, rarely, hospitalization.

Some concern exists that the use of nail polish can contribute to nail dryness and brittleness. This actually is not the case. Nail polish prevents contact of detergents with the nail, acting as a protectant. Furthermore, it decreases nail water vapor loss from 1.6 mg/cm^2/h to 0.4 mg/cm^2/h; this decrease enhances nail moisturization and flexibility. The dryness associated with nail polish is actually due to the nail polish remover, which is usually a harsh acetone-based solvent.

Nail polish can be tested "as is," but it should be allowed to thoroughly dry because the volatile solvent can cause an irritant reaction if not allowed to rapidly evaporate. The toluene-sulfonamide-formaldehyde resin can also be tested alone in 10% petrolatum. Patients who are allergic to this resin may experience no difficulty with hypoallergenic nail polishes, but allergic contact dermatitis is still a possibility. Nail enamels may contain metallic beads to aid in the dispersion of the products before application. These beads contain nickel; therefore, nickel sensitivity can occur with nail enamel usage.

Gel Nails

Gel Nails are a new, more natural looking alternative to acrylic nail extensions. Unlike acrylics, the UV gel used to create the nails has no discomforting odors and no hazardous ingredients. There is no glue involved in

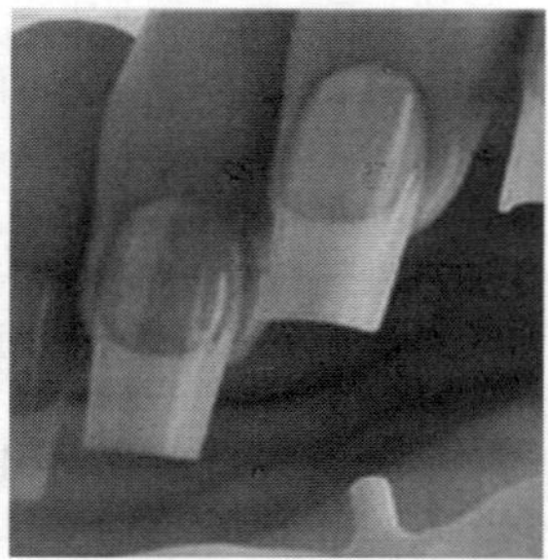

the process, eliminating unnecessary nail damage caused by lifting. Gel nails are created with a layer of base gel, brushed into a mold then cured by UV Lamps: 9 watt, 18 watt, or 36 watt at 110 volts depending on cure rate desired. Gel nails are limitless in their design capabilities, and can even be reinforced with fiberglass for extra strength and repairs purposes. They are also available in a choice of natural colors to enhance the natural look of the nail bed. In short, gel nails are an extremely natural-looking enhancement: thin, clear, flexible, non-yellowing, nonporous, while resist lifting.

Gel nails can be used for natural nail overlays (actually one of the most difficult techniques, but looks the easiest to achieve), tip overlays, and sculpted onto forms for short extensions, and to help encapsulate a damaged free-edge as it grows out. Gel nails are the future of the nail industry. No other service can give clients both a natural nail look and feel combined with the convenience and durability of acrylic nails. Many clients have tired of acrylic nails in this past decade because of the disadvantages that can be associated with them (such as the odors, lifting and nail damage, and components such as ethyl methacrylate, etc).

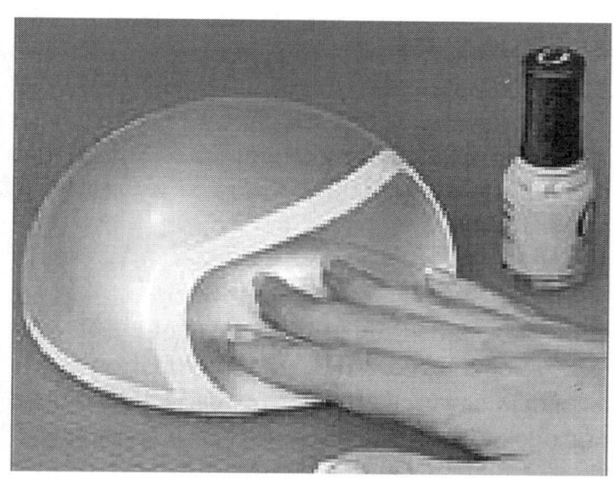

Some major benefits of gel nails are that they are odorless, they have a thin, flexible, natural feeling on the nail, and they are clear and extremely lightweight. Traditional acrylics harden in about 10 minutes, but they can take 24 hours or more to fully cure. Ultra violet gel nails harden in about 15–30 seconds and cure in about 2 minutes.

NAIL HARDENER

Formulation

Nail hardeners are used to increase the strength of brittle nails caused by nail plate dehydration due to excessive contact with solvents, detergents, and water. Originally, nail hardeners were formulated as 10% or greater solutions of formaldehyde; however, the Food and Drug Administration recalled these products following reports of onycholysis, subungual hyperkeratosis, reversible subungual hemorrhage, bluish discoloration of the nail plate, and allergic contact dermatitis in the dermatologic literature.

Free formaldehyde in concentrations of 1–2% is still permitted, but acetates, toluene, nitrocellulose, acrylic, and polyamide resins are now used to structurally reinforce the nail plate. Some products actually contain 1% nylon fibers and are known as fibered nail hardeners. Other additives purported to strengthen the nail include hydrolyzed proteins, modified vegetable extracts, glycerin, propylene glycol, and metal salts.

Application

Nail hardeners are essentially a modification of clear nail enamel with different solvent and resin concentrations. They are the first coat of enamel applied to the clean nail plate and function as a base coat.

Dermatologic Considerations

Nail hardeners may contain the same toluene-sulfonamide-formaldehyde resin as nail polish, and the same dermatologic considerations apply.

NAIL ENAMEL REMOVER

Formulation

Nail polish removers are liquids designed to strip the nail polish from the nail plate. They may contain strong solvents, such as acetone, alcohol, ethyl acetate, or butyl acetate. Conditioning nail enamel removers are available containing fatty materials, such as cetyl alcohol, cetyl palmitate, lanolin, butyl and ethyl stearates, acetonitrile, castor oil, or other synthetic oils. These oily substances are thought to act as occlusive nail moisturizers retarding water evaporation; however, their effectiveness is minimal compared with the dehydrating effect of the strong solvents required to dissolve the nail enamel.

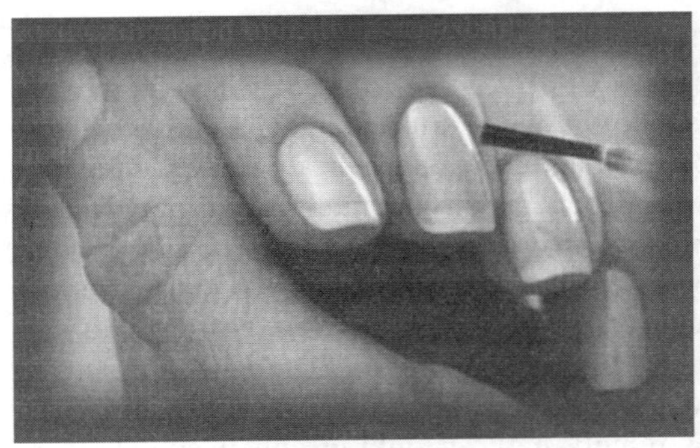

Formula III

Butyl stearate	60.0%
Ethyl acetate	40.0%

Blend butyl and ethyl stearates properly.

Application

Nail enamel remover is applied to a tissue or a cotton ball and wiped across the nail plate to remove old or unwanted nail polish. Several applications and rubbing may be required to remove the polish if several coats have been applied.

Dermatologic Considerations

Nail polish remover can irritate and dry the nail plate and surrounding skin. It can also contribute to nail dryness and brittleness. These problems can be minimized by using the product once a week or less.

Only open patch testing should be attempted with nail polish remover because of its high solvent concentration. It may be tested at a concentration of 10% dissolved in olive oil.

CUTICLE REMOVER

Formulation

The cuticle can be removed mechanically by pushing and trimming or chemically through the use of cuticle removers, which are formulated as liquids or creams that contain an alkali to destroy cuticle keratin. The active agent is usually 2–5% sodium hydroxide or potassium hydroxide with propylene glycol or glycerin added as a humectant. Milder preparations can be made with trisodium phosphate or tetrasodium pyrophosphate, but they are also less effective.

A subset of products within this category is known as cuticle softeners. These products are quaternary ammonium compounds in a 3–5% concentration, and they are sometimes combined with urea. They are designed to soften the cuticular protein and to facilitate mechanical removal.

Application

Cuticle removers dissolve excess cuticular tissue on the nail plate. They are not intended to remove the fibrous cuticular ridge; however, vigorous use can remove the entire cuticle. The product is applied with a cotton ball, and it is left on the nail plate for 10 minutes followed by removal with wiping.

Dermatologic Considerations

Removal or manipulation of the cuticle is not recommended; paronychial inflammation with secondary bacterial infection or yeast colonization can occur. Cuticle removers can also damage the nail plate through softening. Because of the high alkali content of these products, irritant contact dermatitis is common if the product is left on too long. Thus, these products should not be used in closed patch testing. Open patch testing in a 2% aqueous concentration may be used, if necessary.

NAIL MOISTURIZER

Formulation

Nail moisturizers are valuable in patients with dry, brittle, fissured, and/or splitting nails. The healthy nail contains about 16% water, becoming soft with saturation at 30%. The water content of nail keratin is proportional to the relative humidity, being 7% at 20% relative humidity and 30% at 100% relative humidity. The idea of applying a cream or a lotion to the nails to increase the nail water content is somewhat new.

Nail moisturizers are usually creams or lotions that contain occlusive, such as petrolatum, mineral oil, or lanolin. Humectants, such as glycerin, propylene glycol, and proteins, may also be added. Alpha-hydroxy acids, lactic acid, and urea are active ingredients used to increase the water-binding capacity of the nail plate. A well-formulated nail moisturizer should contain substances from all of the aforementioned groups to maximally treat the dehydrated nail plate.

Recent studies demonstrated that daily oral biotin supplementation may be helpful in treating brittle nails; however, no evidence exists that topical biotin added to nail moisturizers has a similar beneficial effect. Furthermore, no evidence exists that topical gelatin, calcium, iron, botanical extracts, and biological extracts are effective in treating nail dehydration.

Application

Nail moisturizers function best if the nails are first soaked for 10–20 minutes in lukewarm water, preferably at bedtime. The moisturizer should then be generously applied under occlusion with a light cotton glove or sock. This procedure should be repeated nightly for at least three months. Certainly, activities contributing to dry nails, such as frequent contact with water, detergents, solvents, or nail polish remover, should be discontinued.

Dermatologic Considerations

Alpha-hydroxy acids, lactic acid, and urea can cause stinging and irritant contact dermatitis in susceptible individuals. Wounds on the hands or fissured cuticles can burn if high concentrations of these substances are used. Most over-the-counter preparations contain 5% or less of these active agents; however, prescription preparations of lactic acid are available in strengths of 12% and 30%.

NAIL ELONGATORS

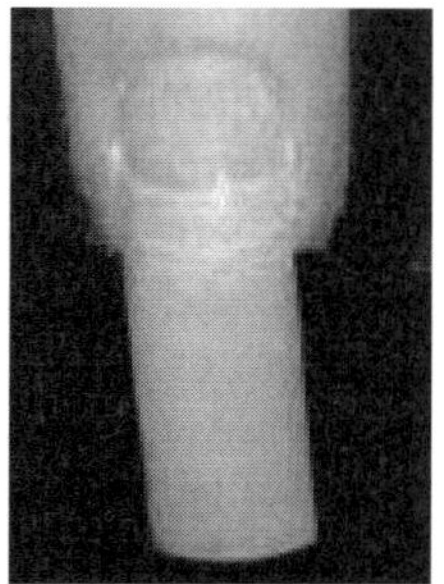

Nail elongators are cosmetic methods of creating the illusion of nail length. This technique can be accomplished through the use of preformed plastics, formed acrylics, or a combination of both methods.

Preformed Artificial Nails

Preformed plastic nails are popular home nail elongators available in press-on and preglued forms and in forms that require the application of glue. The acrylic glue used for adhesion is typically methacrylate based and a possible cause of allergic contact dermatitis. A stronger nail adhesive made from ethyl 2-cyanoacrylate provides better adhesion, but it can cause onycholysis. Traumatic removal of artificial nails may result in onychoschizia and nail pitting.

Sculptured Nails

An increasingly popular method of obtaining long, hard nails is the application of sculptured nails. The word sculptured is used because the custom-made artificial nail is sculpted on a template attached to the natural nail plate. The sculpted nail fits perfectly and, if applied well, can be hard to differentiate from a natural nail.

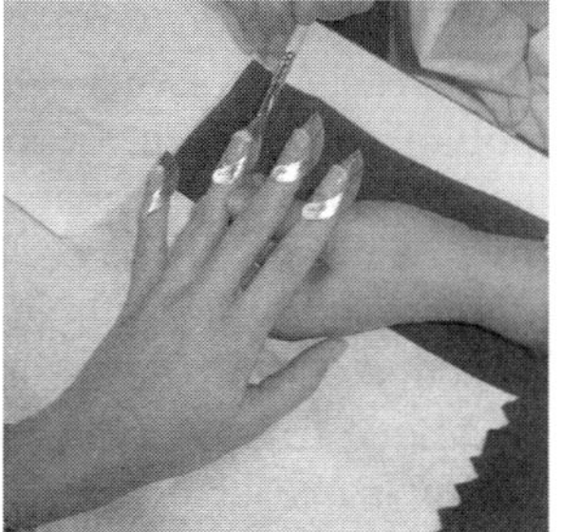

The application of sculptured nails is an involved process requiring approximately two hours to sculpt 10 fingernails. The basic process is as follows:

- All nail polish and oils are removed from the nail.
- The nail is roughened with a coarse emery board, pumice stone or grinding drill to create an optimal surface for adhesion of the sculpted nail.
- An antifungal, antibacterial liquid, such as decolorized iodine, is applied to the entire nail plate to minimize onychomycosis and paronychia.
- The loose edges of the cuticle are trimmed, removed, or pushed back, depending on the operator.
- A flexible template is fit beneath the natural nail plate, upon which the elongated sculpted nail is built.

The acrylic is mixed and applied with a paintbrush to cover the entire natural nail plate, and it is extended onto the template to the desired nail length. A clear acrylic is used over the natural nail plate attached to the nail bed through which the natural pink color can be seen. A white acrylic is used from the free edge of the nail plate distally. This combination perfectly simulates a natural nail, so enamel need not be worn. It also allows both men and women to use sculptured nails.

The final nail sculpture is sanded to a high shine.

Nail polish, jewels, decals, and decorative metal strips may be added, depending on the fashion tastes of the individual.

Originally, methyl methacrylate was the monomer used to fashion the nail, but it has been removed from the market because of its sensitizing potential. Currently, liquid ethyl or isobutyl methacrylate is used as the monomer, and it is mixed with powdered polymethyl methacrylate polymer. The product is allowed to polymerize in the presence of a benzoyl peroxide accelerator, and a formable acrylic, which hardens in 7–9 minutes, is made. Usually, hydroquinone, monomethyl ether of hydroquinone, or pyrogallol is added to slow down polymerization.

Many individuals are not aware that the finished nail sculptures require more care than natural fingernails. With continued wear of the sculptured nail, the acrylic loosens from the natural nail, especially around the edges. These loose edges must be clipped, and new acrylic must be applied approximately every 3 weeks to prevent development of an environment for infection. The sculptured nail grows out with the natural nail plate, and more polymers must be added proximally, depending on the growth rate of the nail. This procedure is known as filling. If necessary, the sculptured nails can be removed by soaking them in acetone.

Allergic contact dermatitis remains an issue even though methyl methacrylate is no longer used; isobutyl, ethyl, and tetrahydrofurfuryl methacrylate are still strong sensitizers. However, the polymerized, cured acrylic is not sensitizing only the liquid monomer. Therefore, a careful operator who avoids skin contact with the uncured acrylic can avoid sensitizing the patient. Patch testing should be performed with methyl methacrylate monomer, 10% in olive oil, and methacrylate acid esters, 1% and 5% in olive oil and petrolatum, in individuals with suspected sensitization.

Sculptured Nails with Tips

Another technique of elongating nails is to combine custom-made sculptured nails with preformed artificial tips, which involves gluing a preformed plastic piece to the tip of the nail and then applying a smaller amount of the liquid acrylic to the remaining exposed natural nail. The natural nail can be visualized beneath the artificial nail. This technique is the most popular form of nail elongation used today because operator time is greatly reduced because most of the nail is preformed.

Photo Bonded Nails

A variation on sculptured nails is known as photo bonded nails. These nails are also formed from a cured acrylic sculpted on the natural nail, but, instead of allowing the acrylic to cure (dry) at room temperature, the nails are placed under magnesium light for 1–2 minutes. This technique is similar to restorative dental bonding. Photo-onycholysis and paresthesias have been reported as a result of this technique, which is currently the least popular of all the nail elongation methods.

NAIL TREATMENT PRODUCTS

Nail treatment pro- ducts are actually a subgroup of nail polishes with added ingredients intended to produce some therapeutic benefit. Therapeutic benefits include decreased nail breakage, increased nail growth, and prevention of fungal infections. Products designed to strengthen nails may contain substances, such as iron or calcium, but these products are essentially nail enamels that are applied before the pigmented nail polish (base coat) or after the pigmented nail polish (top coat). Each successive coat of polish thickens the nail plate and, to some extent, makes the nail less subject to breakage. Nail growth polishes may actually contain some fibered material, such as silk proteins, to add further nail support.

Perhaps the most interesting nail treatment products to the dermatologist are those liquids designed for nail fungal infections. The active ingredient in most of these over-the-counter formulations is 1% tolnaftate in liquid form that is applied beneath the nail plate.

Nail cosmetics are a constant source of concern for the dermatologist because they can promote disease, create deformity, and provide a source for both allergic and irritant contact dermatitis. In some regards, the nails are the healthiest when unadorned and unmanipulated. However, patients persist because nail adornment is fashionable, a wearable art form, and a manner of personal expression. Therefore, the dermatologist must become familiar with new nail cosmetics and develop the ability to recognize possible cosmetic causes of nail disease. Perhaps the major problem with nail polishes—from the consumer's point of view-is the length of the drying time. Various methods of producing fast-drying polish have recently been patented, and these methods, along with others that are still being developed, may result in marketable products. Of all the different types of cosmetics, nail polish is the one that is most likely to continue to be positively affected by advancements and developments in the chemistry field.

EVALUATIONS OF NAIL POLISH

Extreme attention to quality control is essential throughout the manufacturing process. Not only does quality control increase safety in the process, but it is the only way that a manufacturer can be assured of consumer confidence and loyalty. A single bottle of poor quality polish can lose a customer forever. Regardless of quality control, however, no single nail polish is perfect; the polish always represents a chemical compromise between what is desired and what the manufacturer is able to produce.

The nail polish is tested throughout the manufacturing process for several important factors

- Abrasion
- Color matching,
- Drying time,
- Gloss,
- Non-volatile contents
- Hardness,
- Application properties,
- Water resistance,
- Smoothness of flow, etc.

Subjective testing, where the mixture or final product is examined or applied, is ongoing. Objective, laboratory testing of samples, though more time consuming, is also necessary to ensure a usable product. Laboratory tests are both complicated and unforgiving, but no manufacturer would do without them.

Abrasion: abrasive resistance is determined by means of Taber Abrader.

Nonvolatile Content

There are many methods for the determination of the non volatiles present in the lacquer. Generally accepted is the 'dish method'. Place 1 to 0.2g of the sample in a tare, flat dish about 8 cm. in diameter. Spread the sample evenly with a tare wire and place in an oven at 105 ± 2 °C. At intervals break up any skins formed with the wire. After 1 hr. remove, cool, and weigh the dish. Reheat for 1 additional hour in the same manner and reweigh the dish. Use the greater weight loss in calculating the non volatile content of the sample.

Drying Time

Apply a film of the sample with a 0.006 in. Bird applicator under controlled temperature and humidity conditions, at 25 °C and 50 °C and 50% relative humidity, to a completely nonporous surface, such as a plate of glass or melamine coated paper. Measure with a stopwatch the time required to form a dry-to-dry much film.

Nail Product Ingredient Safety

- Consumers should read labels of nail products carefully and heed any warnings.
- Infections and allergic reactions can occur with some nail products.
- Nail products also can be dangerous if they get in the eyes.
- Some can easily catch fire if exposed to the flame of the pilot light of a stove, a lit cigarette, or other heat source, such as the heating element of a curling iron.
- Some ingredients in nail products may be harmful if ingested.

Avaliable Market Products

- Blue and blues
- Chanel nail polish
- Essie nail polish
- Estee lauder nail polish
- Lakme color bar and nail polish
- Maxishine (Oriflame)
- MAYBELLINE sweet nothings nail paint
- Moisture shine sparkles
- Mood nail polish
- Opi nail polish
- Petite nail polish
- Ravlon nail polish
- So Laque nail polishes (by Bourjois)
- Swabplus nail polish remover
- H_2O + hand and nail creme'

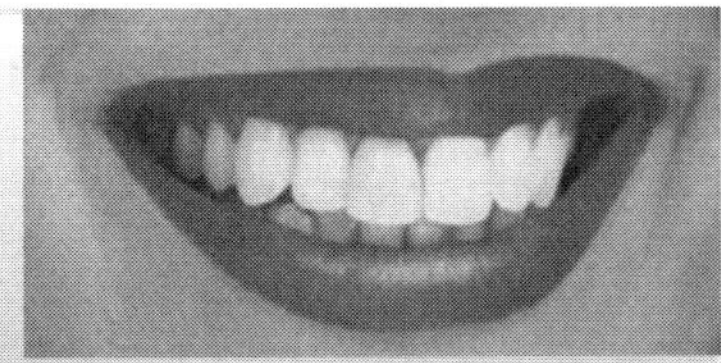

Cosmetics for the Teeth

Beauty is everywhere a welcome guest.
— Goethe

Teeth (singular, tooth) are structures found in the jaws of many vertebrates. The primary function of teeth is to tear and chew food, and in some animals, particularly carnivores, for fighting and/or defense. The roots of the teeth are covered by gums. Adult teeth naturally darken with age as the pulp within the tooth shrinks and dentin is deposited in its place. The shape of the teeth is related to the animal's diet. For example, plant matter is hard to digest, so herbivores have many molars for chewing. Carnivores need canines to kill and tear meat.

ANATOMY OF TEETH

While humans develop two sets of teeth throughout life (diphyodont), some animals develop only one set (monophyodont) or develop many (polyphyodont). Sharks, for example, grow a new set of teeth every two weeks to replace worn teeth. Rodent teeth grow and wear away continually through the animal's gnawing, maintaining approximately constant length.

Humans are diphyodont, meaning that they develop two sets of teeth throughout life. The first set (the "baby," "milk," "primary" or "deciduous" set) normally starts to appear at about six months of age, although some babies are born with one or more visible teeth, known as neonatal teeth. Normal eruption of teeth starting at about six months is known as teething and can be quite painful for an infant. Human children have 20

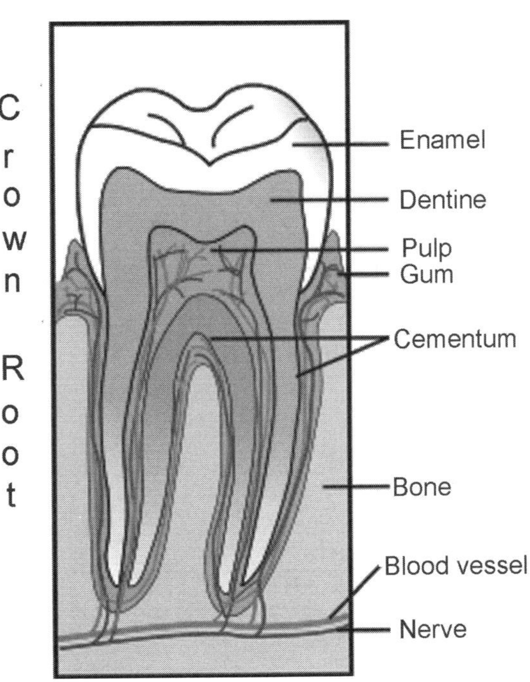

deciduous teeth evenly distributed across the mouth's quadrants. Each quadrant of five teeth has a:

- Central incisor
- Lateral incisor
- Cuspid (canine)
- First molar
- Second molar.

The second, permanent set of teeth consists of 32 teeth. Twenty-eight of them appear between the ages of about 6 and 12 years. Secondary teeth do not push deciduous teeth out of their sockets; instead, a group of cells (odontoclasts) forms in front of tip of second tooth and dissolves the base of first tooth. Finally, the first tooth is held in place only by tissues of gum. Deciduous molars are replaced by premolars. The third molars (the wisdom teeth) are the final teeth to erupt, usually around age 20. However, it is common for the wisdom teeth not to erupt at all (they are congenitally missing); this is often the case in small jaws without room to support the extra teeth. It is

Adult Teeth
Upper Teeth
1. Central incisor
2. Lateral incisor
3. Canine (cuspid)
4. First premolar (first biscuspid)
5. Second premolar (second bicuspid)
6. First molar
7. Second molar
8. Third molar (wisdom tooth)

Lower Teeth

9. Third Molar (wisdom tooth)
10. Second molar
11. First molar
12. Second premolar (second bicuspid)
13. First premolar (first bicuspid)
14. Canine (cuspid)
15. Lateral incisor
16. Central incisor

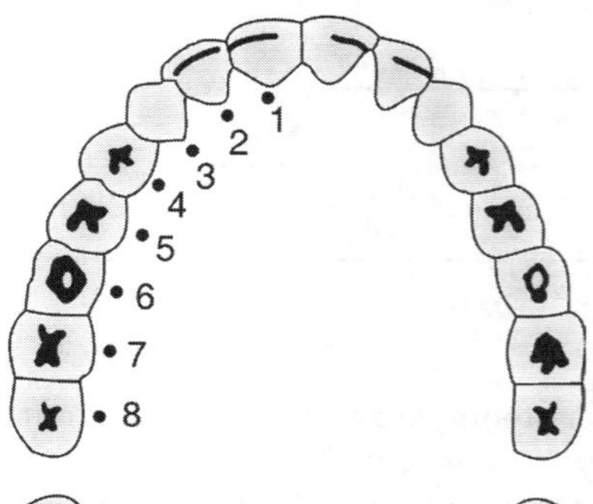

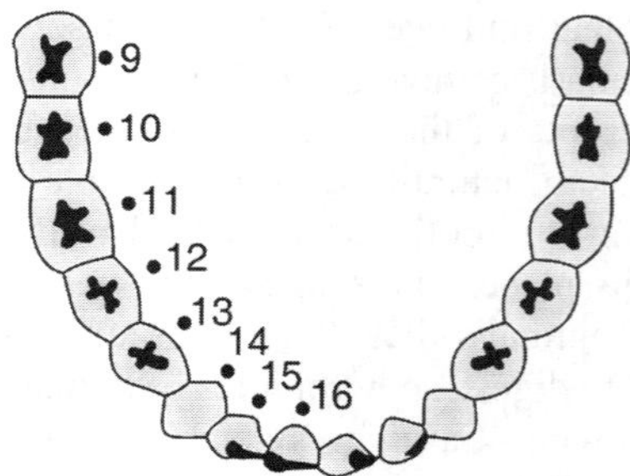

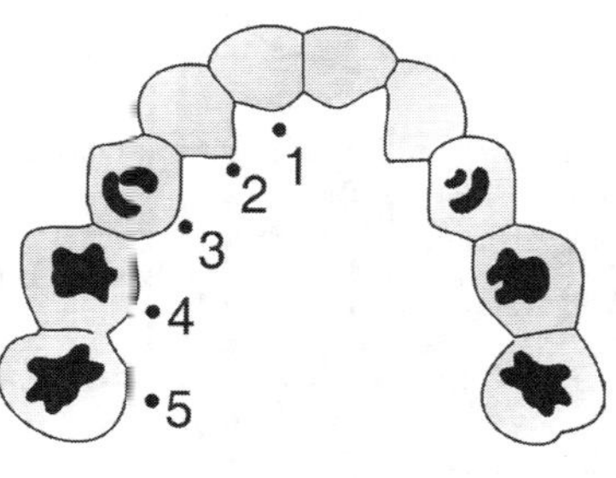

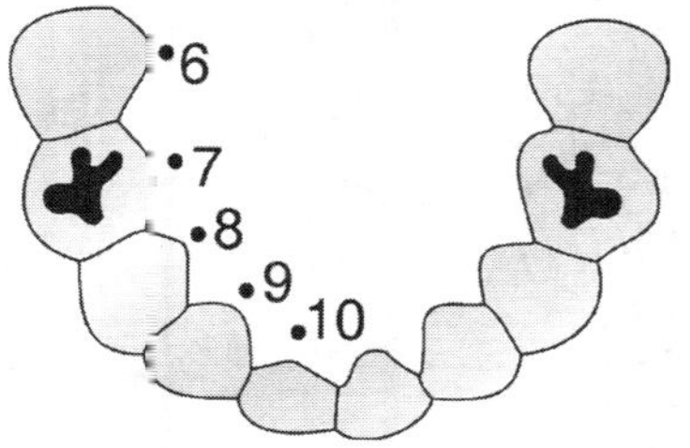

Baby Teeth
Upper Teeth
1. Central incisor
2. Lateral incisor
3. Canine (cuspid)
4. First molar
5. Second molar
6. Second molar
7. First molar
8. Canine (cuspid)
9. Lateral incisor
10. Central incisor

possible, though rare, for a person to have fourth molars, and there have been instances where fifth molars have been present in the dentition.

Permanent teeth are evenly distributed across the mouth's quadrants. Each quadrant of eight teeth has a:

- central incisor (upper jaw: maxillary central incisor; lower jaw: mandibular central incisor)
- lateral incisor (upper jaw: maxillary lateral incisor; lower jaw: mandibular lateral incisor)
- cuspid (canine) (upper jaw: maxillary canine; lower jaw: mandibular canine)
- first premolar (bicuspid) (upper jaw: maxillary first premolar; lower jaw: mandibular first premolar)
- second premolar (bicuspid) (upper jaw: maxillary second premolar; lower jaw: mandibular second premolar)
- first molar (upper jaw: maxillary first molar; lower jaw: mandibular first molar)
- second molar (upper jaw: maxillary second molar; lower jaw: mandibular second molar)
- third molar (wisdom teeth) (upper jaw: maxillary third molar; lower jaw: mandibular third molar).

The permanent set may last for life if cared for properly through a regular program of dental hygiene, including regular brushing and professional cleaning by a dentist or hygienist. Teeth that are susceptible to decay may be sealed for additional protection.

Teeth are attached to the underlying bone of the jaw via the periodontal ligament, though the teeth themselves are not made of bone. The white part of the tooth, which can be seen in the mouth, is the enamel. Immediately one to three mm below the enamel is a slightly softer, yellow tissue called dentin. Dentin is supported by the pulp (commonly called 'the nerve', although it contains many other structures which are not nerves), which lies in the center of the tooth. The teeth's composition is specialized to resist the harsh environment of the oral cavity and withstand the large forces imposed upon them by mastication, or chewing.

Enamel is the hardest and most highly mineralized substance of the body and with dentin, cementum, and dental pulp is one of the four major tissues which make up the tooth. It is the normally visible dental tissue of a tooth and must be supported by underlying dentin. Ninety-six percent of enamel consists of mineral, with water and organic material composing the rest. The normal color of enamel varies from light yellow to grayish white. At the edges of teeth where there is no dentin underlying the enamel, the color sometimes has a slightly blue tone. Since enamel is semi translucent, the color of dentin and any restorative dental material underneath the enamel strongly affects the appearance of a tooth. Enamel varies in thickness over the surface of the tooth and is often thickest at the cusp, up to 2.5 mm, and thinnest at its border, which is seen clinically as the cementoenamel junction (CEJ).

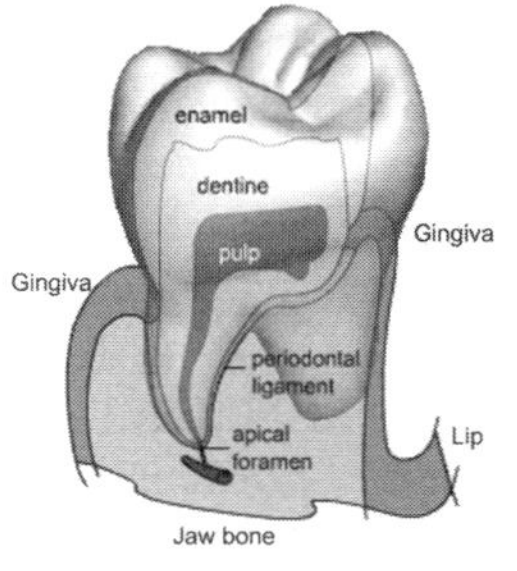

Enamel's primary mineral is hydroxyapatite, which is a crystalline calcium phosphate. The large amount of minerals in enamel accounts not only for its strength but also for its brittleness. Dentin, which is less mineralized and less brittle, compensates for enamel and is necessary as a support. Unlike dentin and bone, enamel does not contain collagen. Instead, it has two unique classes of proteins called amelogenins and enamelins. While the role of these proteins is not fully understood, it is believed that they aid in the development of enamel by serving as a framework support, among other functions.

Dentin is the substance between enamel or cementum and the pulp chamber. It is secreted by the odontoblasts of the dental pulp. The formation of dentin is known as dentinogenesis. The porous, yellow-hued material is made up of 70% inorganic materials, 20% organic materials, and 10% water. Because it is softer than enamel, it decays more rapidly and is subject to severe cavities if not properly treated, but dentin still acts as a protective layer and supports the crown of the tooth.

Dentin is a mineralized connective tissue with an organic matrix of collagenous proteins. Dentin has microscopic channels, called dentinal tubules, which radiate outward through the dentin from the pulp cavity to the exterior cementum or enamel border. These canals have different configurations in different species and their diameter ranges between 0.8 and 2.2 μm. Although they may have tiny side-branches, they do not intersect with each other. Their length is dictated by the radius of the tooth. The three dimensional configuration of the dentinal tubules is under genetic control.

Cementum is a specialized bony substance covering the root of a tooth. It is approximately 45% inorganic material (mainly hydroxyapatite), 33% organic material (mainly collagen) and 22% water. Cementum is excreted by cementoblasts within the root of the tooth and is thickest at the root apex. Its coloration is yellowish and it is softer than either dentin or enamel. The principle role of cementum is to serve as a medium by which the periodontal ligaments can attach to the tooth for stability. At the cementoenamel junction, the cementum is acellular due to its lack of cellular components, and this type covers approximately 1/3 – 1/2 of the root. The more permeable form of cementum, cellular cementum, covers 1/3 – 1/2 of the root apex, where it binds to the dentin.

Plaque is a biofilm consisting of large amounts of various bacteria which forms on teeth. If not removed regularly, it can lead to dental cavities (caries) or periodontal problems (such as gingivitis). Given time, plaque can mineralize along the gingiva, forming tartar. The microorganisms that form the biofilm are almost entirely bacteria (mainly streptococcus and anaerobes), with the composition varying by location in the mouth. *Streptococcus mutans* is the most important bacteria associated with dental caries.

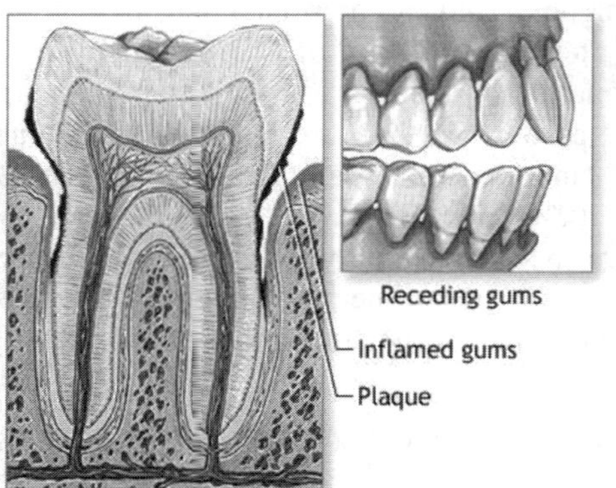

Certain bacteria in the mouth live off the remains of foods, especially sugars and starches. In the absence of oxygen they produce lactic acid, which dissolves the calcium and phosphorus in the enamel. This process, known as "**demineralization**", leads to tooth destruction. Saliva gradually neutralizes the acids which cause the pH of the tooth surface to rise above the critical pH. This causes 'demineralization', the return of the dissolved minerals to the enamel. If there is sufficient time between the intakes of foods (two to three hours) then the impact is limited and the teeth can repair themselves. Nonetheless, in the presence of plaque, saliva is unable to penetrate through the plaque to neutralize the acid produced by the bacteria.

Dental caries, also described as "**tooth decay**" or "**dental cavities**", is an infectious disease which damages the structures of teeth. The disease can lead to pain, tooth loss, infection, and, in severe cases, death.

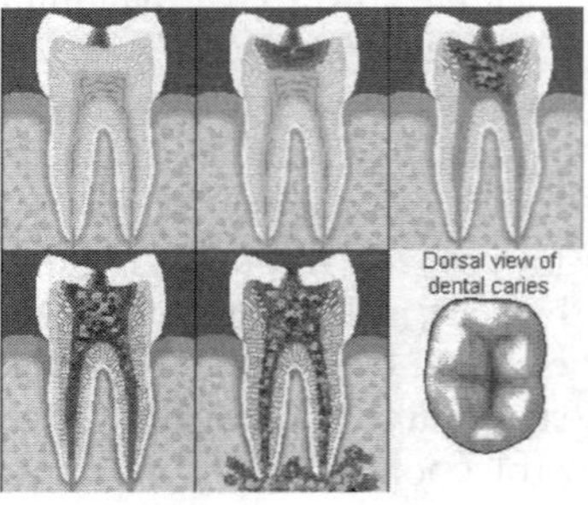

Tooth decay is caused by certain types of acid-producing bacteria which cause the most damage in the presence of fermentable carbohydrates such as sucrose, fructose, and glucose. The resulting acidic levels in the mouth affect teeth because a tooth's special mineral content causes it to be sensitive to low pH. Depending on the extent of tooth destruction,

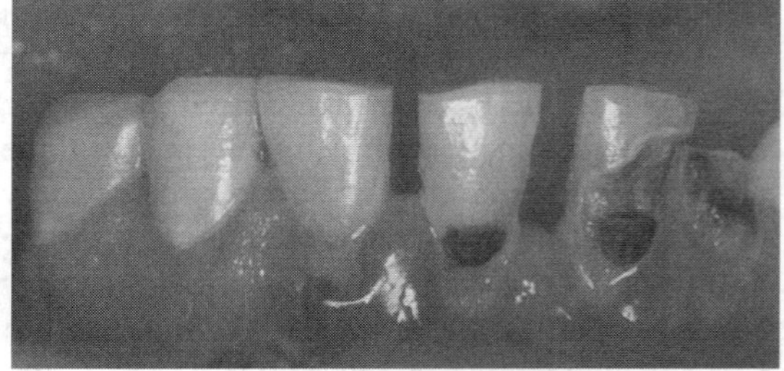

various treatments can be used to restore teeth to proper form, function, and aesthetics, but there is no known method to regenerate large amounts of tooth structure. Instead, dental health organizations advocate preventative and prophylactic measures, such as regular oral hygiene and dietary modifications, to avoid dental caries.

Oral hygiene is the practice of keeping the mouth clean, and is means of prevention from dental caries, gingivitis, periodontal disease, bad breath, and other dental disorders. It consists of both professional and personal care. Regular cleanings, usually done by dentists and dental hygienists, are recommended to remove tartar (mineralized plaque) that may develop even with careful brushing and flossing. Professional cleaning includes tooth scaling, using various instruments or devices to loosen and remove deposits from the teeth.

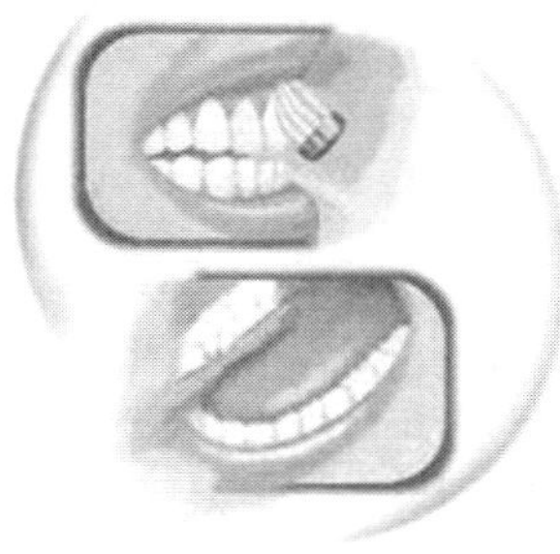

Regular brushing is recommended by healthcare professionals twice a day in order to prevent formation of plaque and tartar. A toothbrush is able to remove plaque on most surfaces of the teeth except for areas between teeth. As a result, flossing is also considered a necessity to maintain oral hygiene. When used correctly, dental floss removes plaque from areas which could otherwise develop caries. The purpose of cleaning teeth is to remove plaque, which consists mostly of bacteria. Electric toothbrushes are not considered more effective than the manual variety. The most important advantages of electric toothbrushes is the ability to aid people with dexterity difficulties, such as those associated with rheumatoid arthritis.

In addition, fluoride therapy is often recommended to protect against dental caries. It has been demonstrated that water fluoridation and fluoride supplements decrease the incidence of dental caries. Fluoride helps prevent dental decay by binding to the hydroxyapatite crystals in enamel. The incorporated fluoride makes enamel more resistant to demineralization and, thus, resistant to decay. Topical fluoride is also recommended to protect the surface of the teeth. This may include a fluoride toothpaste or mouthwash. Many dentists include application of topical fluoride solutions as part of routine visits.

Abnormalities of the Dentition

- Abnormalities with number of teeth.
- Abnormalities with size of teeth.
- Amelogenesis imperfecta – a condition in which the tooth's primary surface, the enamel, does not form properly or at all.
- Anodontia – total lack of tooth development.
- Dental fluorosis – white spotted, yellow, brown, black and sometimes pitted teeth from over-ingesting fluoride.

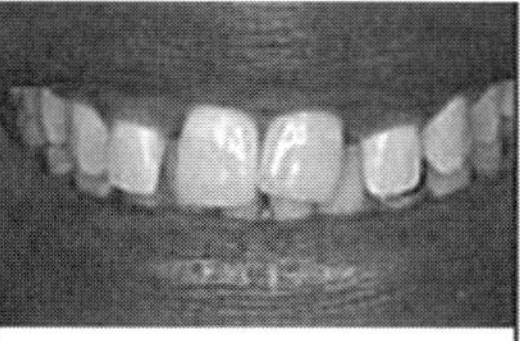

Full smile showing misalignment, diastemas, and poor gingival architecture

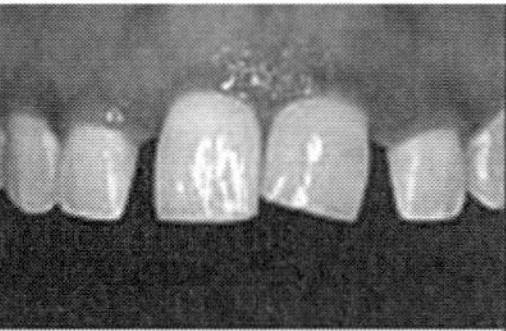

Retracted view of maxillary anteriors showing misalignment, diastemas, and poor gingival architecture

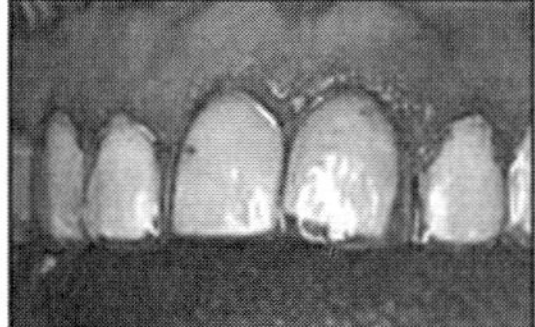

Retracted view with surgical guide in place following initial gingivectomy

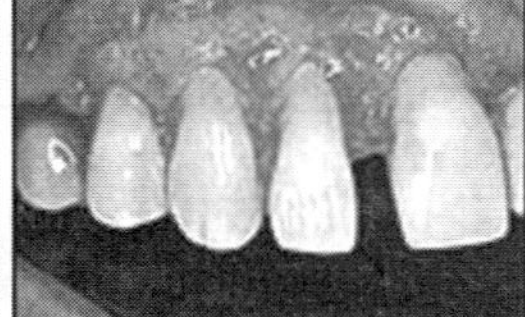

Retracted view of right side with flap reflected immediately after osseous reshaping with surgical handpiece and burs

- Dentinogenesis imperfecta – a similar condition to above, but affects the underlying layer of the tooth.
- Deossification – loss of bone tissue or removal of the mineral elements of bone.
- Hypercalcification – excess of calcium.
- Hyperdontia – presence of a higher-than-normal number of teeth.
- Hypocalcification – reduction of calcium.
- Hypodontia – missing teeth.

- Macrodontia – having abnormally large teeth
- Microdontia – having abnormally small teeth.
- Supernumerary roots – presence of a higher-than-normal number of roots on a tooth. Most common in maxillary bicuspids.

Abnormalities in Shape of Teeth

- Concrescence – mass of hard material deposits on teeth.
- Cusp of Carabelli – abnormally pointed tips of teeth.

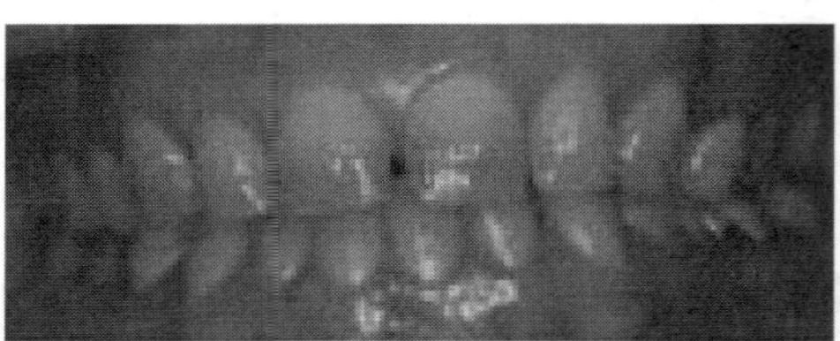
Before

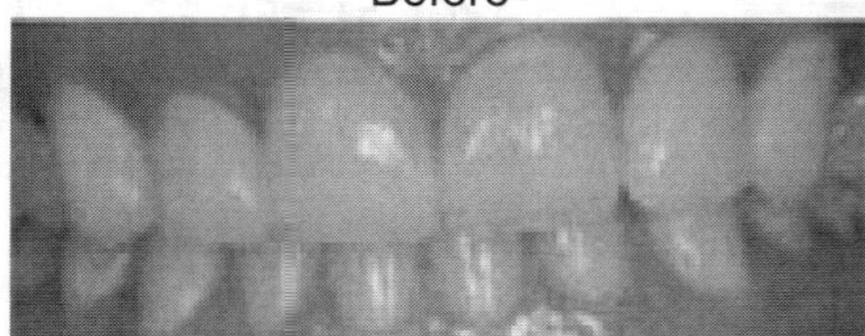
After

- Dens evaginatus – cusp-like elevation of enamel.
- Dens in dente – also called Dens invaginatus it is a malformed tooth caused by invagination of the crown before it is calcified, giving the appearance of a "tooth within a tooth.".
- Dilaceration – trauma or abnormal angulation or curve in the tooth during formation causing damage to the root structure.
- Ectopic enamel – not in normal position or abnormal enamel.
- Hypercementosis – problem caused in cementum of teeth.
- Supernumerary roots – presence of a higher-than-expected number of roots on a tooth.
- Talon cusp—also known as "eagle's talon" is an extra cusp (Pointed end) on an anterior tooth.
- Taurodontism—a condition found in teeth where the body of the tooth and pulp chamber is enlarged.
- Tooth Fusion – the union of two adjacent tooth germs by dentin during formation.
- Tooth Gemination – paired tooth.

DENTIFRICES

Introduction

Dentifrices are hygienic preparations intended to cleanse the teeth of food debris, prevent calculus and plaque formation, polish to impart luster to teeth and to leave a refreshing feeling in mouth.

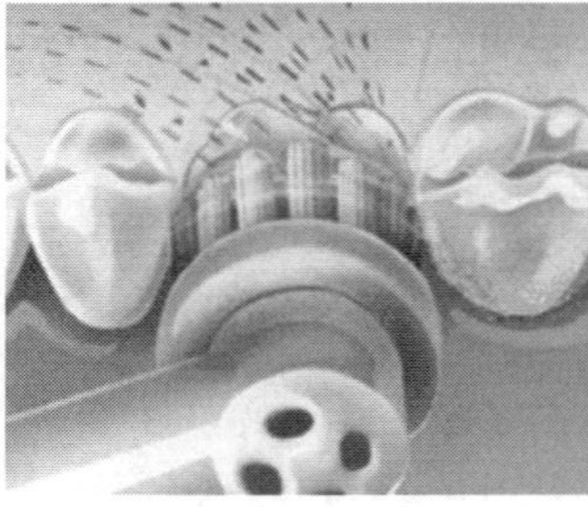

Dentifrices are preparations intended for use with a toothbrush for the purpose of cleaning the accessible surfaces of the teeth. These have been prepared in paste, powder, and to a lesser extent in liquid and block forms. In addition to enhancing personal appearance by maintaining cleaner teeth, brushing with a dentifrice reduces the incidence of tooth decay, helps maintain healthy gingiva, and reduces the intensity of mouth odors. Good dental health increases the possibility of good general health, a leading secondary result of cleaning teeth.

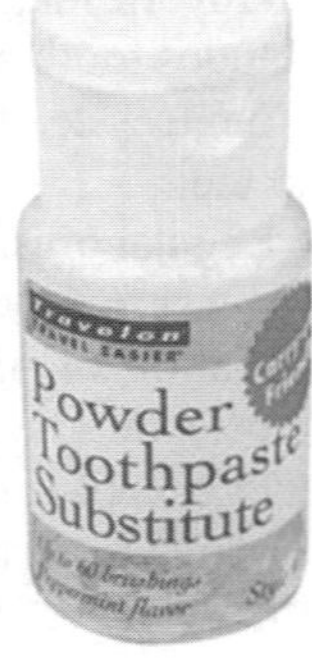

Early writings of the Greeks, Romans, Buddhists, and Hebrews refer to the use of tooth-picks, chew-sticks and sponges in hygienic teeth cleansing rituals. Dried animal parts, herbs, honey, and minerals were suggested as dentifrice ingredients. For many years, materials capable of destroying teeth, irritating the oral mucosa, or detrimental to health were used. Examples include; sulphuric acid, acetic acid, lead ores, and impure, excessively abrasive minerals. Today there has been a grand development of evaluation of methods which have been

designed to ensure the safety and effectiveness of dentifrices.

During the past decade, considerable research effort has been expended in an attempt to develop dentifrices capable of effecting greater reduction in incidence of tooth decay or gingival disorders than possible a tooth- cleaning dentifrice. So, dentifrices may also be defined as, "tooth-cleaning agents have incorporated in them, some drug or chemical by the virtue of it's bactericidal, bacteriostatic, enzyme inhibiting or acid neutralizing qualities reduce the incidence of dental caries or aid in the control of periodontal disease".

Types of Dentifrices

1. Tooth powder
2. Tooth pastes.

Ideal Properties of Good Dentifrices

- It must adequately clean teeth of stain, food debris and plaque.
- It must remove bad odour of mouth cavity and should leave sensation of cleanliness and freshness in mouth.
- It should be non-toxic.
- It should be pleasant and provide good flavour in oral cavity as an after effect.
- It must improve gums health.
- It should be easy to use.
- It must reduce the teeth decay.
- It should not damage tooth enamel because of its abrasiveness.
- It should be economic.

Tooth powder is a powder containing abrasives, detergents, sweetening agents and flavoring agent, etc. meant for cleaning of teeth.

Formulation of Tooth Powder

1. Abrasive and polishing agents.
2. Detergents and foaming agents.
3. Sweetening agents.
4. Flavoring agents.

Abrasive and polishing agents: These agents are included in dentifrices due to their action of abrasiveness, polishing properties and for debris removers. Most commonly used polishing agents are precipitated calcium carbonate, dicalcium phosphate, calcium pyrrophosphate, tribasic calcium phosphate, magnesium trisilicate, sodium metaphosphate, hydrated alumina (aluminum hydroxide), etc.

Detergents and foaming agents: They act by wetting the teeth and provide detergency in order to enhance action of polishing agents. Some of the commonly used detergents and foaming agents are sodium lauryl sulphate, sodium lauryl sarcosinate, diethyl sodium lauryl sulphosuccinate, magnesium lauryl sulphate and soaps like sodium palmitate, etc.

Sweetening agents: Saccharin sodium and xylitol in the concentration of 0.005 to 0.25% is the most commonly used sweetening agent to impart sweet taste to dentifrices.

Flavoring agents: They are able to provide good flavour in oral cavity after its cleansing and maintaining it for a longer period of time. Usually 0.5 to 2.0% Peppermint oil, anise oil, clove oil, spearmint oil, winter green oil, cinnamon oil, etc. are used as flavors.

Tooth paste is simply a tooth powder wetted to make a paste.

Formulation of Tooth Paste

1. Abrasive and polishing agents.
2. Detergents and foaming agents.
3. Humectants.
4. Binding agents.
5. Sweetening agents.
6. Flavoring agents.
7. Preservatives.
8. Colors.
9. Miscellaneous.
   - Anti-caries agents
   - Anti-bacterial agents
   - Desensitizing agents.

Formulation of tooth pastes is similar to tooth powders except that pastes contain humectants, binding agents and preservatives which are not generally added to tooth powders.

- **Humectants:** Glycerine, propylene glycol, sorbitol are usually added in concentration ranging from 10–30%.

- **Sorbitol,** also known as glucitol, is a sugar alcohol the body metabolises slowly. It is obtained by reduction of glucose changing the aldehyde group to an additional hydroxyl group hence the name sugar alcohol. Sorbitol is often used in modern cosmetics as a humectant and thickener. Some transparent gels can only be made with sorbitol as it has a refractive index sufficiently high for transparent formulations. It is also used as a humectant in some cigarettes.
- **Xylitol,** also called wood sugar or birch sugar, "tooth friently" sugar is a five-carbon sugar alcohol that is used as a sugar substitute. Xylitol is a naturally occurring sweetener found in the fibers of many fruits and vegetables, including various berries, corn husks, oats, and mushrooms. It can be extracted from corn fiber, birch, raspberries, plums, and corn. Xylitol is roughly as sweet as sucrose but contains 40% less food energy. Xylitol also actively aid in repairing minor cavities caused by dental caries.

- **Binding agents:** These are important ingredient of tooth pastes; they are added to keep solids and liquids united. Usually 2 to 5% Gum acacia, gum tragacanth, gum karaya, methyl cellulose, agar, Veegum®, sodium alginate, bentonite, etc. are used as binders in tooth pastes.
- **Sweetening agents:** Generally saccharine sodium is used in tooth pastes to impart sweet taste.

Sodium saccharine

- **Flavoring agents:** Peppermint oil, anise oil, clove oil, spearmint oil, cinnamon oil, etc.
- **Preservatives:** Preservatives are necessary in case of tooth paste in order to preserve preparation containing carbohydrate and moisture enough for bacterial growth. Commonly used preservatives are methyl paraben and propyl paraben.
- **Colors:** FD & C approved colors and opacifiers like titanium dioxide, zinc oxide are used.
- **Anti-caries agents:** Sodium fluoride, sodium lauryl sarcosinate, etc. are used as anti-cariogenic agents.
- **Anti bacterial agents:** Triclosan is now a day most widely used anti-bacterial agent for tooth pastes. Triclosan [chemically 5-chloro-2-(2,4-dichlorophenoxy) phenol] is a potent wide spectrum antibacterial and antifungal agent.
- **Desensitizing agents:** Potassium nitrate, sodium chloride, etc. used to reduce the sensitivity of teeth to hot and cold.

FUNCTIONS OF A DENTIFRICE

The main function of a dentifrice is to clean the accessible surfaces of teeth. As a result of such cleaning the effects like incidence of tooth decay and mouth odours have been minimized.

The secondary functions besides the main function are too of the same importance. So all effects that have been noted following the conventional use of dentifrice is taken as the function of dentifrice.

Tooth Polishing

When food debris, bacterial plaques, and tooth surface are removed, the teeth may be clean but exhibit a low lustre. Aesthetically, highly polished teeth are desirable. Polished teeth are less receptive to the retention of dental plaque and debris so can remain cleaner longer.

Dentifrices are necessary to polish teeth and so should be formulated with a two fold purpose in mind:

i. To clean the tooth surface by removal of dental plaque and stain.
ii. To polish the enamel.

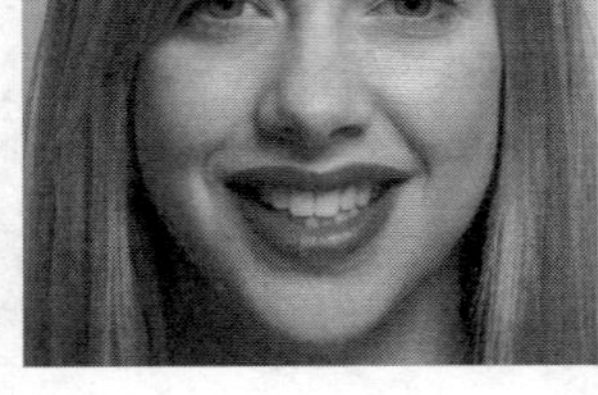

There are several methods for measuring the polishing of human teeth.

i. Qualitative estimations of the degree of polish of teeth consist mainly of visual inspection and comparison.

ii. Quantitative method: In this method, light reflected from a test specimen is measured by means of a photo-electric tube and amplifying circuit.

A tooth paste utilizing hydrated alumina as a polishing agent polish teeth faster and to a higher degree than five commercial pastes or calcium carbonate powder.

Dentifrices containing insoluble sodium metaphosphate and calcium phosphate, or calcium phosphate and calcium carbonate are rated second and third.

Dentifrices containing dicalcium phosphate or calcium carbonate are rated 4th and 5th, respectively.

The combination of insoluble sodium metaphosphate and tri-calcium phosphate is the most effective polishing agent for the enamel surface. It raises the lustre of a dulled tooth and does not dull a highly polished tooth.

Microscopic examination of the enamel surface brushed with the polishing agent revealed no abrasive action.

Reduction of the Incidence of Tooth Decay

Dental caries, i.e. tooth decay, may be defined as a disease of the calcified tissues of the teeth caused by acids formed by the action of micro-organisms on carbohydrates and characterised by a decalcification of the inorganic portion of the tooth accompanied or followed by a disintegration of the organic substance.

Dental caries generally begins on the surface of the tooth. Since the carious lesion begins in a relatively small circumscribed area, bacteria must be localized and held in close contact with enamel before the lesion, called dental caries, is produced.

So the removal of these bacterial plaques as well as mechanical removal of the bacterial media can be considered a method of caries control.

SURFACE ACTIVE DETERGENTS

Half of dentifrices are of foaming type and foaming was result of presence of soap. Manufacture claims that foaming dentifrices have more effective cleansing due to lower surface tension of soap containing dentifrices. Consumers also preferred foaming over non-foaming dentifrices.

Soap is effective as a detergent only in alkaline solution. Some dentifrices had a pH as high as 11, an alkalinity apt to prove irritation to gums. Also soap forms insoluble calcium and magnesium soaps, with calcium salt found in saliva and hard water.

Flavouring materials such as iso-eugenol and vanillin are also not stable in presence of alkali so it cannot be utilised in soap containing dentifrices. Also soap does not have pleasant taste and become rancid so needs high quantity of flavour which can itself be irritating. Finally soap cannot be effectively used effectively with certain polishing agent as sodium metaphosphate.

After development of several synthetic detergents suitable to use in dentifrices, all leading manufacturers adopted the detergent as a replacement of soap. Similar to soap, detergent lower surface tension, penetrate and loosen surface deposits and emulsify or suspend the debris which are removed by dentifrices from teeth surface.

Newer detergent, sodium lauryl sulphate is neutral in reaction, cleanses in acid or alkaline solution, does not precipitates in hard water and saliva, so is suitable for use in tooth paste.

During selection of detergent, stability, taste and compatibility with other ingredient should be taken in account. After these parameters cleaning effectiveness is considered. Although laboratory test are helpful in evaluation of cleaning effectiveness of detergent but one has to relay on results of extensive controlled clinical studies.

Among the test which are utilised in evaluation of these properties of detergents are:

a. Effect on clearing activity of epithelium of oesophagus of flogs.
b. Irrigation test on rabbit eye.
c. Local irrigation produced by subcutaneous injection.
d. Local irritation as determined by capillary permeability.
e. Acute toxicity as necropsies.
f. Chronic toxicity including post-mortem examination of vital organ.
g. Irritation of oral mucosa of animal.
h. Irritation of oral mucosa of human.

When all foregoing test indicate safety of a detergent the complete dentifrices along with detergent at desired concentration are evaluated in controlled clinical studies by expert dentists to determine freedom from irritation and sensitization character.

Detergents described in accepted dental remedies are:

Dioctyl sodium sulfosuccinate, NF: It is available as white, waxlike plastic solid with characteristic odour suggestive of octyl alcohol. Freely soluble in alcohol and glycerol and is compatible with hard water. Marketed under name aerosol OT.

Sodium alkyl sulfoacetate—sodium lauryl sulfonate: It is a mixture consisting principally of compound of general formula $ROCOCH_2SO_3Na$. R is alkyl group and predominantly by lauryl. It accrues as white powder having mild coconut like odour and taste like acrid in solution. Aqueous solution is neutral or slightly alkaline, marketed under name lathanol LAL.

Sodium lauryl sulphate: It is mixture of compound of general formula $ROSO_3Na$. R is alkyl group and predominately lauryl. It occur as white neutral or slightly alkaline. Daily use of dentifrices twice a day for 30 days on young dogs showed no symptoms of irritation on oral mucosa. Clinical studies on dentifrices containing sodium lauryl sulphate confirm the complete safety of sodium lauryl sulphate at a concentration 2% or less.

Sulfocolaurate: Sulfocolourate is a mixture to contain 91% of lauric acid esters of potassium salt of sulfoacetic acid acidified with B-amino ethyl alcohol and 7 to 8% of potassium chloride.

It is a white crystalline powder, sparingly soluble in water, quite soluble in warm water at body temperature or higher. Recommended use in dentifrices at concentration 2% as foaming agent.

Sodium salt of sulphated monoglycerids: Most of metal salts are soluble including calcium and magnesium in detergent. Solution containing this detergent has extremely low surface tension. It has excellent foaming, emulsifying and deterging properties.

Detergent can be prepared by mixing coconut oil, glycerol and sulphuric acid in desired order at a temperature which make them to react. Detergent can also be formed by reacting fatty acid glycerol with aid of heat and an alkaline catalyst thereafter treating the product with sulphuric acid. Then resulting mixture is neutralised.

Sodium Coconut Monoglyceride Sulfonate: Surface active agent possesses all of the attributes of outstanding value for a dentifrices, i.e. stability and inertness to other ingredients, freedom from toxicity, whiteness of colour, neutrality of reaction and case in preparation. Also sulfonate has mildness of flavour, freedom from taste reaction with food and has good foaming activity.

Sodium N-lauroyl Sarcosinate: Synthesized by condensing lauroyl chloride and sodium salt of sarcosine in presence of alkali.

HUMECTANTS

During preparation of tooth paste, the abrasive are mixed with a liquid phase which include humectants, its function is to retain moisture when paste is exposed to air, preventing the paste from hardening.

Glycerol, sorbitol and propylene glycol commonly used humectants. Glycerol and sorbitol are sweat while propylene glycol has a slightly acrid taste. Under favourable condition humectants may develop bacterial or mould growth. They should be add with preservative particularly if it is to be stored for any period prior to use in dentifrices formulation. Benzoic acid has been recommended for this use.

Laboratory and clinical data shows that sorbitol has no deleterious effects on oral tissue and systemic toxicity is same as that of sugar, dextrose and sucrose. It is not readily fermented as sugar.

BINDER

To avoid separation of simple admixture of solid and liquid phase of tooth paste, binder is added particularly during storage. All binder is hydrophilic colloids, and forms viscous liquid phase after absorbing water and swells. Binder stabilizes the mass against separation of liquid phase.

Starch in form of glycerite is employed as common binder. Other binders include gum arabic, ghatti, gum karaya, tragacanth and also seaweed colloids. Water dispersible cellulose prepared synthetically is also employed for this purpose.

In addition to organic binder, bentonite, hychated aluminium silicate and veegum also have been recommended as binders.

Choice of binder depends on many factors as on composition of dentifrices and desired case of dispersion of tooth paste in mouth. It is important to establish the effectiveness of the binders in stabilizing the paste for expected shelf life. After selection of binder, specification for binder and dentifrices are established for uniform viscosity of paste batch after batch.

Gum karaya: Gum Karaya obtained from gummy oxidation from *Sterculia urns* are tears of variable size or in broken irregular pieces with crystalline appearance. It possesses acetic odour and mucilaginous and slightly acetic taste. It is also available as pinkish grey or white to brown powder. It swells in water but insoluble in chloroform and alcohol. Aqueous dispersion of gum is acidic.

Tragacanth USP - Gum Tragacanth: Dried gummy extraction of *Astragalus gummifer* ungrounded tragacanth are flattened, camellated frequently curved pieces, white to yellow, translucent and horny with short fracture. Purer form is odourless and has mucilaginous taste. Powder for is white show starch grain under microscope. Upon addition of water it swells and forms a cloudy gelatinous mass. It is insoluble in alcohol.

Sodium Alginate NF: It is purified carbohydrate obtained from brown seaweed by use of dilute alkali. Consist of mainly sodium salt of alginic acid. It is white to light cream coloured amorphous, odourless, tasteless powder. It dissolves in water forming a viscous colloidal solution. Addition of calcium ions is available. Aqueous solution of sodium alginate is also not compatible with acid; also it is not stable below a pH of 3.3. Alginate is frequently attacked by microorganism. To avoid this effect preservatives are employed. Formaldehyde is particularly effective.

Irish Moss Extract—Chondrus extract NF: It is dried refined hydrocolloidal aqueous extract prepared from chondrus, bleached or unbleached. Chondrus is dried unbleached plant of *Chondrus crispus*.

Chondrus extract occurs as coarse fine powder, odourless with mucilaginous taste. Aqueous solution is alkaline to litmus. It contains 70–75% calcium salt of sulphuric acid ester of colloidal carbohydrate complex with high proportion of galactose group. Sodium, potassium and magnesium lotion are also present. It is most widely used binder in United States.

Methyl cellulose: Methyl cellulose is methyl ether of cellulose. It is a greyish white fibrous powder which swells in water and forms a clear viscous colloidal solution of neutral reaction. Mucilage of methyl cellulose is stable for long time. Mucilages of methyl cellulose is used as wetting agent.

Sodium carboxy methyl cellulose: It is sodium salt of polycarboxymethyl either of cellulose. It occurs as white granular odourless powder prepared by action of monochloroacetic acid on alkali cellulose and conversion of resulting product into sodium salt. Preservative of recommended purified grades are said to be physiologically inert so suitable for pharmaceutical use for human. Preservative like benzoic acid, phenol, chlorinated phenol are added to reduced bacterial and mould contamination.

FLAVOURS

Taste is an important factor in consumer acceptance of a paste. Therefore, development of appropriate flavour is both an art and a science. It is art as

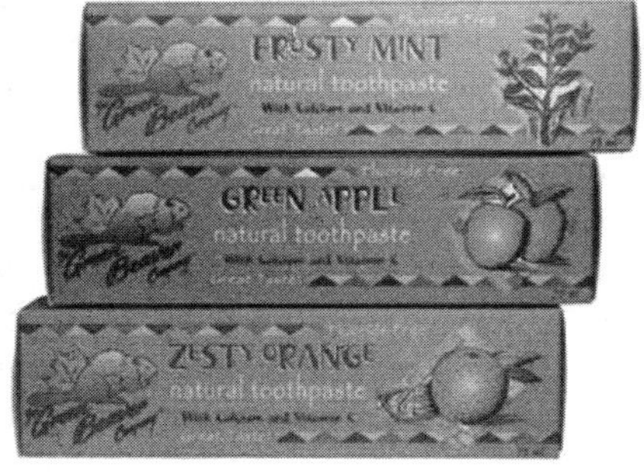

it requires blending of number of component so resulting mixture should have well round, smooth flavour and lasting pleasant aftertaste. It is science in that selected flavour should be compatible with others ingredients and should not change during shelf life.

Spearmint, peppermint, wintergreen, sassafras and cinnamon mint flavour are very popular among consumers. Principle flavour ingredients are not used alone but are added with other essential oils to modify the flavour and create distinctive notes. Essential oil used to modify primary flavour is anise, clove, caraway, coriander, pimento, eucalyptus, nutmeg and thyme. Citrus oils, eugenol, eucalyptus, anethole, irone, orris and menthol can also be used.

In developing flavour, trial blends are incorporated in base. The formulation can be fudged while fresh, freshly flavoured paste is harsher than aged paste.

Polishing materials are generally absorptive material so it is necessary to re-evaluate the preparation after a specified time to see changes because of selective absorption of some flavouring agent. In addition to physical changes, chemical changes also affect flavour stability. Oxidation of terpenes, hydrolysis of esters and alkaline alteration of phenol comes under this heading of chemical modification of flavour.

Concentration of flavour in any preparation depends on type of flavour of base, lastly consumer acceptance. Generally concentration of flavour ranges from 0.5–2%. While selecting of flavour, one should be aware of undesirable effect on mucous membrane of mouth.

Allergic reactions are possible in some flavours. All synthetic flavours contain sweating agent. Saccharine is commonly used sweetener generally present in concentration 0.05 to 0.25% of dentifrices. Not all sweeteners are safe as dulcin is not safe to use. So it is important to establish safety of selected sweetener prior to use in dentifrices.

Loss of flavour, either due to excessive exposure to heat during heating period or due to evaporation during decoration step. If decoration is must, it should be minimum. Care should be taken during manufacturing for chemical changes like oxidation, hydrolytic or other changes.

MISCELLANEOUS DENTIFRICE INGREDIENTS

It has been reported that dentifrices contain astringents (zinc chloride), preservative (dichlorophene benzoate, formaldehyde), oxidising agent (potassium chlorate, urea peroxide).

Care should be taken while selecting other ingredient for specific purpose.

LIQUID DENTIFRICES

Liquid dentifrices have been formulated with and without abrasives. Since the need for an abrasive to ensure cleansing of teeth has been established, the ability of a liquid dentifrice to cleanse teeth should be carefully considered before it is offered to the public.

FORMULAS AND MANUFACTURE

Tooth Paste

CLEANSING TOOTH PASTE

Formula I: Chalk as abrasive and soap as surface active agent.

Chalk	: 39%
Water	: 32%
Glycerol	: 20.0%
Soap	: 6.3%
Gum	: 0.4%
Saccharin	: 0.1%
Flavouring oils	: 1.2%

Formula II

Calcium carbonate	: 45.7%
Starch	: 7.0%
Glycerol	28.2%
Water	: 14.4%
Sodium benzoate	: 2.2%
Flavour	: 1.3%
Na salt of H_2SO_4 ester of a fatty acid monoglyceride	: 1.2%

Formula III

Dicalcium phosphate*	: 42.50%
Sodium coconut# -monoglyceride sulphonate	: 2.00%
Water	: 28.35%
Glycerol	: 25.00%
Irish moss extract**	: 1.45%

Saccharin : 0.10%
Flavour (mint) : 0.60%
* = Abrasive, # = Surfactant, ** = Binder

Formula IV

Insoluble sodium* metaphosphate : 26.6%
Dicalcium phosphate : 26.6%
Gum : 1.4%
Flavour : 1.6%
Purified alkyl sulphate (SLS)# : 1.1%
Glycerol and water : 42.7%
* = abrasive, # = Surfactant

Formula V

Aluminium hydroxide gel* : 42.50%
Alumina : 02.00%
Glycerol : 28.00%
Water : 24.50%
SLS# : 1.00%
Flavour : 0.70%
Methyl p-hydroxy bencoate : 0.10%
Tragacanth$: 0.50%
Saccharin : 0.05%
Phosphoric acid : 9.5%
Ammoniated tooth paste
* = Polishing agent; $ = Binder; # = Surfactant

Formula VI

Tricalcium phosphate : 26.67%
Glycerol : 45.40%
Water : 15.40%
Flavouring (Peppermint) : 0.58%
Gum tragacanth : 0.96%
Saccharin : 0.10%
Surface active agent (Nacconol LAL) : 2.89%
Diammonium phosphate : 5.00%
Urea, 100 mesh : 3.00%

It is a formula representing 'low' urea ammoniated tooth paste Tricalcium phosphate is used as polishing agent

Formula VII

High urea ammoniated paste, carboxy methyl cellulose is binder.

Tricalcium phosphate : 26.67%
Glycerol : 45.40%
Water : 15.40%
Flavour (Peppermint) : 0.58%
Gum tragacanth : 0.96%
Saccharin : 0.10%
Surface active agent : 2.89%
Diammonium phosphate : 5.00%
Urea, 100 mesh : 3.00%

Chlorophyll tooth paste: These contain water soluble chlorophyll derivative

Formula VIII

Calcium carbonate : 50.05%
Sodium copper chlorophyll in : 0.3%
Tetrasodium pyrophosphate : 0.25%
Sodium lauryl sulphate : 2.5%
Oil of spearmint : 1.0%
Gum tragacanth : 1.0%
Glycerol : 20.0%
Saccharin : 0.1%
Water : 24.85%

Formula IX

Dicalcium phosphate dihydrate : 50.0%
Sod. Copper chlorophyllin : 0.1%
Tetrasodium pyrophosphate : 2.0%
Carboxy methyl cellulose : 1.5%
Water : 46.4%

Formula X

Sodium copper chlorophyllin, 100%, : 0.1%
Insoluble sodium polymetaphosphate# : 27.0%
Dicalcium phosphate : 27.0%
Gum tragacanth* : 1.3%
Saccharine : 0.2%
Flavour : 1.0%
Sodium lauryl sulphate$: 20.0%
Distilled water : 22.4%
= Abrasive, * = Binder, $ = Surfactant

Formula XI Antienzyme toothpaste:

Calcium carbonate* : 12.1%
Dicalcium phosphate dehydrate* : 36.2%
Sodium N-lauroyl sarcosinate# : 2.0%
Glycerol : 30.6%
Water : 15.3%
Irish moss$: 1.0%
Sweetening agent, flavour, preservative : 2.8%
* = Abrasive $ = Binder, # = Surfactant

Formula XII *Fluoride tooth paste:*
Contain stannous fluoride calcium pyrophosphate as abrasive.

Heat treated orthophosphate : 42.0%
Detergent : 2.0%
Humectant : 25.0%
Water : 29.2%
Binder : 1.4%
Stannous fluoride : 0.4%
Flavour : 9.5%

Formula XIII

Calcium pyrophosphate : 42.2%
Synthetic detergent : 1.3%
Glycerol : 25.0%
Water : 29.0%
Gum tragacanth : 1.4%
Sodium fluoride : 0.2%
Flavour : 0.9%

Formula–XIV
Microcrystalline hydroxide and aluminium hydroxide: Abrasive

Microcrystalline aluminium hydroxide : 38.2 parts
Aluminium hydroxide (#32.5) : 5.0 parts
Sodium fluoride : 0.1 parts
Sodium alginate : 1.0 parts
Glycerol : 15.3 parts
Sorko (70% aqueous sorbitol) : 15.3 parts
Water : 20.8 parts
Colour : 0.004
Saccharin, soluble : 0.25 parts
Flavour : 1.2 parts
Sodium lauryl sulphate : 2.2 parts
Orthophosphoric acid to pH 7.3 : 0.6 parts

Total 99.954 parts

Penicillin tooth paste: Penicillin readily decomposes in aqueous media so formulation containing such antibiotic is of no use.

TOOTH POWDER

Cleansing Tooth Powder

Formula XV

Hard Soap, powdered # : 50 g
Calcium carbonate precipitated * : 935 g
Saccharin, soluble : 2 g
Oil of peppermint : 4 cc
Oil of cinnamon : 2 cc
Methyl salicylate : 8 cc

* = Polishing agent, # = S.A.A.

Formula XVI

Insoluble sodium metaphosphate : 76.8%
Tricalcium phosphate : 20.0%
Alkyl sulphate (Sodium lauryl sulphate) : 1.0%
Flavour : 2.0%
Saccharin : 0.2%

Formula XVII

Microcrystalline aluminium hydroxide : 50 parts
Sodium lauryl sulphonate : 2 parts
Saccharin : 0.1 parts
Gum Tragacanth : 0.1% parts

Total 52.2 Parts

Formula XVIII: Soap as surfactant

Microcrystalline aluminium hydroxide : 90.0 parts
Soap : 5.0 parts
Methyl salicylate : 3.5 parts
Saccharin : 0.1 parts

Total 98.6 parts

Ammoniated Tooth Powder

Formula XIX

Low ammoniated tooth powder Contain : 3% of urea
Calcium carbonate : Abrasive
Dibasic ammonium phosphate : 50.0 g
Carbamide : 30.0 g
Bentonite : 50.0 g
Saccharine, soluble : 02.0 g
Calcium carbonate, ppr. : 866.0 g
Oil of peppermint : 0.2.0 g
Oil of wintergreen : 6.0 cc
Duponol (Sod. lauryl sulphate) : 10.0g

Formula XX
High urea ammoniated tooth powder.

Carbamide : 22.5%
Diammonium phosphate : 5.0%
Calcium carbonate : 51.67%
Calcium phosphate : 10.55%
Bentonite : 5.0%
Sodium chloride : 3.0%

Saccharine : 0.28%
Flavour : 1.5%
Sulfocolaurate : 0.5%

Contain 22.5 and 5% of urea and diammonium phosphate respectively. Calcium carbonate and calcium phosphate used as abrasive.

Chlorophyll tooth powder:
Water soluble chlorophyllins are present.

Formula XXI

Dicalcium phosphate dihydrate* : 91.8%
Sodium copper chlorophyllin : 0.2%
Sodium lauryl sulphate# : 3.0%
Sodium tripolyphosphate : 3.0%
Oil of peppermint : 2.0%
* = Abrasive; # = Detergent

Formula XXII

Trimagnesium phosphate : 17.0%
Insoluble sodium polymeta phosphate* : 80.0%
Saccharin : 0.2%
Sodium magnesium chlorophyllin : 0.1%
Flavour : 1.4%
Sodium lauryl sulphate# : 1.3%
Penicillin tooth powder : Present
Penicillin : Present
* = Abrasive; # = Detergent

Formula XXIII

Contain 500 units potassium penicillin per gram.

Calcium carbonate, ppr : 96.27%
Tricalcium phosphate : 0.02%
Sulfocolaurate : 2.03%
Menthol : 0.20%
Methyl salicylate : 0.80%
Oil of peppermint : 0.20%
Saccharin, soluble : 0.30%
Potassium penicillin : 500 units/gram

Formula XXIV

Calcium carbonate : 74.0%
Sodium bicarbonate : 2.0%
Tricalcium phosphate : 15.0%
Neutral soap : 6.5%
Saccharin : 0.3%
Essential oils : 2.2%
Penicillin : 100,000 units.

Formula XXV: Anti-enzyme powder:

Sodium N-lauryl sarcosinate : 30%
Dicalcium phosphate dihydrate* : 94.2%
Saccharin, soluble : 0.3%
Flavour : 2.5%
* = Surfactant

Formula XXVI: *Fluoride tooth powder:*

Microcrystalline aluminium hydroxide : 91.25%
Aluminium hydroxide (#32.5)* : 5.0%
Flavour : 1.2%
Saccharin, soluble : 0.25%
Sodium fluoride : 0.1%
Sodium lauryl sulfoacetate : 22%
* = Abrasive

LIQUID DENTIFRICES

Formula XXVII

Soap powdered : 7.2 g
Saccharin : 0.24 g
Amaranth (solution) : 1.0cc
Oil of cinnamon : 0.6cc
Oil of peppermint : 0.6 cc
Oil of clove : 1.2 cc
Alcohol : 90.0 cc
Water, distilled ad : 120.0 cc

MANUFACTURE OF DENTIFRICES

Tooth Powder

Primary objective during manufacture of tooth powder is homogeneous distribution of all ingredients without contamination by foreign substance. It is convenient to premix those ingredients present in small amount prior to admixture with the remainder of the component.

Mixing operation is generally performed in ribbon like mixture. Flavour can be mixed with part of polishing agent, screened and resulting mixture added to the bulk of powder. Homogenous mixture of tooth powder is screened prior to storage in filling hopper. Some powder tend to agglomerate and cause resistance in flow to overcome this problem, granulating the powder by drying slurries with finely divided polishing agent, a detergent and binder. The dried product is then comminuted and mixed with flavouring agent. For free flow property, the size of granules of tooth powder should be such that substantially all are retained on a 100 mesh screen but pass through 40 mesh screen.

Liquid Dentifrices

Equipments required for manufacturing are mixing, storage tank, filtration unit and filters. Glass lined or stainless steel tank are commonly used manufacturing process that involve simple solution of all ingredient. Hydrophilic colloids are added to increase viscosity. The material is homogeneously dispersed in portion of solvent prior to mixing with other ingredient.

Flavours are dissolved in solvent to which then glycerol and aqueous solution of colouring matter, saccharine and detergent are added.

Tooth Paste

Manufactured by two general methods:
In first method the binder which is wetted by humectant is dispersed in liquid portion containing saccharine and preservative and allow swelling to form homogenous mass or gel. Swelling is accelerated by heating and agitation. Homogenous gel is then pumped into mixer and solid abrasive is added to homogenous mass. Flavours and detergents are added at last. Excessive aeration should be avoided. The paste is then milled, cleavated and tubed.

In second method, binder is premixed with solid abrasive and introduced with aqueous solution of humectant, preservative and saccharine into suitable mixture. After mixing flavours and detergents are added. In this method heat is not employed. One must establish that binder can be adequately swelled during the mixing process for uniform paste of derived consistency.

PACKAGING DENTIFRICES

Tin is a desirable metal for collapsible tubes used for packaging tooth paste because of low tendency to corrode. Due to high price during post World War II period limited its use.

Aluminium tubes are light, strong and take decoration as well. These are well adopted for formulation of dicalcium phosphate. Aluminium is more reactive than tin or lead hence subjected to corrosion. Dentifrices packed in aluminium tube should be adequately judge for acceptability of tube for specific product. Corrosion of tubes can be minimized by waxing or lacquering or by protecting aluminium internally and externally by a heat sealed film such as polyethylene, pliofilin or saran. Sodium silicate like inhibition can also be included in paste.

Closures of tube are also important consideration, earlier metal caps were used but plastic cap has taken their position currently. Such caps are either valve-seat type or liner type.

Tooth powder is generally packed in metal cans with dispersing top, closed with metal or plastic caps. Cans are maked up of tin-plated or chemically treated steel or may be coated with suitable lacquer.

EVALUATION OF DENTIFRICES

Dentifrices are evaluated for:

- Abrasiveness
- Particle size
- Cleansing property
- Foaming property
- pH
- Consistency.

MOUTHWASH

Mouth rinse or mouthwash is a liquid dentifrices and product used for oral hygiene. Antiseptic and anti-plaque mouth rinse claims to kill the germs that cause plaque, gingivitis, and bad breath. Anti-cavity mouth rinse uses fluoride to protect against tooth decay. However, it is generally agreed that the use of mouthwash does not eliminate the need for both brushing and flossing. Mouthwashes are available in concentrated and ready to use form. Mouthwashes are mainly alcoholic or hydroalcoholic solutions. The main function of mouthwashes is deodorants and antiseptics.

Mouthwashes are available as function and composition wise. Mouthwashes can be available as solid, powder, or concentrated products which can be easily diluted with water. Now a day's several lozenges or chewing gums type mouthwashes or fresheners have also been introduced.

Anthony van Leeuwenhoek discovered living motile organisms in 17th century in deposits on the dental plaque. He experimented with samples by adding vinegar or brandy and found that this resulted in the immediate immobilization or killing of the organisms in suspended in water. Next he tried rinsing the mouth of himself and somebody else with a rather foul mouth with vinegar or brandy and found that living organisms remained in the dental plaque. He concluded - correctly - that the mouthwash either did not reach, or was not presenting long enough, to kill the plaque organisms.

Common use involves rinsing the mouth with about 20 ml (2/3 fl oz) of mouthwash two times a day after brushing. The mouthwash is typically swished or gargled for about half a minute and then spat out.

Various Types of Mouthsash are Available

1. **Mouthwash containing antiseptic or anti-bacterial substances**: These are used to reduce bacterial population of the mouth. Substances commonly used are phenol and its derivatives, tymol, hexachlorophene, hydroxy benzoates, salicylic acid, quaternary ammonium compounds, boric acid, formalin, tannic acid, etc. 
2. **Fluoride containing mouthwashes** are used to reinforce fluoride layer of teeth enamel.
3. **Mouthwash containing astringents or minerals** is used for repairing early caries and lesions in mouth, to protect and shrink inflamed mucous surfaces, precipitate protein of saliva and diminish accumulated mucous secretions by precipitation. Various substances like zinc chloride, zinc phenosulphate, alum (aluminum sulphate), citric acid, tannic acid, lactic acid, acetic acid, etc. are used in concentration of 0.05 to 0.5%.
4. **Mouthwashes containing both anti-plaque and anti-caries properties**, e.g. chlorhexidine mouthwash.

Ideal mouthwashes should have following properties:

i. It should have good and quick antiseptic action at the dilution required.
ii. It should have sweet taste.
iii. It should be non-toxic, non-irritant to oral cavity and mucous membrane.
iv. It should impart attractive odour.
v. It should be economic.

Composition

Active ingredients in commercial brands of mouthwash can include

- Chlorhexidine gluconate
- *Drug extracts:* Used as stimulants, astringents and flavouring agents, e.g. tincture of myrrh, tincture of cinchona, benzoin tincture, quillia ticture, etc.
- Hydrogen peroxide
- Fluoride
- *Astringents:* Zinc chloride, zinc acetate, aluminium potassium sulphate, tannic acid, citric acid, lactic acid, acetic acid.
- *Essential oils as deodorizing agents:* Eucalyptol, Menthol, Thymol (isopropyl metacresol), Methyl salicylate, anethole, chlorophillin
- Cetylpyridinium chloride
- Hexetidine
- *Quaternary ammonium compounds:* Benzalkonium chloride
- Methylparaben
- Enzymes and calcium.

Ingredients also include

- *Vehicle:* Commonly water, sometimes alcohol or combination, also glycerine can be used.
- Sweeteners such as sorbitol and sodium saccharine
- *Colours:* Generally vegetable dyes are used, e.g. saffron, carmine, erythrosine and phloxine, etc.
- A significant amount of alcohol (around 4–20%).

Because of the alcohol content, it is possible to fail a breathalyzer test after rinsing. Many newer brands are alcohol-free.

Formula XXVIII

Sodium lauryl sulphate	1.0%
Alcohol	5.0%
Flavouring oils	2.0%
Distilled Water ad	100.0%

Formula XXIX: Anti-bacterial mouthwash

Thymol	0.03%
Ethanol	3.5%
Borax	2.0%
Sodium bicarbonate	1.0%
Glycerin	8.0%
Distilled water ad	100.0%
Flavour	q.s.

Formula XXX: Astringents mouthwash

Zinc chloride	0.12%
Saccharine sodium	0.1%
Menthol	0.05%
Ethanol	20.0%
Distilled water ad	100.0%
Flavour	q.s.

Formula XXXI

Hexylresorcinol	: 0.0800%
Glycerol	: 20.000%
Ascorbic acid	: 0.100%
Clove oil	: 0.020%
Methyl salicylate	: 0.0200%
Ethyl alcohol	: 18.000%
Water	: 61.7800%
FD&C Blue No. 2	: .0005%

A salt mouthwash is a home treatment for mouth infections and/or injuries, or post extraction, and is made by dissolving a teaspoon of salt in a cup of warm water.

Plain (diluted) hydrogen peroxide is another common mouthwash.

H–O–O–H Hydrogen peroxide

GARGLES

Gargling means to treat mucous membrane of pharynx and tonsils with quantity of fluid taken into mouth and without swallowing it, air is expeled from pharynx and globules of air passes through fluid.

Purpose: Soothing astrigent and antiseptic for sore throat, canker soars, gingivitis and other oral inflammations.

Types:

Antiseptic gargles:	H_2O_2 solution, saturated boric acid solution, etc.
Soothing gargles:	Warm saline solution, hot milk or water, mild alkaline or mucilaginous solution, etc.
Astrigent gargles:	Tannic acid or drug containing it, solutions, diluted alcohol, tincture of iron solutions, glycerine solutions, silver nitrate solutions, etc.

Method:

1. Take deep inspiration.
2. Take liquid into mouth.
3. Throw back the head, and allow liquid to flow into pharynx.
4. Swallow once with mouth open, then slowly expel the breath through the liquid for about half a minute, the mouth remaining open.
5. The head should be thrown quickly forward and down and liquid thrown out.

Mouth-washes are also made of the same liquids but the object is to treat mucous membrane of the mouth and gums.

RINSES

Whether it's to mask bad breath, fight cavities or prevent the buildup of plaque, the sticky material that contains germs and can lead to oral diseases, mouth rinses serve a variety of purposes. Though they may leave your mouth with a clean, fresh taste, some rinses can be harmful, concealing the bad breath and unpleasant taste that are signs of periodontal diseases which cause inflammation and degeneration of the supporting structures of the teeth and tooth decay. Your dentist will tell you. Most mouth rinses just do not wash.

Rinses are generally classified by the U.S. Food and Drug Administration (FDA) as either cosmetic or therapeutic, or a combination of the two. Cos metic rinses are commercial over-the-counter (OTC) products that help remove oral debris before or after brushing, temporarily suppress bad breath, diminish bacteria in

the mouth and refresh the mouth with a pleasant taste. Therapeutic rinses have the benefits of their cosmetic counterparts, but also contain an added active ingredient that helps protect against some oral diseases. Therapeutic rinses are regulated by the FDA and are voluntarily approved by the American Dental Association (ADA).

Therapeutic rinses also can be categorized into types according to use:

- Antiplaque/ antigingivitis rinses and
- Anticavity fluoride rinses.

Various Types of Rinses

1. **Antiplaque/Antigingivitis Rinses**
   A. Therapeutic Antiseptics
      1. Chlorhexidine products
      2. Phenol products
      3. Sanguinaria products
   B. Cosmetic antiplaque rinses
2. **Therapeutic Anticavity Fluoride Rinses**
3. **Cosmetic Breath Freshening Mouth Rinses**
4. **Others**
   A. Topical antibiotic rinses
   B. Enzyme rinses
   C. Artificial saliva rinses
   D. Rinses that control tartar (the hard, crusted calcium deposits that form on teeth).

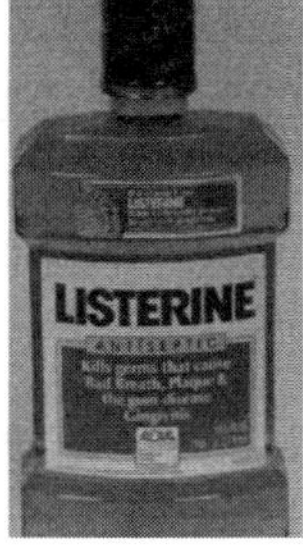

Most rinses are effective oral antiseptics that fresh the mouth and curb bad breathe for up to three hours. Their success in preventing tooth decay, gingivitis and periodontal disease is limited, however.

Rinses are not considered substitutes for regular dental examinations and proper home care. Dentists stress a regimen of brushing with fluoride toothpaste followed by flossing, twice a day. If done consistently and properly, the brushing and flossing, along with routine trips to the dentist, should be sufficient in fighting, tooth decay and periodontal disease.

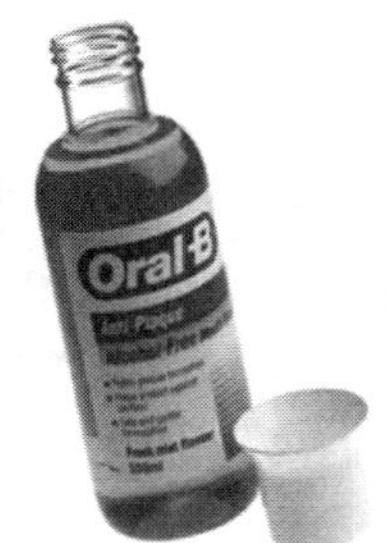

Most over-the-counter antiplaque rinses and antiseptics are not much more effective against plaque and periodontal disease than rinsing with plain water. Most dentists are skeptical about the value of these antiplaque products, and studies point to only a 20 to 25 percent effectiveness, at best, in reducing the plaque that causes gingivitis.

Anticavity rinses with fluoride, however, have been clinically proven to fight up to 50 percent more of the bacteria that cause cavities. Nevertheless, many dentists consider the use of fluoride.

$$Na^+ \ {}^-O - \overset{\overset{\displaystyle O}{\|}}{\underset{\underset{\displaystyle F}{|}}{P}} - O^- \ Na^+$$

Eucalyptol is a natural organic compound which is a colorless liquid. It is cyclic ether and a monoterpene.

Eucalyptol is also known by a variety of synonyms: 1,8-cineol, limonene oxide, cajeputol, 1,8-epoxy-p-menthane, 1,8-oxido-p-menthane, eucalyptol, eucalyptole, 1,3,3-trimethyl-2-oxabicyclo[2,2,2]octane, cineol, cineole.

Hexetidine (Latin: Hexetidinum) is an anti-bacterial and anti-fungal agent commonly used in both veterinary and human medicine. It is a local-anesthetic, astringent and deodorant and has antiplac effects.

Hexetidine is the medicinal ingredient in Sterisol, which is labelled for "the symptomatic treatment of 'strep' throat, tonsillitis, pharyngitis, laryngitis, gingivitis, ulcerative stomatitis, oral thrush (Mouth ulcer or canker sores) and Vincent's angina; postoperative hygiene following tonsillectomy, throat or oral surgery."

Methyl salicylate (chemical formula C_6H_4 (HO) $COOCH_3$; also known as salicylic acid methyl ester, oil of wintergreen, betula oil, and methyl-2-hydroxybenzoate) is a natural product of many species of plants. Some of the plants producing it are called wintergreens, hence the common name.

Menthol is a covalent organic compound made synthetically or obtained from peppermint or other mint oils. It is a waxy, crystalline substance, clear or white in color, which is solid at room temperature and melts slightly above. The main form of menthol occurring in nature is (-)-menthol, which is assigned the (1R, 2S, 5R) configuration. Menthol has local anesthetic and counterirritant qualities, and it is widely used to relieve minor throat irritation.

Chlorhexidine Gluconate is a chemical antiseptic, to combat both gram positive and gram negative microbes. It is both bacteriostatic and bacteriocidal. Chlorhexidine compound demonstrated by Dr. Harald Loe (professor at the Royal Dental College in Aarhus, Denmark) in 1960s could prevent the build-up of dental plaque. The reason for chlorhexidine effectiveness is that it strongly adheres to surfaces in the mouth and thus remains present in effective concentrations for many hours.

Since then commercial interest in mouthwashes has been intense and several products claim effectiveness in reducing the build-up in dental plaque and the associated severity of gingivitis (inflammation of the gums). The mechanism of action is believed to be membrane disruption, and not ATPase inactivation as previously thought.

Products containing Chlorhexidine Gluconate in high concentrations must be kept away from eyes (corneal ulcers) and the inner ear (deafness), although it is used in minute concentrations in some contact lens solutions. In some countries it is available by prescription only. It is often used as an active ingredient in mouthwash designed to kill dental plaque and other oral bacteria. Chlorhexidine gluconate can thus be used to improve bad breath or helitosis. Chlorhexidine Gluconate-based products are usually utilized to combat or prevent gum diseases such as gingivitis.

Chlorhexidine is deactivated by anionic compounds, including the anionic surfactants commonly used as detergents in toothpastes and mouthwashes. For this reason, chlorhexidine mouth rinses should be used at least 30 minutes after other dental products. For best effectiveness, food, drink, smoking, and mouth rinses should be avoided for at least one hour after use. Chlorhexidine Gluconate is also used in non-dental applications. It is used for general skin cleansing, a surgical scrub and a pre-operative skin preparation. Due to other chemicals listed as inactive ingredients, the cleanser solution is not suitable for use as mouthwash.

Cetylpyridinium chloride is a cationic quaternary ammonium compound in some types of mouthwash such as nasal sprays. It is an antiseptic that kills bacteria and other microorganisms. It has been shown to be effective in preventing dental plaque and reducing gingivitis. It has also been used as an ingredient in certain pesticides.

COSMETIC DENTISTRY

Today there is less use of amalgam (silver) for fillings and more tooth colored material (composite) being used.

A cosmetic restoration is defined as "reproducing a tooth for esthetic reasons, without the use of metal".

A cosmetic restoration may be any of the following:

- A porcelain veneer that is bonded to the front surface of your teeth.

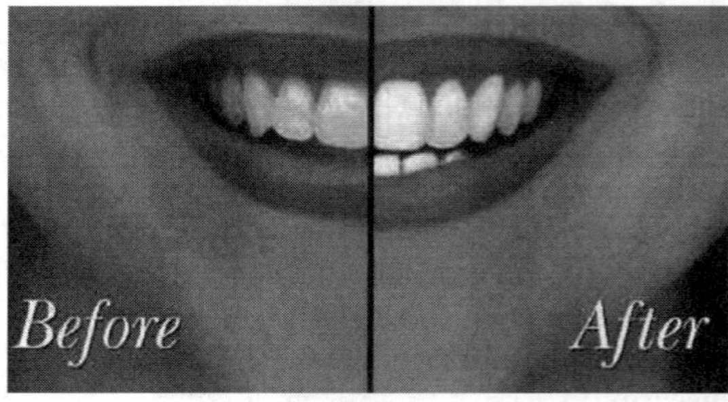

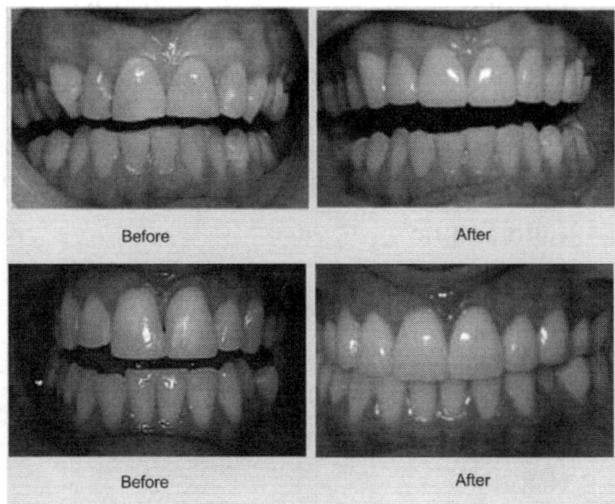

- An all ceramic cap (full crown) that has no metal.
- An inlay (a tooth-colored filling)
- An onlay (a tooth-colored filling)
- Crowns and bridges can sometimes alter lip support or cheek contour.

Other benefits of cosmetic dentistry include:

- Brighter teeth (Teeth Whitening Gel)
- Better smile
- Greater self-esteem
- Higher employment rate
- More successful personal relationships
- Materials that are presently being used
- Porcelain
- Composite
- Polymer glass.

EVALUATIONS

Adequate quality control of all raw materials is imperative. There is no more positive invitation to product variation than lack of control of the component ingredients.

Formula changes of seemingly minor consequences may have unexpected and unfortunate results on a well balanced composition; further more, these effects may not appear until the product has been stored for some time.

Accelerated stability tests at elevated temperatures are generally valuable but cannot always predict behavior under more normal conditions.

Most flavors and therapeutic ingredients are usually of a complex, organic nature and subject to chemical change.

Antiseptic property: This property determined *in vitro* and *in vivo*, to determine effectiveness against number of microorganisms. By this optimum time required to keep mouthwash in mouth can also be determined.

Stability effect: Activity of antiseptic and other ingredients decreased with time. This can be determined by normal stability study or accelerated stability study according to ICH guidelines.

Deodorizing activity: This can be analyzed by chemical analysis, and surface tension effects, level of odour determined by gas chromatography or by professional odour testers.

Tests for therapeutic activity: Tests like effects on dental caries, effects on oral soft tissue problems, cleaning and astringent effects, etc.

Dental Health

TIPS TO KEEP YOUR TEETH WHITE

Here are some tips to help maintain your pearly whites.

- Avoid the consumption of or exposure to products that stain your teeth. If you do choose to consume beverages that stain, consider using a straw so that the liquid bypasses your front teeth.
- Brush or rinse immediately after consuming stain-causing beverages or foods.
- Follow good oral hygiene practices. Brush your teeth at least twice daily and floss at least once daily to remove plaque. Use whitening toothpaste (once or twice a week only) to remove surface stains and prevent yellowing. Use regular toothpaste the rest of the time.

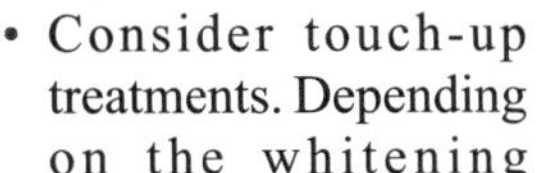

- Consider touch-up treatments. Depending on the whitening method used, you may need a touch-up every 6 months or after a year or two. If you smoke or drink lots of stain-causing beverages, you may need a touch up more often.

Popular Toothpaste Brands Include

- Aim toothpaste
- Aquafresh
- Close-Up
- Colgate
- Crest, Crest Plus Scope® and Crest Sensitivity®
- Dabur
- Fresh Breath
- Gleem
- Glister (Amway)
- Himalaya Toothpaste
- Mentadent
- Oral-B
- Pepsodent
- Sensodyne (for sensitive teeth)
- Signal
- The Natural Dentist

Lozenges, throat sprays and anti-snore throat sprays

- Breathe Right®
- Rite Aid CVS,
- Walgreens Oasis®
- Aquawhite Teeth Whitening Gel.

Moisturizing Mouth Spray and Breath Spray

- Ayr No-Drip Sinus®
- BreathRX®
- SinoFresh®

Popular Mouthwash and Mouth Rinse Brands

- Act
- Anticavity Mouth rinse
- BetaCell®,
- BreathRX®, and
- Chloraseptic
- Clear Choice
- Close-Up Anti-Plaque
- Corsodyl
- Crest Pro-Health®,
- Dr. Fresh®,
- Fluorigard
- Hexidine
- Lavoris, Scope
- Listerine (Parke Devis)
- Listermint with Fluoride
- Oasis®,
- Peridex
- Plax
- Reach ACT®,
- Rembrandt Mouth Refreshing Rinse
- Scope®,
- Signal
- Swish®,
- Viadent®,
- Clohex (Eroup)
- A.M.-P.M. (Warren)
- Rexidin (Warren)
- Colget Total plax.

Popular Gargles

- Pidin (Elazo well come)
- Ponvidine gugle.

Part

5

Perfumes and Soaps

24. Prefumes
25. Soaps

24 Perfumes

> *The best perfume in the world is our good deeds.*
> — Abraham Lincoln

Fragrance has long played an important role in personal grooming. In ancient times, when daily bathing was not common, dabs of fragrance here and there helped a person smell more pleasing. Nowadays, we still use fragrances in the form of perfumes, deodorants, lotions, hair products, soaps and cosmetics to please, attract and entice. Perfume is a mixture of fragrant essential oils and aroma compounds, fixatives, and solvents used to give the human body, animals, objects, and living spaces a pleasant smell. The word perfume used today derives from the Latin "per fumum", meaning through smoke. Perfumery, or the art of making perfumes, began in ancient Mesopotamia and Egypt but was developed and further refined by the Romans and Persians. Although perfume and perfumery also existed in East Asia, much of its fragrances are incense based.

The world's first recorded chemist is considered to be a person named Tapputi, a perfume maker who was mentioned in a cuneiform tablet from the second millennium BC in Mesopotamia.

The U.S. Code of Federal Regulations describes a "natural flavorant" as:

"The essential oil, oleoresin, essence or extractive, protein hydrolysate, distillate, or any product of roasting, heating or enzymolysis, which contains the flavoring constituents derived from a spice, fruit or fruit juice, vegetable or vegetable juice, edible yeast, herb, bark, bud, root, leaf or any other edible portions of a plant, meat, seafood, poultry, eggs, dairy products, or fermentation products thereof, whose primary function in food is flavoring rather than nutritional."

UK Food Law Defines A Natural Flavor As

" A flavouring substance (or flavouring substances) which is (or are) obtained, by physical, enzymatic or microbiological processes, from material of vegetable or animal origin where material is either raw or has been subjected to a process normally used in preparing food for human consumption and to no process other than one normally so used. "

Perfume consists mostly of chemicals called volatile organic compounds, or VOCs. An aroma compound, also known as odorant, aroma, fragrance or flavor, is a chemical compound that has a smell or odor. A chemical compound has a smell or odor when two conditions are met: the compound needs to be volatile, so it can be transported to the olfactory system in the upper part of the nose, and it needs to be in a sufficiently high concentration to be able to interact with one or more of the olfactory receptors.

Aroma compounds can be found in food, wine, spices, perfumes, fragrance oils, and essential oils. For example, many form biochemical during ripening of fruits and other crops. In wines, most form as byproducts of fermentation. Odorants can also be added to a dangerous odorless substance, like natural gas or hydrogen, as a warning. As well many of the aroma compounds plays a significant role in the production of flavorants, which are used in the food service industry to flavor, improve and increase the appeal of their products.

Advantages

- Enjoy the wafting fragrance of a well-perfumed person passing by the side.
- Smell more pleasing.

Disadvantages of Perfumes

- Some VOCs, such as formaldehyde, ethanol and d-limonene, cause eye, nose and throat irritation, difficulty in breathing, allergy symptoms and headaches. Formaldehyde is considered a probable carcinogen by the U.S. Environmental Protection Agency (EPA). Perfume can be a trigger for asthmatics and migraine and sinus headache sufferers.
- Chemicals may irritate others, especially in tight spaces, like an elevator.
- Contact dermatitis, an allergic reaction in the skin can occur.
- Diethyl phthalate (DEP), an irritant and suspected hormone disrupter that is absorbed through the skin and can accumulate in human fat tissue.

There are three principal types of flavorings used in foods, under definitions agreed in the E.U. and Australia:

- *Natural flavouring substances:* Flavouring substances obtained from plant or animal raw materials, by physical, microbiological or enzymatic processes. They can be either used in their natural state or processed for human consumption, but cannot contain any nature-identical or artificial flavouring substances.
- *Nature-identical flavouring substances:* Flavouring substances that are obtained by synthesis or isolated through chemical processes, which are chemically identical to flavouring substances naturally present in products intended for human consumption. They cannot contain any artificial flavouring substances.
- *Artificial flavouring substances:* Flavouring substances not identified in a natural product intended for human consumption, whether or not the product is processed.

Natural perfumes are complex creations based on the artistic vision of the perfumer or nose. Natural perfumers' materials include:

- Absolutes,
- Concretes,
- Essential oils,
- Macerations,
- Pressed oils,
- Resins and
- Tinctures.

The perfumes can take anything from two months to several years or more to complete.

To the perfumer, apart from stay within security guidelines, the health payback of the perfume is of no real significance.

The greatest importance of perfumes are of:

- The artist's vision,
- The concept or theme,
- The quality of the materials used, and
- The structure and composition of the perfume.

These factors are what makes natural perfumery an art. Modern day natural perfumers are learning from the past while at the same time taking a giant leap into the future. If we look back at pre-1900's we will see that the perfumers then did not have the benefit of the same enormous choice of usual aromatics that are on furnish today. These days, natural perfumers have an immense choice of essences to use in their works, such as

- Ambrette seed oil.
- Coco absolute.

- Coffee absolute
- Pink lotus absolute and so forth.

Aromatherapy blends are based first on the health giving benefits of the chosen oils and second on how well the oils complement each other. In aromatherapy blends, more than three, and usually not more than six oils are used. This is a sensual aromatherapy blend which will smell pleasing and will help to uplift the emotions and improve the sense of well-being of the user. It is not a perfume.

Aroma compounds are classified on the basis of functional groups as:

1. Alcohols
   - Menthol (peppermint).
   - Benzyl alcohol (oxidises to benzaldehyde, almond).
   - Furaneol (strawberry).
   - Ethyl maltol (sugary, cooked fruit).
   - cis-3-Hexen-1-ol (fresh cut grass).
   - 1-Hexanol (herbaceous, woody).
2. Aldehydes
   - Acetaldehyde (pungent).
   - Benzaldehyde(marzipan, almond).
   - Cinnamaldehyde (cinnamon).
   - cis-3-Hexenal (green tomatoes).
   - Citral (lemongrass, lemon oil).
   - Furfural (burnt oats).
   - Hexanal (green, grassy).
   - Neral (citrus, lemongrass).
   - Vanillin (vanilla).
3. Amines
   - Trimethylamine (fish).
   - Indole (jasmine flowery, feces).
   - Skatole (bad breath, feces).
   - Cadaverine (rotting flesh).
   - Putrescine (rotting flesh).
   - Pyridine (very unpleasant).
   - Alkylpyrazines.
   - Methoxypyrazines.
   - Substituted pyrazines: 2-ethoxy-3 isopropylpyrazine, 2-methoxy-3-sec-butylpyrazine, 2-methoxy-3-methylpyrazine (toasted seeds of fenugreek, cumin, and coriander).
4. Esters
   - Isoamyl acetate (banana)
   - Ethyl acetate (fruity, solvent)
   - Hexyl acetate (apple, floral, fruity)
   - Methyl butanoate (methyl butyrate) (apple, fruity)
   - Pentyl pentanoate (apple, pineapple)
   - Fructone (fruity, apple-like)
   - Ethyl butanoate (fruity) (ethyl butyrate)
   - Strawberry aldehyde (strawberry)
   - Ethyl decanoate (as ethyl caproate)
   - Ethyl hexanoate (ethyl caproate)
   - Ethyl octanoate (ethyl caprylate)
   - Methyl salicylate (oil of wintergreen)
   - Pentyl butanoate (pear, apricot)
   - Sotolon (maple syrup, curry, fenugreek).
5. Ethers
   - Anethole (liquorice, anise seed, ouzo, fennel)
   - Anisole (anise seed)
   - Eugenol (clove oil)
   - 2,4,6-Trichloroanisole (corktaint).
6. Ketones
   - Dihydrojasmone (fruity woody floral)
   - Oct-1-en-3-one (blood, metallic, mushroom-like)
   - 2-Acetyl-1-pyrroline (fresh bread, jasmine rice)
   - 6-Acetyl-2,3,4,5-tetrahydropyridine (fresh bread, tortillas, pop corn).
7. Lactones
   - gamma-Decalactone intense peach flavor
   - gamma-Nonalactone coconut odor, popular in suntan lotions
   - delta-Octalactone creamy note
   - Jasmine lactone powerful fatty fruity peach and apricot
   - Massoia lactone powerful creamy coconut.
8. Terpenes
   - Camphor (Cinnamomum camphora)
   - Citronellol (rose)
   - Linalool (floral, citrus, coriander)
   - Nerol (sweet rose)
   - Nerolidol (wood, fresh bark)
   - alpha-Terpineol (lilac)
   - Thujone (juniper, common sage, Nootka cypress, and wormwood)
   - Thymol (Thyme-like).
9. Thiols

- Ethanethiol, formerly called Ethyl mercaptan (Durian or leek, added to natural gas)

- Grapefruit mercaptan (grapefruit)
- Methanethiol, formerly called methyl mercaptan (added to natural gas).

10. Other compounds

- Methylphosphine and dimethylphosphine (garlic-metallic, two of the most potent odorants known)
- Nerolin (orange flowers)
- Tetrahydrothiophene (added to natural gas).

Aroma Compound Receptors

Animals which are capable of smell detect aroma compounds with olfactory receptors. Olfactory receptors are cell membrane receptors on the surface of sensory neurons in the olfactory system which detects air-borne aroma compounds. In mammals, olfactory receptors are expressed on the surface of the olfactory epithelium in the nasal cavity.

Flavor is the sensory impression of a food or other substance, and is determined mainly by the chemical senses of taste and smell. The "trigeminal senses", which detect chemical irritants in the mouth and throat, may also occasionally determine flavor. The flavor of the food, as such, can be altered with natural or artificial flavorants, which affect these senses. Flavorant is defined as a substance that gives another substance flavor, altering the characteristics of the solute, causing it to become sweet, sour, tangy, etc. Of the three chemical senses, smell is the main determinant of a food item's flavor. While the taste of food is limited to sweet, sour, bitter, salty, and savory.

Regulations on Natural Flavoring

The European Union's guidelines for natural flavorants are slightly different. Certain artificial flavorants are given an **E** number, which may be included on food labels.

Method of Preparation of Perfumes

The most popular method for extraction is steam distillation, but as technological advances are being made more efficient and economical method are being developed. Essential oils can be extracted using a variety of methods.

- *Steam distillation:* The plant material is placed into a still where pressurized steam passes through the plant material, which ruptures the oil gland and releases the oil. The essential oil vapor and the steam then passes out of the still into water-cooled pipe where the vapor are condensed back to liquids. At this stage, the volatile oils are separated from the water content. This process generates two products: the volatile oil, which contains oil soluble molecule, and a hydrosol, which contains water soluble molecule, e.g. Rose water.
- *Maceration:* The plant material is soaked in vegetable oil, heated and strained. This process actually creates more of infused oil, rather than an "essential oil." These produced oils can be used for massage.
- *Expression:* This method is also known as cold pressing; this process is used to extract volatile oils from citrus fruits such as lemon, orange, grape fruit, etc. The rinds are separated from the fruit, are ground, and are then pressed. The result is a watery mixture of essential oil and liquid which will separate.
- *Extraction:* In this process, a hydrocarbon solvent is added to the plant materials, which help in dissolving the volatile oil. The solution is filtered and concentrated by distillation. A substance containing resin (resin oil) or a combination of wax and essential oil (concentrate) remains. Pure alcohol is used to extract the volatile oil from the concentrate. After evaporation of alcohol, the oil is left behind.
- *Enflurage:* The enflurage method is called pomade. This method for producing essential oils is not used much any more, as it is an expensive and time-consuming process. Pomade was obtained by the use of layers of fat onto which the petals of plants such as Tuberose and Jasmine were laid out and left to dry. The fat collected the essential oils, which were later extracted. This process has now been replaced by solvent extraction.

Chemical Odor

- Allyl hexanoate: Pineapple
- Benzaldehyde: Bitter almond

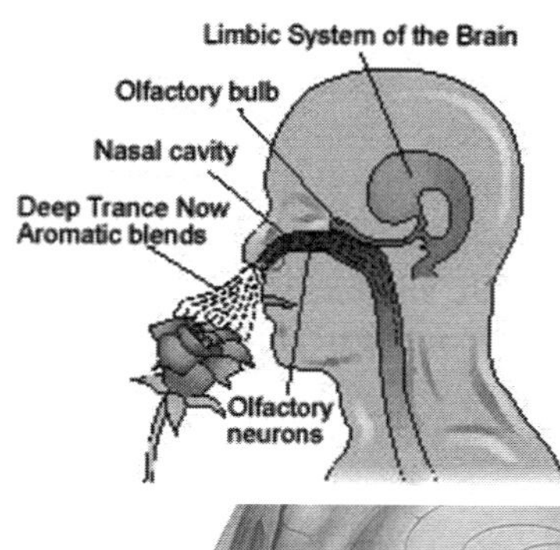

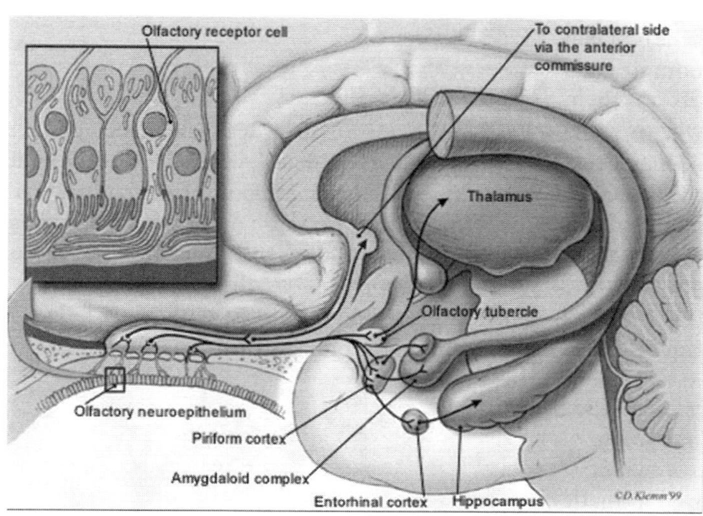

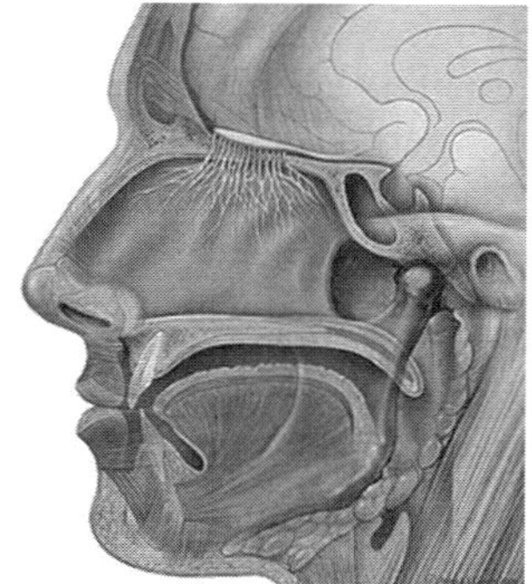

- Cinnamic aldehyde: Cinnamon
- Diacetyl: Buttery
- Ethyl- (E, Z) -2,4-decadienoate: Pear
- Ethyl maltol Sugar: Cotton candy
- Ethyl propionate: Fruity
- Isoamyl acetate: Banana
- Limonene: Orange
- Methyl salicylate: Wintergreen.

Flavors from food products are usually the result of a combination of natural flavors, which set up the basic smell profile of a food product while artificial flavors modify the smell to accent it.

Determination

Few standards are available or being prepared for sensory analysis of flavors. In chemical analysis of flavors, solid phase extraction (SPE), solid phase microextraction (SPME), and headspace gas chromatography [GC] are applied to extract and separate the flavor compounds in the sample. The determination is typically done by various mass spectrometric techniques [MS].

Concentration

Perfume types reflect the concentration of aromatic compounds in a solvent, which in fine fragrance is typically ethanol or a mix of water and ethanol. Various sources differ considerably in the definitions of perfume types. The concentration by percent/volume of perfume oil is as follows:

- *Perfume extract (Extrait):* 15–40% (IFRA: typical 20%) aromatic compounds
- *Eau de Perfume (EdP):* 10–20% (typical ~15%) aromatic compounds. Sometimes listed as "eau de perfume".
- *Eau de Toilette (EdT):* 5–15% (typical ~10%) aromatic compounds

- *Eau de Cologne (EdC):* Chypre citrus type perfumes with 3–8% (typical ~5%) aromatic compounds
- *Splash and After shave:* 1–3% aromatic compounds.

Perfume oil is necessarily diluted with a solvent because undiluted oils (natural or synthetic) contain high concentrations of chemical components (natural or otherwise) that will likely result in allergic reactions and possibly injury when applied directly to skin or clothing. As well, the scents in pure perfume oils are far too concentrated to smell pleasant. By far the most common solvent for perfume oil dilution is ethanol or a mixture of ethanol and water. Perfume oil can also be diluted by means of neutral-smelling liquid oils such as fractionated coconut oil, or liquid waxes such as jojoba oil.

The intensity and longevity of perfume bases on the concentration, intensity and longevity of the used aromatic compounds (natural essential oils / perfume oils): As the percentage of aromatic compounds increases, so does the intensity and longevity of the scent created. Different perfumeries or perfume houses assign different amounts of oils to each of their perfumes. Therefore, although the oil concentration of a perfume in Eau de Parfum (EdP) dilution will necessarily be higher than the same perfume in Eau de Toilette (EdT) from within the same range, the actual amounts can vary between perfume houses. An EdT from one house may be stronger than an EdP from another.

Men's fragrances are rarely as EdP or perfume extracts. As well, women's fragrances are rarely sold in EdC concentrations. Although this gender specific naming trend is common for assigning fragrance concentrations, it does not directly have anything to do with whether a fragrance was intended for men or women.

Furthermore, some fragrances with the same product name but having a different concentration name may not only differ in their dilutions, but actually use different perfume oil mixtures altogether. For instance, in order to make the EdT version of a fragrance brighter and fresher than its EdP, the EdT oil may be "tweaked" to contain slightly more top notes or fewer base notes.

In some cases, words such as "extrême", "intense" or "concentrée", that might indicate aromatic concentration are sometimes completely different fragrances that relates only because of a similar perfume accord. An example of this would be Chanel's Pour Monsieur and Pour Monsieur Concentrée.

Fragrance Notes

Perfume is described in a musical metaphor as having three sets of 'notes', making the harmonious scent accord. The notes unfold over time, with the immediate impression of the top note leading to the deeper middle notes, and the base notes gradually appearing as the final stage. These notes are created carefully with knowledge of the evaporation process of the perfume.

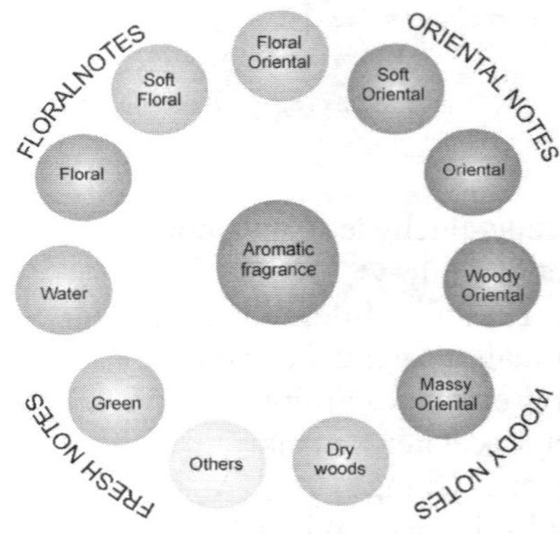

Top notes: The scents that are perceived immediately on application of a perfume. Top notes consist of small, light molecules that evaporate quickly. They form a person's initial impression of a perfume and thus are very important in the selling of a perfume. Also called the head notes.

Middle notes: The scent of a perfume that emerges just prior to when the top notes dissipate. The middle note compounds form the "heart" or main body of a perfume and act to mask the often unpleasant initial impression of base notes, which become more pleasant with time. They are also called the "heart notes".

Base notes: The scent of a perfume that appears close to the departure of the middle notes. The base and middle notes together are the main theme of a perfume. Base notes bring depth and solidity to a perfume. Compounds of this class of scents are typically rich and "deep" and are usually not perceived until 30 minutes after application.

The scents in the top and middle notes are influenced by the base notes; as well the scents of the base notes will be altered by the type of fragrance materials used as middle notes. Manufactures of perfumes usually publish perfume notes and typically they present it as fragrance pyramid, with the components listed in imaginative and abstract terms.

Olfactive Families

Grouping perfumes, like any taxonomy, can never be a completely objective or final process. Many fragrances contain aspects of different families. Even a perfume designated as "single flower", however subtle, will have undertones of other aromatics. "True" unitary scents can rarely be found in perfumes as it requires the perfume to exist only as a singular aromatic material.

Classification by olfactive family is a starting point for a description of a perfume, but it cannot by itself denote the specific characteristic of that perfume.

Traditional

The traditional classification which emerged around 1900 comprised the following categories:

i. *Single floral:* Fragrances that are dominated by a scent from one particular flower; in French called a soliflore. (e.g. Serge Lutens' Sa Majeste La Rose, which is dominated by rose.)
ii. *Floral bouquet:* Containing the combination of several flowers in a scent.

- *Amber:* A large fragrance class featuring the sweet slightly animalic scents of ambergris or labdanum, often combined with vanilla, flowers and woods, can be enhanced by camphorous oils and incense resins, which bring to mind Victorian era imagery of the Middle East and Far East.
- *Wood:* Fragrances that are dominated by woody scents, typically of agarwood, sandalwood and cedar. Patchouli, with its camphoraceous smell, is commonly found in these perfumes.
- *Leather:* A family of fragrances which features the scents of honey, tobacco, wood and wood tars in its middle or base notes and a scent that alludes to leather.
- *Chypre:* Meaning Cyprus in French, this includes fragrances built on a similar accord consisting of bergamot, oakmoss, patchouli, and labdanum. This family of fragrances is named after a perfume by François Coty. A notable example is Mitsouko (a popular name for girls in Japanese) by Guerlain.
- *Fougère:* Meaning Fern in French, built on a base of lavender, coumarin and oakmoss. Houbigant's Fougère Royale pioneered the use of this base. Many men's fragrances belong to this family of fragrances, which is characterized by its sharp herbaceous and woody scent.

Modern

Since 1945, due to great advances in the technology of perfume creation (i.e. compound design and synthesis) as well as the natural development of styles and tastes; new categories have emerged to describe modern scents:

i. *Bright floral:* Combining the traditional single floral and floral bouquet categories.
ii. *Green:* A lighter and more modern interpretation of the Chypre type.
iii. *Oceanic/Ozone:* The newest category in perfume history, appearing in 1991 with Christian Dior's Dune. A very clean, modern smell leading to many of the modern androgynous perfumes.
iv. *Citrus or Fruity:* An old fragrance family that until recently consisted mainly of "freshening" eau de colognes due to the low tenacity of citrus scents. Development of newer fragrance compounds has allowed for the creation of primarily citrus fragrances. Gourmand: scents with "edible" or "dessert"-like qualities. These often contain notes like vanilla and tonka bean, as well as synthetic components designed to resemble food flavors. An example is Thierry Mugler's Angel.

The five standard families consist of

- Floral,
- Oriental,
- Woody,

- Fougère, and
- Fresh,

With the former four families being more "classic" while the latter consisting of newer bright and clean smelling citrus and oceanic fragrances that have arrived due to improvements in fragrance technology. With the exception of the Fougère family, each of the families are in turn divided into sub-groups and arranged around a wheel.

AROMATICS SOURCES

I. Plant sources

Plants have long been used in perfumery as a source of essential oils and aroma compounds. These aromatics are usually secondary metabolites produced by plants as protection against herbivores, infections, as well as to attract pollinators. Plants are by far the largest source of fragrant compounds used in perfumery. The sources of these compounds may be derived from various parts of a plant. A plant can offer more than one source of aromatics, for instance the aerial portions and seeds of coriander have remarkably different odors from each other. Orange leaves, blossoms, and fruit zest are the respective sources of petit grain, neroli, and orange oils.

- *Bark:* Commonly used barks include cinnamon and cascarilla.
- *Flowers and blossoms:* Undoubtedly the largest source of aromatics. Includes the flowers of several species of rose and jasmine, as well as osmanthus, mimosa, tuberose, as well as the blossoms of citrus, unopened flower buds of the clove, vanilla orchid and ylang-ylang trees.
- *Fruits:* Notable exceptions include litsea cubeba, vanilla, and juniper berry. The most commonly used fruits yield their aromatics from the rind; they include citrus such as oranges, lemons, and limes, although grapefruit rind is still used for aromatics.
- *Leaves and twigs:* Commonly used for perfumery are citrus leaves, sage, lavender leaf, rosemary, violets and patchouli.
- *Resins:* Commonly used resins in perfumery include labdanum, frankincense/olibanum, myrrh, Peru balsam, gum benzoin. Pine and fir resins are a particularly valued source of terpenes used in the organic synthesis of many other synthetic or naturally occurring aromatic compounds. Some of what is called amber and copal in perfumery today is the resinous secretion of fossil conifers.
- *Roots, rhizomes and bulbs:* Commonly used terrestrial portions in perfumery include iris rhizomes, vetiver roots, and various rhizomes of the ginger family.
- *Seeds:* Commonly used seeds include cocoa, anise, cardamom, nutmeg, coriander, caraway, mace, and tonka bean.
- *Woods:* Highly important in providing the base notes to a perfume, wood oils and distillates are indispensable in perfumery. Commonly used woods include sandalwood, rosewood, agarwood, birch, cedar, juniper, and pine. These are used in the form of macerations or dry-distilled (rectified) forms.

II. Animal Sources

- *Ambergris:* Ambergris [by the Sperm Whale] is commonly referred to as "amber" in perfumery.
- *Castoreum:* Obtained from the odorous sacs of the North American beaver.
- *Civet:* Also called Civet Musk, this is obtained from the odorous sacs of the civets, animals in the family Viverridae, related to the Mongoose.
- *Honeycomb:* From the honeycomb of the Honeybee. Both beeswax and honey can be solvent extracted to produce beeswax absolute. Beeswax is extracted with ethanol and the ethanol evaporated to produce beeswax absolute.
- *Musk:* Originally derived from the musk sacs from the Asian musk deer, it has now been replaced by the use of synthetic musks which usually are called "white musk".

III. Other Natural Sources

- *Lichens:* Commonly used lichens include oakmoss and treemoss thalli.

- *"Seaweed":* Distillates are sometimes used as essential oil in perfumes. An example of commonly used seaweed is bladder wrack [*Fucus vesiculosus*].

IV. Synthetic Sources

Many modern perfumes contain synthetic odorants. Synthetics can provide fragrances which are not found in nature. For instance, Synthetic aromatics are often used as an alternate source of compounds that are not easily obtained from natural sources. For example,

- Linalool and coumarin are both naturally occurring compounds that can be inexpensively synthesized from terpenes.
- Calone, a compound of synthetic origin, imparts a fresh ozonous metallic marine scent that is widely used in contemporary perfumes.
- Orchid scents (typically salicylates) are usually not obtained directly from the plant itself but are instead synthetically created to match the fragrant compounds found in various orchids.

Technique

Paper blotters are commonly used by perfumers to sample and smell perfumes and odorants.

Basic Framework

Perfume oils usually contain tens to hundreds of ingredients and these are typically organized in a perfume for the specific role they will play. These ingredients can be roughly grouped into four groups:

- *Primary scents:* Can consist of one or a few main ingredients for a certain concept, such as "rose". Alternatively, multiple ingredients can be used together to create an "abstract" primary scent that does not bear a resemblance to a natural ingredient. For instance, jasmine and rose scents are commonly blends for abstract floral fragrances. Cola flavourant is a good example of an abstract primary scent.
- *Modifiers:* These ingredients alter the primary scent to give the perfume a certain desired character: for instance, fruit esters may be included in a floral primary to create a fruity floral; calone and citrus scents can be added to create a "fresher" floral. The cherry scent in cherry cola can be considered a modifier.
- *Blenders:* A large group of ingredients that smooth out the transitions of a perfume between different "layers" or bases. Common blending ingredients include linalool and hydroxycitronellal.
- *Fixatives:* Used to support the primary scent by bolstering it. Many resins and wood scents, and amber bases are used as fixatives.

The perfume's fragrance oils are then blended with ethyl alcohol and water, aged in tanks for a minimum of two weeks and filtered through processing equipment to remove any sediment and particles before the solution can be filled into the perfume bottles.

There are different ways of using essential oils therapeutically.

- *Inhalation:* Steam inhalation is an excellent way to treat sinus problems, coughs, colds, sore throats, and nasal allergies as hay fever; and for cleansing the skin. There are different ways of inhalation.

Straight from the bottle treats headache, memory problems, nausea, etc.

Oil burners kill airborne bacteria thereby prevents colds spreading to others, insomnia, stress, etc.)

Drops on a tissue or cotton ball treat colds, coughs, migraine, etc.

Drop in a bowl of hot water with 3–4 drop of essential oil treats respiratory infection, colds catarrh (runny nose) etc.

- *Bath:* Bathing with pure essential oils is one of life's greatest pleasures. The warmth of a bath not only relaxes a person, it also enables the skin to absorb the essential oil better. They are of different types
  - Hot or warm bath treats colds, muscle cramp, stiffness etc. (use 10 drops maximum).
  - Foot bath treats athlete's foot, aching feet, arthritis, etc. (use 5 drops maximum.
  - Shallow bath treats thrush,-piles, etc.
- *Massage:* The most pleasant, relaxing and therapeutic way of using essential oils is through aromatherapy massage. Actually, friction produced by massage helps the penetration of oils through skin. Oil used in massage includes almond oil, sesame, and olive oil.

Aromatherapy: Today, where stress and depression are a major cause of a health imbalance, then a full aromatherapy body massage is also the best complementary therapy treatment available. It is also an excellent prophylactic treatment to ensure continuation of good health.

Massage therapy has been shown to be highly beneficial. It can affect the autonomic nervous system, and calm the "fight or flight" response, reducing the level of harmful stress hormones in the body. It is an excellent way of reducing stress. Massage stimulates the blood circulation, increasing the supply of nutrients and oxygen to cells. Massage stimulates lymphatic flow, improving tissue drainage and improving the immune system. During the massage, much of the volatile oils will be inhaled and some absorption is likely through the mucous of the nose and mouth.

Different ways of massage therapy includes:

Massage diluted oil into affected area treats varicose veins, strain, constipation, muscle aches, etc.

Massage diluted oils all over body treats stress, insomnia, anxiety, etc.

- *Internal use:* Some essential oils such as oil of peppermint and cinnamon can be used to make teas or mouthwashes, or mixed with a glass of honey and water.
- *Directly to the skin:* There are only four essential oils, lavender, sandalwood, tea tree, chamomile that may safely be applied directly onto the skin. The oil should be applied only onto an affected area, i.e. a cut or burn.
- *Skin/Hair tonics:* Use this when oil is not suitable, for e.g. on an oily scalp, or to dry out a cut. Use few drops of essential oil into a teaspoon of isopropyl alcohol.

Health and Environmental Issues

Perfume ingredients [For instance, acetophenone, ethyl acetate and acetone while present in many perfumes are also known or potential respiratory allergens], regardless of natural or synthetic origins, may all cause health or environmental problems like asthma, carcinogenicity, pollution when used or abused in substantial quantities. Although the areas are under active research, much remains to be learned about the effects of fragrance on human health and the environment. The demands for aromatic materials like sandalwood, agarwood, musk has led to the endangerment of these species as well as illegal trafficking and harvesting.

Safety Regulation

The perfume industry in the US is not directly regulated by the FDA, instead the FDA controls the safety of perfumes through their ingredients and require that they be tested to the extent that they are Generally recognized as safe (GRAS). Due to the need for protection of trade secrets, companies rarely give the full listing of ingredients regardless of their effects on health. In Europe, the mandatory listing of any of a number of chemicals thought to be hazardous has just begun. Many old perfumes of chypres and fougeres classes, which require the use of oakmoss extract are being reformulated because of these new regulations.

Storage of Perfume

Fragrance compounds in perfumes will degrade or break down if improperly stored in the presence of heat, light and oxygen. Proper preservation of perfumes involves keeping them away from sources of heat and storing them where they will not be exposed to light. An open bottle will keep its aroma intact for several years, as long as it is well stored. However the presence of oxygen in the head space of the bottle and environmental factors will in the long run alter the smell of the fragrance.

Perfumes are best preserved when kept in light-tight aluminum bottles or in their original packaging when not in use, and refrigerated at a relatively low temperatures between 3–7°C. Although it is difficult to completely remove oxygen from the headspace of a stored flask of fragrance, opting for spray dispensers instead of rollers and "open" bottles will minimize oxygen exposure. Sprays also have the advantage of isolating fragrance inside a bottle and preventing it from mixing with dust, skin, and detritus, which would degrade and alter the quality of a perfume.

Lists of Perfumes

Famous perfumes classified by year of creation Bottles of some notable commercial perfumes: (clockwise from top left) Bois de Violette, Serge Lutens, 1992; Angel, Thierry Mugler, 1994; Shalimar, Guerlain, 1925; Beyond Paradise, Estée Lauder, 2003; No. 5, Chanel, 1921 (Pre-1950 bottle); Cabochard, Parfums Grès, 1959 (original bottle); Bellodgia, Caron, 1927; Arpège, Lanvin, 1927 (original bottle); Nombre Noir, Shiseido, 1981; Mitsouko, Guerlain, 1919; Pour Un Homme, Caron, 1934.

25 Soaps

Dirt cannot wash dirt; hate cannot cure hate.
— Swami Vivekananda

Soap is perhaps the first manufactured substance with which we come into contact in our lives and it remains a daily necessity thereafter. Soap is the simplest and regularly used cosmetic all over the world. It's a cosmetic which is used by peoples from all walks of life. Soap has become a necessary commodity in day-to-day life. Soaps are specificially excluded from cosmetic in food, drug and cosmetics Act (FDCA) and no cosmetic drug regulations are applicable to soaps.

The earliest recorded evidence of the production of soap-like materials dates back to around 2800 BC in Ancient Babylon. A formula for soap consisting of water, alkali and cassia oil was written on a Babylonian clay tablet around 2200 BC. The Romans used a type of clay found near Rome called **"sapo"** from which the word soap is derived. True soaps made from vegetable oils (such as olive oil), aromatic oils (such as thyme oil) and lye (sodium hydroxide) was first produced in the medieval Islamic world. Castile soap was later produced in Europe from the 16th century. Manufactured bar soaps first became available in the late nineteenth century, and advertising campaigns in Europe and the United States helped to increase popular awareness of the relationship between cleanliness and health.

Saponins are widely distributed in the plant kingdom and such plants as *Saponaria officinalis*, *Quillaia saponaria*, *Gypsophila spp*. and *Sapindus spp*. contain useful amounts which might be used for cleaning purposes. Cleansing and conditioning the skin and hair are two important aspects of maintaining the outward health and appearance of the body. An understanding of the mechanism of action and chemistry of products designed to improve the ability of the skin and hair to function is medically important. Soap itself has found numerous applications in pharmacy, such as pill making, dentifrices, lotions and liniments, enemas, plasters, suppositories and poultices, in addition to veterinary applications.

Probably the first soap was white hard soap: Jabon de Castilla, or **Castile soap**, also known as Sapo hispaniensis or Sapo castilliensis. Originally an important product for the Castile region of central Spain, Castile eventually became the generic name for hard, white, olive oil soaps. In Britain early production of soap was usually based on rendered animal fat, such as tallow from beef or mutton. It is recorded that a type of black soft soap was known as "**Bristol soap**". Another harder type soap is, "Bristol grey soap". Until the Industrial Revolution, soap-making was done on a small scale and the product was rough. Andrew Pears started making a high-quality, transparent soap in 1789 in London. With his grandson, Francis Pears, they opened a factory in Isleworth in 1862.

William Gossage produced low-price good quality soap from the 1850s. Robert Spear Hudson began manufacturing a soap powder in 1837, initially by grinding the soap with a mortar and pestle. William Hesketh Lever and his brother, James, bought a small soap works in Warrington in 1885 and founded what is still one of the largest soap businesses, now called Unilever. These soap businesses were among the first to employ large scale advertising campaigns.

Most cleansing is accomplished with a product known as soap, which is obtained through the chemical reaction between a fat and an alkali, resulting in a fatty acid salt with detergent properties. Modern refinements include adjustments in the alkaline pH to decrease skin irritation and to incorporate substances that prevent precipitation of calcium fatty acid salts in hard water, known as soap scum.

Nevertheless, modern soap is basically a blend of tallow and nut oil, or the fatty acids derived from these products, in a ratio of 4:1. Increasing this ratio results in superfatted soaps designed to leave an oily film on the skin.

Bar and liquid cleansers can be divided into three basic types, as follows:

1. True soaps composed of long chain fatty acid alkali salts with a pH of 9–10;
2. Combars composed of alkaline soaps to which surface active agents have been added, also with a pH of 9–10; and
3. Syndet, or synthetic detergent, bars composed of synthetic detergents and fillers that contain less than 10% soap and that have an adjusted pH of 5.5–7.

The purpose in developing new synthetic detergents is to provide a product that is less irritating to the skin than traditional soaps are.

Common detergents in bar-type cleansers are:

- Sodium cocoate,
- Sodium tallowate,
- Sodium cocoglyceryl ether sulfonate.
- Sodium cocoyl isethionate,
- Sodium dodecyl benzene sulfonate, and
- Sodium isethionate,
- Sodium palm kernelate,
- Sodium palmitate,
- Sodium stearate,
- Triethanolamine stearate,

Detergents in liquid formulations are:

- Sodium laureth sulfate,
- Cocoamido propyl betaine,
- Lauric acid diethenolamine (lauramide DEA),
- Sodium cocoyl isethionate, and
- Disodium laureth sulfosuccinate.

The normal pH of the skin is acidic, between 4.5 and 6.5. Applying alkali soap theoretically raises the pH of the skin, making it feel dry and uncomfortable.

Special Additives

Additives to soap also are responsible for its characteristic appearance, feel, and smell.

- Lanolin and paraffin may be added to a moisturizing syndet soap to create a superfatted soap.
- Sucrose and glycerin can be added to create a transparent bar.
- Olive oil added as fat, e.g. in castile soap.
- Medicated soaps may contain benzoyl peroxide, sulfur, or resorcinol antibacterials, such as triclocarban or triclosan. Triclocarban is excellent for eradicating gram-positive organisms, but triclosan eliminates both gram-positive and gram-negative bacteria. These soaps have a pH of 9-10 and may cause skin irritation.
- Moisturizing syndet bar soaps contain sodium lauryl isethionate with a pH adjusted to 5–7 by using lactic or citric acid. These products are less irritating to the skin and are sometimes labeled beauty bars. Most bar soaps marketed by cosmetic companies are of this type.
- Titanium dioxide is added in concentrations as high as 0.3% to opacify the bar and to increase its optical whiteness.
- Pigments, such as aluminum lakes, can color the bar without producing colored foam, which is considered an undesirable characteristic.
- Foam builders, such as sodium carboxymethylcellulose and other cellulose derivatives, can make the lather feel creamy.
- Perfume in concentrations of 2% or more also can be added to ensure that the soap bar smells pleasant until it is completely used up.

Soap, consisting of sodium (soda ash) or potassium (potash) salts of fatty acids is obtained by reacting fat with lye in a process known as saponification. The fats are hydrolyzed by the base, yielding alkali salts of fatty acids (crude soap) and glycerol.

Permitted Colours

- Citrus red No. 2
- Pigment orange No. 5 (Iragalite Red-CVPB Paste)
- Rhodamine B-500
- Iragalite carmine SP powder (Pigment red-5)
- Monolight Red 4R-HV-Paste (Pigment Red-7)
- Phthalocyanine blue
- Oil Red no. 1.

Mechanism of Working of Soap

Soaps are useful for cleaning because soap molecules attach readily to both nonpolar molecules (such as grease or oil) and polar molecules (such as water). Although grease will normally adhere to skin or clothing, the soap molecules can attach to it as a "handle" and make it easier to rinse away. Applied to a soiled surface, soapy water effectively holds particles in suspension so the whole of it can be rinsed off with clean water.

$$\text{(Fatty end)}:CH_3\text{–}(CH_2)_n\text{–}COONa: \text{(water soluble end)}$$

The hydrocarbon ("fatty") portion dissolves dirt and oils, while the ionic end makes it soluble in water. Therefore, it allows water to remove normally-insoluble matter by emulsification.

Saponification is the hydrolysis of an ester under basic conditions to form an alcohol and the salt of a carboxylic acid. Saponification is commonly used to refer to the reaction of a metallic alkali (base) with a fat or oil to form soap. Saponifiable substances are those that can be converted into soap.

$$CH_2\text{–}OOC\text{–}R\text{–}CH\text{–}OOC\text{–}R\text{–}CH_2\text{-}OOC\text{–}R \text{ (fat)} + 3\ NaOH \text{ (or KOH)}$$

$$\text{Both heated} \rightarrow$$

$$CH_2\text{–}OH\text{–}CH\text{–}OH\text{–}CH_2\text{-}OH \text{ (glycerol)} + 3\ R\text{–}CO_2\text{–}Na \text{ (soap)}$$

Where $R= (CH_2)_{14}CH_3$ in the example (right)

Sodium hydroxide (NaOH) is a caustic base. If NaOH is used a hard soap is formed, whereas when potassium hydroxide (KOH) is used, a soft soap is formed. Vegetable oils and animal fats are fatty esters in the form of triglycerides. The alkali breaks the ester bond and releases the fatty acid and glycerol. If necessary, soaps may be precipitated by salting it out with saturated sodium chloride.

Manufacturing Methods

Traditional soap manufacturing methods involved the boiling of oils and fats with caustic solution in open pans, followed by the addition of salt or brine in the "salting out" process, in which the soap separated from the lye. The skilled operator would control the process by "tallowing". From the way the soap slid from a heated hand trowel he could judge whether more brine or caustic was required and when the batch was ready for "settling". By successive washing in brine the lye was separated from the soap and the glycerin recovered. The soap was dried and cut into bars for supply to the wholesale and retail trade. Soap is derived from either vegetable or animal fats. Sodium tallowate, a common ingredient in many soaps, is derived from rendered beef fat. Soap can also be made of vegetable oils, such as palm oil, and the product is typically softer. If soap is made from pure olive oil it may be called Castile soap or Marseille soap. Castile is also sometimes applied to soaps with a mix of oils, but a high percentage of olive oil. An array of oils and butters are used in the process such as olive, coconut, palm, cocoa butter, hemp oil and shea butter to provide different qualities, e.g., olive oil provides mildness in soap; coconut oil provides lots of lather; while coconut and palm oils provide hardness. Sometimes castor oil can also be used as an ebullient. Most common, though, is a combination of coconut, palm, and olive oils.

Process

There are two process of soap making, viz. cold process and hot process. In cold-process and hot-process soap-making, heat may be required for saponification.

Cold Process

Cold-process soap-making takes place at a temperature sufficiently above room temperature to ensure the liquefaction of the fat being used, and requires that the lye and fat be kept warm after mixing to ensure that the soap is completely saponified.

A cold-process soap-maker first looks up the saponification value of the fats being used on a saponification chart, which is then used to calculate the appropriate amount of lye. Excess untreated lye in the soap will result in a very high pH and can burn or irritate skin. Not enough lye and the soap are greasy. Most soap makers formulate their recipes with a 4–10% discount of lye so that all of the lye is reacted and that excess fat is left for skin conditioning benefits.

The lye is dissolved in water. Then oils are heated, or melted if they are solid at room temperature. Once both substances have cooled to approximately 100–110°F (37–43°C), and are no more than 10°F (~5.5°C) apart, they may be combined. This lye-fat mixture is stirred until "trace" (modern-day amateur soap-makers often use a stick blender to speed this process). There are varying levels of trace. Depending on how your additives will affect trace, they may be added at light trace, medium trace or heavy trace. After much stirring, the mixture turns to the consistency of a thin pudding.

Essential oils, fragrance oils, botanicals, herbs, oatmeal or other additives are added at light trace, just as the mixture starts to thicken.

The batch is then poured into molds, kept warm with towels or blankets, and left to continue saponification for 18 to 48 hours. Milk soaps are the exception. They do not require insulation. Insulation may cause the milk to burn. During this time, it is normal for the soap to go through a "gel phase" where the opaque soap will turn somewhat transparent for several hours before turning opaque again. The soap will continue to give off heat for many hours after trace.

After the insulation period the soap is firm enough to be removed from the mold and cut into bars. At this time, it is safe to use the soap since saponification is complete. However, cold-process soaps are typically cured and hardened on a drying rack for 2–6 weeks (depending on initial water content) before use. If using caustic soda it is recommended that the soap is left to cure for at least 4 weeks.

Unlike cold-processed soap, hot-processed soap can be used right away because lye and fat saponify more quickly at the higher temperatures used in hot-process soap-making.

Hot Process

Hot-process was used when the purity of lye was unreliable, and can use natural lye solutions such as potash. The main benefit of hot processing is that the exact concentration of the lye solution does not need to be known to perform the process with adequate success.

Cold-process requires exact measurement of lye to fat using saponification charts to ensure that the finished product is mild and skin-friendly. Saponification charts can also be used in hot-process soap-making, but are not as necessary as in cold-process.

In the hot-process method, lye and fat are boiled together at 80–100 °C until saponification occurs, which the soap-maker can determine by taste (the bright, distinctive taste of lye disappears once all the lye is saponified) or by eye (the experienced eye can tell when gel stage and full saponification have occurred).

After saponification has occurred, the soap is sometimes precipitated from the solution by adding salt, and the excess liquid drained off.

The hot, soft soap is then spooned into a mold.

Purification and Finishing

The common process of purifying soap involves removal of sodium chloride, sodium hydroxide, and glycerol. These components are removed by boiling the crude soap curds in water and re-precipitating the soap with salt.

Most of the water is then removed from the soap. This was traditionally done on a chill roll which produced the soap flakes commonly used in the 1940s and 1950s. This process was superseded by spray dryers and then by vacuum dryers.

The dry soap (approximately 6–12% moisture) is then compacted into small pellets. These pellets are now ready for soap finishing, the process of converting raw soap pellets into a salable product, usually bars.

Soap pellets are combined with fragrances and other materials and blended to homogeneity in an amalgamator (mixer). The mass is then discharged from the mixer into a refiner which, by means of an auger, forces the soap through a fine wire screen. From the refiner the soap passes over a roller mill (French milling or hard milling) in a manner similar to calendering paper or plastic or to making chocolate liquor. The soap is then passed through one or more additional refiners to further plasticize the soap mass. Immediately before extrusion it passes through a vacuum chamber to remove any entrapped air. It is then extruded into a long log or plank, cut to convenient lengths, passed through a metal detector and then stamped into shape in refrigerated tools. The pressed bars are packaged in many ways.

Sand or pumice may be added to produce a scouring soap. This process is most common in creating soaps used for human hygiene. The scouring agents serve to remove dead skin cells from the surface being cleaned. This process is called exfoliation. Many newer materials are used for exfoliating soaps which are effective but do not have the sharp edges and poor size distribution of pumice.

Glycerin Soap

Glycerin Soaps are soaps that contain glycerin, a component of fat or oil. The soap is most notably different in that it is translucent.

Production

Glycerin soap is made through melting and continuously heating soap until the mixture has reached a clear jelly-like consistency. The cooking process requires that the soap be partially dissolved with high proof alcohol. With home and hand-made soaps that still contains glycerin left over from saponification, the grating, melting and cooking of the soap can proceed without the addition of anything into the mixture, however sugar or additional glycerin is sometimes added. Glycerin soap can also be produced without remelting soap through directly cooking raw home-made soap.

In industrial soap-making, the glycerin is then usually removed. As such, to produce glycerin soap a good quantity of glycerin must be added back into the mixture prior to melting.

Lipid-free Cleansers

Lipid-free cleansers are liquid products that clean without fats. They are applied to dry or moistened skin, rubbed to produce lather, and rinsed or wiped away. These products may contain water, glycerin, cetyl alcohol, stearyl alcohol, sodium laurel sulfate, and (occasionally) propylene glycol.

Lipid-free cleansers leave behind a thin, moisturizing film and can be used effectively to remove facial cosmetics and dirt in persons with sensitive or dermatitic skin. Lipid-free cleansers cause less cutaneous irritation in photoaged skin than other cleansers. However, propylene glycol can cause stinging, and sodium laurel sulfate is a detergent.

Cleansing Creams

Cleansing creams are applied to the face both to clean and to moisturize. They are composed of water, mineral oil, petrolatum, and waxes. The classic cream for facial cleansing is known as cold cream. Cold creams combine the effect of a lipid solvent, such as beeswax and mineral oil, with detergent action from borax, also known as decahydrate of sodium tetraborate. These products are popular to remove cosmetics and to provide cleansing for patients with dry skin.

Body Washes

Body washes are a special subset of liquid synthetic detergents that combine mild skin cleansing with moisturizing and emollient qualities. They are applied with a puff that does not support bacterial growth to break the emulsion through the incorporation of generous amounts of air and water. High amounts of petrolatum can be incorporated in body wash emulsions to improve skin dryness and hydration.

Other Kinds of Soaps

- Olive Oil (Castile) Soap
- Citrus Honey Bock Soap
- Clay Soap Sampler
- Coconut Milk Soap
- Coffee Soap—For Cooks and Coffee Lovers
- Fisherman's Soap
- Gardeners Soap
- Goat's Milk Soap
- Good Old Fashioned Hard White Soap
- Green Tea and Lemongrass Soap
- Grocery Store Soap
- Shaving Soap—olive oil shaving soap.

Cleansing Implements

- Particulate abrasive scrubs
- Woven mesh sponges (luffa or loofah)
- Woven disposable face cloths
- Woven cleansing pouches
- Mechanized face brush.

Market Preparations

(From various soap manufacturing companies like Hindustan Lever Ltd., Godrej, Nirma, R&C, Proctor and Gamble, Amway, etc.)

- Lux
- Dove
- Pears
- Lifebuoy
- Dettol
- Santoor
- Nirma
- Hamaam
- Margo
- Persona
- Vivel
- Camay
- Wipro shikakai
- Moti
- OK
- Rexona.
- Mysore
- Medicare
- Palmolive
- Neem

SAMPLE
LOGO
① Facial Moisturizer
② NET WT./POIDS NET
1 OZ./30 ml ℮
④ Apply to areas of the face where needed.
⑨ CAUTION: AVOID EYE AREA
INGREDIENTS:
③ ALCOHOL DENAT. • FRAGRANCE (PARFUM) • WATER (AQUA PURIFICATA) PURIFIED • HYDROXYCITRONELLAL • ALPHA-ISOMETHYL IONONE • HYDROXY-ISOHEXYL 3-CYCLOHEXENE CARBOX-ALDEHYDE • BUTYLPHENYL METHYL-PROPIONAL • EVERNIA PRUNASTRI (OAKMOSS) EXTRACT • HEXYL CINNAMAL • EUGENOL • LINALOOL • BENZYL BENZOATE • CITRONELLOL • GERANIOL • BENZYL SALICYLATE • LIMONENE • FARNESOL • COUMARIN • ISOEUGENOL • CINNAMIC ALCOHOL • BENZYL ALCOHOL • BENZYL CINNAMATE • AMYL CINNAMAL
<ILN X0000>
⑩
⑥ © COSMETIC CO.
N.Y., N.Y. 10022
LONDON W1K 3BQ
PARIS • MILANO
⑤ MADE IN U.S.A.
⑦
⑧

PHARMACY

Part 6

Legal Aspects of Cosmetics

26. Cosmetics Labeling Packaging and Drugs and Cosmetics Act 1940

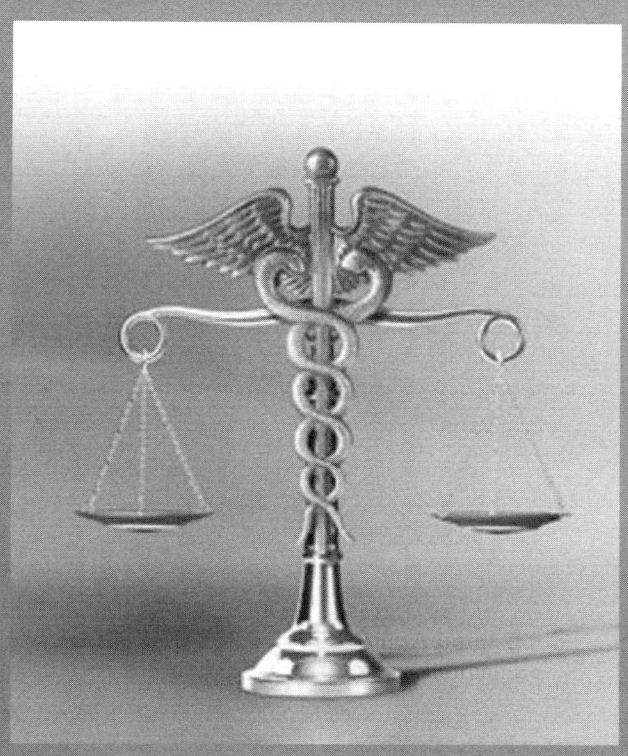

26 Cosmetics Labelling, Packaging and Drugs and Cosmetics Act, 1940

Beauty is the daughter of love.
— Havelock

Recently, more and more people are becoming regular users of cosmetic preparations. Therefore for legal control, its manufacturing has been governed under the Drug and Cosmetics Act, 1940 and defined as follows:

"Any article intended to be rubbed, poured, sprinkled or sprayed on or introduced into or otherwise applied to the human body or over other part thereof for cleaning, beautifying, promoting attractiveness or altering the appearance."

Proper labeling is an important aspect of putting a cosmetic product on the market. FDA regulates cosmetic labeling under the authority of the Federal Food, Drug, and Cosmetic Act (FD & C Act) and the Fair Packaging and Labeling Act (FPLA). These laws and their related regulations are intended to protect consumers from health hazards and deceptive practices and to help consumers make informed decisions regarding product purchase.

It is illegal to introduce a misbranded cosmetic into interstate commerce, and such products are subject to regulatory action. Some of the ways a cosmetic can become misbranded are:

- Its labeling is false or misleading,
- Its label fails to provide required information,
- Its required label information is not properly displayed, and
- Its labeling violates requirements of the Poison Prevention Packaging Act of 1970.

FDA does not have the resources or authority under the law for pre-market approval of cosmetic product labeling. It is the manufacturer's and/or distributor's responsibility to ensure that products are labeled properly. Failure to comply with labeling requirements may result in a misbranded product.

SOME LABELING TERMS

Before proceeding with a discussion of labeling requirements, it is helpful to know what some labeling terms mean:

Labeling: This term refers to all labels and other written, printed, or graphic matter on or accompanying a product.

Principal Display Panel (PDP): This is the part of the label most likely displayed or examined under customary conditions of display for sale.

Hypoallergenic: It means only that the manufactures feels that the product is less likely to cause an allergic reaction. Maufacturers of these cosmetics are not liable to submit substantiation of their hypoallergenic claims to FDA.

Natural: Products containing natural ingredients can cause allergic reactions. The term implies that ingredients are extracted directly from plants or animal products as opposed to being produced synthetically.

Alcohol free: This term on the label only implies that the cosmetic product doesn't contain ethyl or grain alcohol; however it may contain other fatty alcohols, such as cetyl, stearyl, cetearyl, or lanolin.

Fragrance free or Unscented: Fragrance is mainly added in order to mask the odor of the other ingredients. Fragrance free label doesn't mean that fragrance is not added but instead they imply that they have no perceptible odor.

Information Panel: Generally, this term refers to a panel other than the PDP that can accommodate label information where the consumer is likely to see it. Since the information must be prominent and conspicuous, the bottom of the package is generally not acceptable for placement of required information, such as the cosmetic ingredient declaration.

As part of the prohibition against false or misleading information, no cosmetic may be labeled or advertised with statements suggesting that FDA has approved the product. This applies even if the establishment is registered or the product is on file with FDA's Voluntary Cosmetic Registration Program (VCRP), which prohibit the use of participation in the VCRP to suggest official approval. False or misleading statements on labeling make a cosmetic misbranded.

Therapeutic Claims

Be aware that promoting a product with claims that it treats or prevents disease or otherwise affects the structure or any function of the body may cause the product to be considered a drug. FDA has an Import Alert in effect for cosmetics labeled with drug claims.

Labeled for Products if they are both Drugs and Cosmetics

If a product is an over-the-counter (OTC) drug as well as a cosmetic, its labeling must comply with the regulations for both OTC drug and cosmetic ingredient labeling. The drug ingredients must appear according to the OTC drug labeling requirements and the cosmetic ingredients must appear separately, in order of decreasing predominance. Contact the Center for Drug Evaluation and Research (CDER) for further information on drug labeling.

All labeling information that is required by law or regulation must be in English. The only exception to this rule is for products distributed solely in a U.S. territory where a different language is predominant, such as Puerto Rico. If the label or labeling contains any representation in a foreign language, all label information required under the FD&C Act must also appear in that language.

Labeling Information

The following information must appear on the principal display panel:

An identity statement, indicating the nature and use of the product, by means of either the common or usual name, a descriptive name, a fanciful name understood by the public, or an illustration.

An accurate statement of the net quantity of contents, in terms of weight, measure, numerical count or a combination of numerical count and weight or measure.

The following information must appear on an information panel:

1. **Name and place of business:** This may be the manufacturer, packer, or distributor.
2. **Distributor statement:** If the name and address are not those of the manufacturer, the label must say **"Manufactured for..."** or **"Distributed by...".**

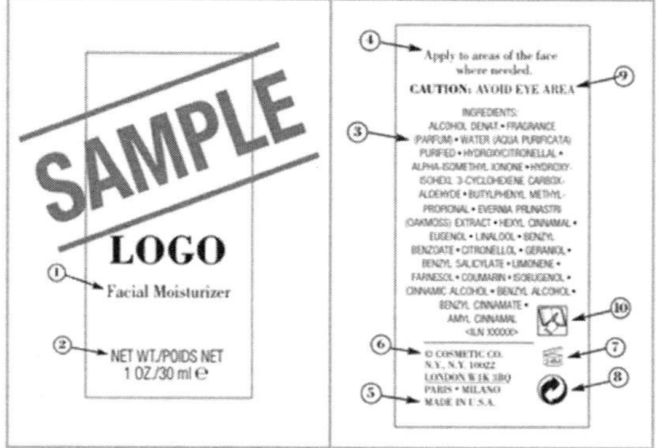

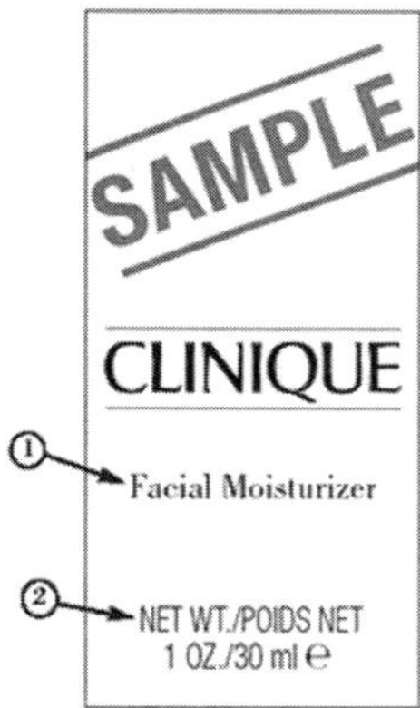

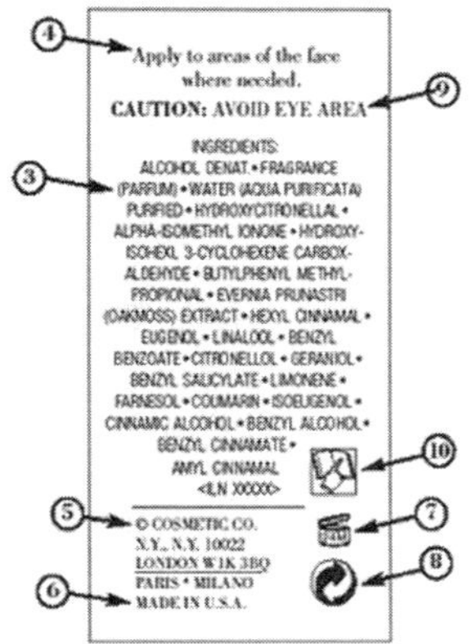

3. **Material facts:** Failure to reveal material facts is one form of misleading labeling and therefore makes a product misbranded. An example is directions for safe use, if a product could be unsafe if used incorrectly.
4. **Warning and caution statements:** These must be prominent and conspicuous. The FD & C Act and related regulations specify warning and caution statements related to specific products. In addition, cosmetics that may be hazardous to consumers must bear appropriate label warnings. An example of such hazardous products is flammable cosmetics.
5. **Ingredients:** If the product is sold on a retail basis to consumers, even it is labeled **"For professional use only"** or words to that effect, the ingredients must appear on an information panel, in descending order of predominance. Remember, if the product is also a drug, its labeling must comply with the regulations for both OTC drug and cosmetic ingredient labeling, as stated above.

REGULATORY REQUIREMENTS FOR LABELING OF COSMETICS MARKETED IN THE UNITED STATES

Cosmetics marketed in the United States, whether manufactured here or imported from abroad, must be in compliance with the provisions of the Federal Food, Drug, and Cosmetic (**FD&C**) Act,

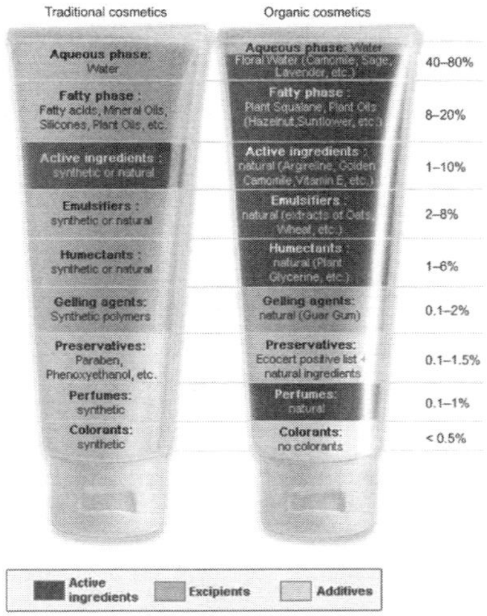

the Fair Packaging and Labeling (**FP&L**) Act, and the regulations published under the authority of these laws.

The regulations published by the Food and Drug Administration (**FDA**) are all codified in **Title 21**, Code of Federal Regulations (**21 CFR**). The regulations applicable to cosmetics are stated at 21 CFR, parts 700 to 740 (**21 CFR 700 to 740**). The color additive regulations applicable to cosmetics are found at **21 CFR 73, 74, 81 and 82.**

The FD&C Act defines **cosmetics** as:

"Articles intended to be applied to the human body for cleansing, beautifying, promoting attractiveness, or altering the appearance without affecting the body's structure or functions. Included in this definition are products such as skin creams, lotions, perfumes, lipsticks, fingernail polishes, eye and facial make-up preparations, shampoos, permanent waves, hair colors, toothpastes, deodorants, and any material intended for use as a component of a cosmetic product. Soap products consisting primarily of an alkali salt of fatty acid and making no label claim other than cleansing of the human body are not considered cosmetics under the law".

COSMETICS THAT ARE ALSO DRUGS

Products that are cosmetics but are also intended to treat or prevent disease, or affect the structure or functions of the human body, are considered also drugs and must comply with both the drug and cosmetic provisions of the law. Examples of products which are drugs as well as cosmetics are anticaries toothpastes (e.g., "fluoride" toothpastes), hormone creams, sun tanning preparations intended to protect against sunburn, antiperspirants that are also deodorants, and antidandruff shampoos.

Most currently marketed cosmetics which are also drugs are over-the-counter (**OTC**) drugs. Several are new drugs for which safety and effectiveness has to be proved to the agency before they can be marketed. A new drug is a drug which is not generally recognized by experts as safe and effective under the conditions of intended use or which has become so recognized but has not been used to a material extent or for a material time under such conditions.

The regulatory requirements for drugs are more extensive than the requirements applicable to cosmetics. For example, the FD&C Act requires that drug manufacturers register every year with the FDA and update their lists of all manufactured drugs twice annually. Additionally, drugs must be manufactured in accordance with current good manufacturing practice (**cGMP**) regulations as codified at **21 CFR 210** and **211**.

ADULTERATED, MISBRANDED OR SPURIOUS COSMETICS

The FD&C Act prohibits the distribution of cosmetics which are adulterated or misbranded. These are mentioned in Chapter III of the Drug and Cosmetic Act which deals with import of the cosmetics.

Adulterated cosmetic: A substance which may make the product harmful to consumers under customary conditions of use; if it contains a-

a. Filthy,

b. Putrid, or

c. Decomposed substance;

If it is manufactured or held under insanitary conditions whereby it may have become contaminated with filth, or may have become harmful to consumers; or if it is not a hair dye and it contains a non-permitted color additive. Coal-tar hair dyes bearing on the label the caution statement prescribed by law and that give "patch-test" instructions are exempted from the adulteration provision even if they are irritating to the skin or are otherwise harmful to the human body. Eyelash and eyebrow dyes are not included in this exemption. All dyes used in eyelash and eyebrow dye products must be approved by the FDA for such use.

Misbranded cosmetics: Its labeling is false or misleading, if it does not bear the required labeling information, or if the container is made or filled in a deceptive manner. For the purpose of this chapter, a cosmetic shall be deemed to be misbranded-

a. if it contains a colour which is not prescribed; or
b. if it is not labeled in the prescribed manner; or
c. if the label or container or anything accompanying the cosmetic bears any statement which is false or misleading in any particular.

Spurious cosmetics: For the purpose of this chapter, a cosmetic shall be deemed to be spurious:

a. if it is imported under a name which belongs to another cosmetics; or
b. if it is an imitation of, or a substitute for, another cosmetic or resemble another cosmetic in a manner likely to deceive or bears upon its label or container the name of another cosmetic, unless it is plainly and conspicuously marked so as to reveal its true character and its lack of identity with such other cosmetic; or
c. if the label or container bears the name of an individual or a company purporting to be the manufacturer of the cosmetic which individual or company is fictitious or does not exist; or
d. if it purports to be the product of a manufacturer of whom it is not truly a product.

COSMETIC LABELING

The cosmetics distributed in the United States must comply with the labeling regulations published by the FDA under the authority of the FD&C Act and the FP&L Act.

Labeling means: all labels and other **written, printed or graphic matter on or accompanying a product**. The label statements required under the authority of the FD&C Act must appear on the inside as well as any outside container or wrapper.

Cosmetic preparations need to be labeled for outdating ('**Expiry date**' or '**Best use before**' date).

FP&L Act requirements, e.g. ingredient labeling and statement of the net quantity of contents on the principal display panel, only apply to the label of the outer container. The labeling requirements are codified at **21 CFR 701 and 740**. Cosmetics bearing false or misleading label statements or otherwise not labeled in accordance with these requirements may be considered misbranded and may be subject to regulatory action.

The principal display panel, i.e., the part of the label most likely displayed or examined under customary conditions of display for sale (21 CFR 701.10), must state:

- The name of the product,
- Identify by descriptive name or illustration the nature or use of the product, and
- Bear an accurate statement of the net quantity of contents of the cosmetic in the package in terms of weight, measure, numerical count, or a combination of numerical count and weight or measure.
- The declaration must be distinct, placed in the bottom area of the panel in line generally parallel to the base on which the package rests, and in a type size commensurate with the size of the container as prescribed by regulation.
- The net quantity of contents statement of a solid, semisolid or viscous cosmetic must be in terms of the avoirdupois pound and ounce, and a statement of liquid measure must be in terms of the U.S. gallon of 231 cubic inches and the quart, pint, and fluid ounce subdivisions thereof. If the net quantity of contents is one pound or one pint or more, it must be expressed in ounces, followed in parenthesis ( ) by a declaration of the largest whole units (i.e., pounds and ounces or quarts and pints and ounces). The net quantity of contents may additionally be stated in terms of the metric system of weights or measures.
- The name and place of business of the firm marketing the product must be stated on an information panel of the label (21 CFR 701.12). The address must state the street

address, city, state, and zip code. If a firm is listed in a current city or telephone directory, the street address may be omitted. If the distributor is not the manufacturer or packer, this fact must be stated on the label by the qualifying phrase "Manufactured for ......" or "Distributed by ......" or similar, appropriate wording.

- The Tariff Act of 1930 requires that all imported articles state on the label the English name of the country of origin.

Class of Cosmetics Labeling Particulars (on both inner and outer labels)

Cosmetics in general Name of the cosmetics and name and address of the manufacturer. Manufacturing license number preceded by letter **B** if the cosmetics is packed in containers having more than 10.0 g.**On the outer labels** Net contents of the package expressed as weight for solids and semi-solids, as volume for liquids or as numerical counts, if the cosmetic is subdivided provided that this statement need not appear if the contained cosmetic is not more than 60 ml/30 g.**On the inner labels only Adequate** directions for safe use; warning, caution, or special directions; names and quantities of ingredients that are hazardous.

Hair dyes containing dyes, colours and pigments With the words: *Caution.* This product contains ingredients which may cause skin irritation in certain cases and so a preliminary test according to the accompanying directions should be made. This product should not be used for dying the eyelashes or eye brows as such use may cause blindness. (Equivalent labeling in local languages is also mandatory).

DECLARATION OF INGREDIENTS

Cosmetics produced or distributed for retail sale to consumers for their personal care are required to bear an ingredient declaration (**21 CFR 701.3**). Cosmetics not customarily distributed for retail sale, e.g. hair preparations or make-up products used by professionals on customers at their establishments and skin cleansing or emollient creams used by persons at their places of work, are exempt from this requirement provided these products are not also sold to consumers at professional establishments or workplaces for their consumption at home.

The ingredient declaration must be conspicuous so that it is likely to be read at the time of purchase. It may appear on any information panel of the package, i.e. the folding carton, box wrapping if the immediate container is so packaged, and may also appear on a firmly affixed tag, tape or card. The letters must not be less than 1/16 of an inch in height (21 CFR 701.3 (b)). If the total package surface available to bear labeling is less than 12 square inches, the letters must not be less than 1/32 of an inch in height (21 CFR 701.3(p)). Off-package ingredient labeling is permitted if the cosmetic is held in tightly compartmented trays or racks, it is not enclosed in a folding carton, and the package surface area is less than 12 square inches (21 CFR 701.3(i)).

The ingredients must be declared in descending order of predominance. Color additives (21 CFR 701.3(f)(3)) and ingredients present at one percent or less (21 CFR 701.3(f)(2)) may be declared without regard for predominance. The ingredients must be identified by the names established or adopted by regulation (21 CFR 701.3(c)); those accepted by the FDA as exempt from public disclosure may be stated as "and other ingredients" (21 CFR 701.3(a)).

Cosmetics which are also drugs must first identify the drug ingredient(s) as "active ingredient(s)" before listing the cosmetic ingredients (21 CFR 701.3(d)).

All label statements required by regulation must be in the English language and must be placed on the label or labeling with such prominence and conspicuousness that they are readily noticed and understood by consumers under customary conditions of purchase (21 CFR 701.2).

LABEL WARNINGS

Cosmetics which may be hazardous to consumers when misused must bear appropriate label warnings and adequate directions for safe use. The statements must be prominent and conspicuous. Some cosmetics must bear label warnings or cautions prescribed by regulation (21 CFR 740). Cosmetics in self-pressurized containers (aerosol products), feminine deodorant sprays, and children's bubble bath products are examples of products requiring such statements.

Although the FD&C Act does not require that cosmetic manufacturers or marketers test their products for safety, the FDA strongly urges cosmetic manufacturers to conduct whatever toxicological or other tests are appropriate to substantiate the safety of their cosmetics. If the safety of a cosmetic is not adequately substantiated, the product may be considered misbranded and may be subject to regulatory action unless the label bears the following statement: Warning-The safety of this product has not been determined. Sec. 21 CFR 740.10.

TAMPER-RESISTANT PACKAGING

Liquid oral hygiene products (e.g. mouthwashes, fresheners) and all cosmetic vaginal products (e.g., douches, tablets) must be packaged in tamper-resistant packages when sold at retail. A package is considered tamper resistant if it has an indicator or barrier to entry (e.g., shrink or tape seal, sealed carton, tube or pouch, aerosol container) which, if breached or missing, alerts a consumer that tampering has occurred. The indicator must be distinctive by design (breakable cap, blister) or appearance (logo, vignette, other illustration) to preclude substitution. The tamper-resistant feature may involve the immediate or outer container or both. The package must also bear a prominently placed statement alerting the consumer to the tamper-resistant feature. This statement must remain unaffected if the tamper-resistant feature is breached or missing. Sec. 21 CFR 700.25.

LAW ENFORCEMENT AUTHORITY

For enforcement of the law, the FDA may conduct examinations and investigations of products, inspect establishments in which products are manufactured or held, and seize adulterated (harmful) or misbranded (incorrectly or deceptively labeled or filled) cosmetics. Adulterated or misbranded foreign products may be refused entry into the United States. To prevent further shipment of an adulterated or misbranded product, the agency may request a federal district court to issue a restraining order against the manufacturer or distributor of the violative cosmetic. The FDA may also initiate criminal action against a person violating the law. Examples of products seized in recent years are nail preparations containing methyl methacrylate or formaldehyde, various eyebrow and eyelash dye products containing prohibited coal-tar dyes, and products contaminated with harmful microorganisms.

The cosmetics marketed in the United States, whether they are manufactured here or are imported from abroad, must comply with the labeling requirements of the Federal Food, Drug, and Cosmetic (FD&C) Act, the Fair Packaging and Labeling (FP&L) Act, and the regulations published by the Food and Drug Administration under the Authority of these two laws.

Alcholic fragrance solutions such as eau-de-cologne	i. The words: Harmful if taken internally.
Containing diethyl phthalate	ii. Contents of Diethyl phathalate in each ml.
Cosmetics for export	i. Specific requirements, if any
	ii. Name and address of manufacturer and name of cosmetic or a code no. approved by the licensing authority.
Soap containing hexachlorophene	Contents hexachlorophene; not to be used on babies
Tooth pastes containing fluorides	i. Contents of fluoride 1 p.p.m. (max 1000 p.p.m.)
	ii. Date of expiry

The FD&C Act was enacted by Congress to protect consumers from unsafe or deceptively labeled or packaged products by prohibiting the movement in interstate commerce of adulterated or misbranded food, drug devices and cosmetics. As defined in section 201(i) of the FD&C Act, a cosmetic is a product, except soap, intended to be applied to the human body for cleansing, beautifying, promoting attractiveness or altering the appearance.

In short, one may say that a cosmetic is a product intended to exert a physical, and not a physiological, effect on the human body.

The raw materials used as ingredients of cosmetic products are by law also cosmetics.

In section 701.20 of Title 21 of the Code of Federal Regulations [21 CFR 701.20], the Food and Drug Administration (FDA) defines the term "**soap**" as a product in which the non-volatile portion consists principally of an alkali salt of fatty acids, i.e. the traditional composition of soap; the product is labeled as soap; and the label statements refer only to cleansing. If cosmetic claims, e.g. moisturizing, deodorizing, skin softening etc., are made on a label, the product is a cosmetic. Synthetic detergent bars are also considered cosmetics, although they may be labeled as "soap."

SCHEDULE M-II

Minimum requirements of space, equipment and machinery for manufacture of cosmetics have been prescribed under Schedule M-II to the Drug and Cosmetic Rules.

Form 31 required for obtain license for manufacture of cosmetics accompanied by a license fee of Rs.2500.00 and inspection fee of Rs.1000.00. For loan license Form 31-A along with fee of Rs.3500.00. The application accompanied by the following documents:

- Layout plan of factory premises;
- A list of equipment and machinery installed;
- A document about the constitution of the firm, i.e. either sole or partnership
- A document about possession of the premises.

The application forms are provided by state drug control dept. / FDA from their office.

Before grant of license (which is issued in **Form 32**) the factory premises inspected by officers of the state regulatory agency, during inspection committee find out all the affidavits and certain whether:

- The applicant has provided adequate space for manufacturing operations, quality control and storage of raw materials, packaging materials and finished products.
- The applicant has provided adequate equipment and machinery for manufacture of cosmetics which it intends to manufacture. (Given under Schedule M-II).
- The applicant has provided adequate testing facilities for raw and finished forms of materials.
- The applicant has whole time services of a person who has either of the qualifications: D. Pharm., Registered Pharmacist or Intermediate in Science as chemistry as one of the subjects.

REQUIREMENT OF FACTORY PREMISES FOR MANUFACTURE OF COSMETICS

General Requirements

a. *Location and surroundings:* The factory shall be located in a sanitary place and hygienic conditions shall be maintained in the premises. Premises shall not be used for residence or be interconnected with residential area. It shall be well ventilated and clean.
b. *Buildings:* The buildings used for the factory shall be constructed so as to permit production under hygienic conditions and not to permit entry of insects, rodents, files, etc. The walls of the room in which manufacturing operations are carried out, shall be up to a height of six feet from the floor, be smooth, waterproof and capable of being kept clean. The flooring shall be smooth, even and washable and shall be such as not to permit retention or accumulation of dust.
c. *Water supply:* The water used in manufacture shall be of potable quality.
d. *Disposal of water:* Suitable arrangements shall be made for disposal of waste water.

e. *Health, clothing and sanitary requirements of the staff:* All workers shall be free from contagious or infectious diseases. They shall be provided with clean uniforms, masks, headgears, and gloves wherever required. Washing facilities shall also be provided.
f. *Medical services:* Adequate facilities for first aid shall be provided.
g. *Working benches* shall be provided for carrying out operations such as filling; labeling, packing, etc. Such benches shall be fitted with smooth, impervious tops capable of being washed.
h. *Adequate facilities* shall be provided for washing and drying of glass containers if the same are to be used for packing the product.

REQUIREMENT OF PLANT AND EQUIPMENT

The following equipment, area and other requirements are recommended for the manufacture of:

a. **Powders:** Face powder, cake make-up, compacts, face packs, masks and rouges, etc.
   Equipment.
   a. Powder mixer of suitable type provided with a dust collector.
   b. Perfume and colour blender.
   c. Sifter with sieves of suitable mesh size.
   d. Ball mill or suitable grinder.
   e. Trays and scoops (stainless steel).
   f. Filling and sealing equipment provided with dust extractor.
   g. For compacts:
      i. A separate mixer,
      ii. Compact pressing machine.
   h. Weighing and measuring devices
      i. Storage tanks.

   An area of 15 square meters is recommended. The section is to be provided with adequate exhaust fans.
b. **Creams, lotions, emulsions, pastes, cleansing milks, shampoos, pomade, brilliantine, shaving creams and hair-oils, etc.**
   a. Mixing and storage tanks of suitable materials.
   b. Heating kettle – steam, gas or electrically heated.
   c. Suitable agitator.
   d. Colloidal mill or homogenizer (wherever necessary).
   e. Triple roller mill (wherever necessary).
   f. Filling and sealing equipment.
   g. Weighing and measuring devices.

   An area of 25 square meters is recommended.
c. **Nail polishes and nail lacquers**
   Equipment:
   a. A suitable mixer.
   b. Storage tanks.
   c. Filling machine – hand operated or power driven.
   d. Weighing and measuring devices.

   An area of 15 square meters is recommended. The section shall be provided with flameproof exhaust system.

   Premises: The following are the special requirements related to Nail Polishes and Nail Lacquers:
   a. It shall be suited in an industrial area.
   b. It shall be separate from other cosmetic-manufacturing areas by metal/brick partition up to ceiling.
   c. Floors, walls, ceiling and doors shall be fireproof.
   d. Smoking, cooking and dwelling shall not be permitted and no naked flame shall be brought in the premises.
   e. All electrical wiring and connections shall be concealed and main electric switch shall be outside the manufacturing area.
   f. All equipment, furniture and light fittings in the section shall be flameproof.
   g. Fire extinguisher like foam and dry powder and sufficient number of buckets containing sand shall be provided.
   h. All doors of the section shall open outwards.

   Storage: All explosive solvents and ingredients shall be stored in metal cupboards or in a separate enclosed area.

   Manufacture
   a. Manufacture of lacquer shall not be undertaken unless the above conditions are complied with.
   b. Workers shall be asked to wear shoes with rubber soles in the section.

   Other requirements: No objection certificate from the local Fire Brigade Authorities shall be furnished.

d. Lipsticks and Lip-gloss, etc.

Equipment

a. Vertical mixer
b. Jacketed kettle – steam, gas or electrically heated.
c. Mixing vessel (stainless steel)
d. Triple roller mill/Ball mill.
e. Moulds with refrigeration facility.
f. Weighing and measuring devices.

An area of 15 square meters is recommended.

e. Depilatories

Equipment:

a. Mixing tanks.
b. Mixer
c. Triple roller mill or homogeniser (where necessary).
d. Filling and sealing equipment.
e. Weighing and measuring devices.
f. Moulds (where necessary)

An area of 10 square meters is recommended.

f. Preparations used for Eyes: Such preparations shall be manufactured under strict hygienic conditions to ensure that these are safe for use.

Eyebrows, Eyelashes, Eyeliners, etc.

Equipment

a. Mixing tanks.
b. A suitable mixer.
c. Homogeniser (where necessary).
d. Filling and sealing equipment.
e. Weighing and measuring devices.

An area of 10 square meters is recommended.

Kajal and Surma

Equipment

a. Base sterilizer
b. Powder sterilizer (dry heat oven).
c. Stainless steel tanks.
d. A suitable mixer.
e. Stainless steel sieves.
f. Filling and sealing arrangements.
g. Weighing and measuring devices.
h. Homogeniser (where necessary).
i. Pestle and Mortar (for Surma).

An area of 10 square meters with a separate area of 5 square meters for base sterilization is recommended.

Other requirements for 1 and 2

a. False ceiling shall be provided wherever required.
b. Manufacturing area shall be made fly proof. An airlock or an air curtain shall be provided.
c. Base used for Kajal shall be sterilized by heating the base at 150 degree C for required time in a separate enclosed area.
d. The vegetable carbon black powder shall be sterilized in a drying oven at 120 degree C for required time.
e. All utensils used for manufacture shall be of stainless steel and shall be washed with detergent water, antiseptic liquid and again with distilled water.
f. Containers employed for 'Kajal' shall be cleaned properly with bactericidal solution and dried.
g. Workers shall put on clean overalls and use hand gloves wherever necessary.

g. Aerosol.

Equipment:

a. Air-compressor (wherever necessary).
b. Mixing tanks.
c. Suitable propellant filling and crimping equipments.
d. Liquid filling unit.
e. Leak testing equipment.
f. Fire extinguisher (wherever necessary)
g. Suitable filtration equipment.
h. Weighing and measuring devices.

An area of 15 square meters is recommended.

Other requirements: No objection certificate from the Local Fire Brigade Authorities shall be furnished.

h. Alcoholic Fragrance Solutions

Equipment

a. Mixing tanks with stirrer.
b. Filtering equipment.
c. Filling and sealing equipment.
d. Weighing and measuring devices.

An area of 15 square meters is recommended.

i. Hair Dyes

Equipment

a. Stainless steel tanks.
b. Mixer.

c. Filling Unit.
d. Weighing and measuring devices.
e. Masks, gloves and goggles.

An area of 15 square meters with proper exhaust is recommended.

j. Tooth powders and toothpastes, etc.

1. Tooth-powder in General.

Equipment

a. Weighing and measuring devices.
b. Dry mixer (powder blender).
c. Stainless steel sieves.
d. Powder filling and sealing equipments.

An area of 15 square meters with proper exhaust is recommended.

2. Toothpastes

Equipment

a. Weighing and measuring devices.
b. Kettle – steam, gas or electrically heated (where necessary).
c. Planetory mixer with de-aerator system.
d. Stainless steel tanks.
e. Tube filling equipment.
f. Crimping machine.

An additional area of 15 square meters with proper exhaust is recommended.

3. Tooth-powder (Black)

Equipment

a. Weighing and measuring devices.
b. Dry mixer powder blender.
c. Stainless steel sieves.
d. Powder filling arrangements.

An area of 15 square meters with proper exhaust is recommended. Areas for manufacturing “Black” and “White” tooth powders should be separate.

k. Toilet Soaps

Equipment

a. Kettles/pans for saponification.
b. Boiler or any other suitable heating arrangement.
c. Suitable stirring arrangement.
d. Storage tanks or trays.
e. Driers.
f. Amalgamator/chipping machine.
g. Mixer
h. Triple roller mill.
i. Granulator.
j. Plodder.
k. Cutter.
l. Pressing, stamping and embossing machine.
m. Weighing and measuring devices.

A minimum area of 100 square meters is recommended for the small-scale manufacture of toilet soaps.

The areas recommended above are for basic manufacturing of different categories of cosmetics. In addition to the separate adequate space for storage of raw materials, finished products, packing materials shall be provided in factory premises.

Note No. I: The above requirements of the Schedule are made subject to modification at the direction of the Licensing Authority, if he is of the opinion that having regard to the nature and extent of the manufacturing operations it is necessary to relax or alter them in the circumstances of a particular case.

Note No. II: The above requirements do not include requirements of machinery, equipments and premises required for preparation of containers and closers of different categories of cosmetics. The Licensing Authority shall have the discretion to examine the suitability and adequacy of the machinery, equipments and premises for the purpose of taking into consideration of the requirements of the license.

Note No. III: Schedule M-II specifies equipments and space required for certain categories of cosmetics only. There are other cosmetics items, viz. Attars, perfumes, etc., which are not covered in the above categories. The Licensing Authority shall, in respect of such items or categories of cosmetics have the discretion to examine the adequacy of factory premises, space, plant and machinery and other requisites having regard to the nature and extent of the manufacturing operations involved and direct the licensee to carry on necessary modification in them.

Note No. IV: Areas for formulations meant for external use and areas for formulations meant for internal use shall be separately provided to avoid mix-up even though they are from the same category of formulations.

MACHINES USED IN COSMETIC MANUFACTURING

Cosmetic Machines: Cosmetic machines which are also available in semi automatic cosmetic machines as follows:

- Colloid Mill
- Cream Manufacturing Plant
- Inline Homogenizer
- Inline Mixer
- Ointment Manufacturing Plant
- Ointment Tube Filling Machine
- Planetary Mixer.

Others

- 36″ sieving unit with SS dust collector cowl.
- 400 lt high intensity powder blender with side mounted chopper blade. DCE dust collector unit and loading platform.
- 400 lt Powder Blending Suite with P 400 A 400 lt high intensity powder blender with side mounted chopper blade. DCE dust collector unit and loading platform.
- 4500 lt Jacketed and insulated SS mixing vessels with counter rotating stirrer and bottom mounted.
- Homogeniser.
- 9000 lt Jacketed and insulated SS mixing vessels with sweep agitator and bottom mounted homogeniser.
- 4500 lt Jacketed and insulated SS mixing vessels with counter rotating stirrer and bottom mounted homogeniser.
- 4500 lt Jacketed and insulated SS premixing vessel with agitator.
- 2500 lt Jacketed and insulated SS premixing vessel with agitator.
- 1000 lt Jacketed and insulated SS premixing vessel with agitator.
- 4500 lt Jacketed and insulated SS mixing vessel with variable speed flameproof agitator.
- 9000 lt insulated SS hot water tank,controls pumps heat exchanger.
- Colloid mills with SS enclosures.
- High speed disperser with hydraulic drive.
- Steam heated drum warming ovens, 8 drum capacities.

Storage Vessels

- 9000 lt SS storage tanks with hopper bottom.
- 9000 lt SS storage tanks on legs.
- 4500 lt SS storage tanks on legs.
- 4500 lt SS storage tanks with mild steel supports.
- 10,000 lt Resin coated mild steel storage tanks.
- 9000 lt Resin coated mild steel storage tanks.
- 8000 lt Mild steel storage tank.

Packaging Plant

- 60 bottle/min Filling Range – for nail polish remover.
- Double sided self adhesive labeler.
- 4 head vacuum fillers.
- 8 head rotary filling machine.
- 6 station indexing capping.
- 6 - IP65 Ink jet coders.
- Auto sleeve sealers and shrink tunnels.

MACHINES USED IN COSMETIC MANUFACTURING

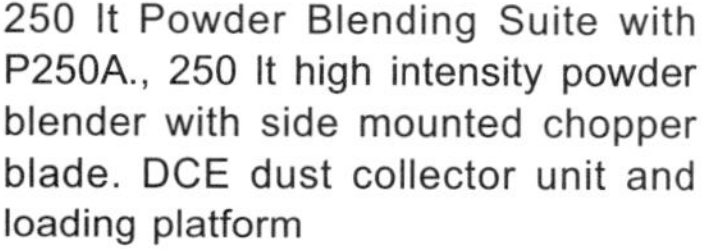

250 lt Powder Blending Suite with P250A., 250 lt high intensity powder blender with side mounted chopper blade. DCE dust collector unit and loading platform

36" sieving unit 2 - Mikrolpul Milling Suites each with Mikropul Ducon Mikropulveriser 2DH pulveriser mill. Filter bag type receiver. DCE Unimaster 152G3 dust collector unit

500 lt Jacketed and insulated SS pressure mixing vessel with counter rotating stirrer, bottom mounted homogeniser and flameproof drive together with 500 lt insulated and jacketed SS premix vessel with Lightnin agitator

250 lt Jacketed and insulated SS pressure mixing vessel with counter rotating stirrer, bottom mounted homogeniser and flameproof drive

380 lt Jacketed and insulated SS mixing vessel with underdriven sweep agitator and flameproof drive together with 300 lt insulated and jacketed SS premix vessel

290 lt Jacketed and insulated SS tipping mixing vessel with post mounted sweep agitator

Automatic CIP plant with 4500 lt insulated SS cold water tank

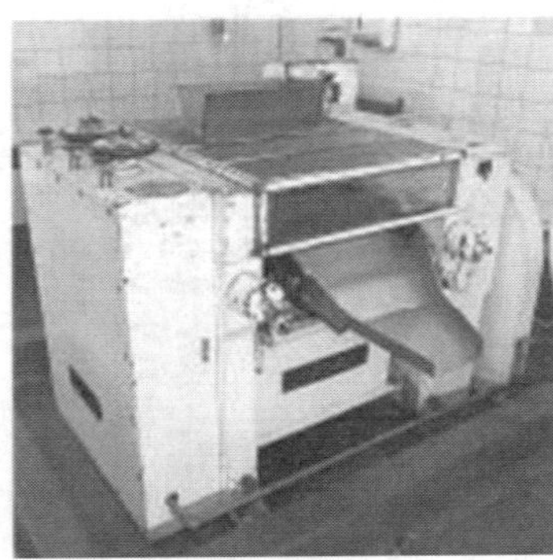

3 roll refiner mill

Downflow booths with air operated drum tipper unit

Wet dust collector with extraction fan

59,000 lt Aluminium alloy plastic granule silo 3m dia × 8.025 m high—600 kg/m³ bulk density on load cells.

10,000 lt SS storage on legs

2000 lt SS storage tanks on legs

Lipstick filling and Assembly Ranges with CL Automation twin premelt unit

6 station indexing capping machine

Auto sleeve sealer and shrink tunnel

Automation assembly and fill Desheathing and capping

6 head inline air operated inline filling machines, 30–1000 ml fill

16 head rotary vacuum filling machine with flameproof drive

6-IP65 Ink jet coder

Double sided self adhesive labelling machine with top steady conveyor

Rotory indexing nail polish filling and capping machine with auto fill (5 ml–15 ml)

24-station rotary indexing mascara filler copper 40/min 3–15 gm fill

SCHEDULE Q OF THE DRUG AND COSMETIC RULES 1945

This Schedule depicts about the coal tar colours used in cosmetics.

Rule 134 of the rules stipulates that no cosmetic shall be imported which contains a coal tar colour other than the one prescribed in Schedule Q to the said rules and coal tar colour used in cosmetic shall not contain more than:

i. 2 ppm of arsenic calculated as arsenic oxide,
ii. 20 ppm of lead calculated as lead,
iii. 100 ppm of heavy metals other than lead calculated as the total of the respective metals.

The rule further stipulates that no cosmetic intended for use on the eyebrow, or the eyelash, or around the eye shall be imported that contains any coal tar dye colour, coal tar base or coal tar dye intermediate.

Table 26.1: List of dyes, colours, and pigments permitted to be used in cosmetics and soaps

[2]Schedule Q
[See rules 134 and 144]
[3][Part I]
[4][List of Dyes, colours and Pigments permitted to be used in Cosmetics and Soaps as given under IS: 4707 (Part-I) 1988 as amended by the Bureau of Indian Standards]

Common name of the color	*Colour Index Number*	*Chemical name of the colour*
1	*2*	*3*
Guinea Green B	42085	Monosodium salt of 4-(N-ethyl-p-sulfobenzylamino)—diphenylmethylone-(1-(N-ethyl-N-p-sulfoniumbenzyl) 2,5-cyclohexadienimine).
Light Green SF	42095	Disodium salt of 4-[4 (N-ethyl-p-sulfo benzylamine-phenyl)-4-sulphoniumphenyl) methylene]-(2 (N-ethyl-N-sulfobenzyl) 2, 5-Cyclohexadienimine.
Tartrazine	19140	Trisodium salt of 3-carboxy-5-hydroxy-1-p-sulfophenyl-4-p-sulfophenylazo-pyrazole.
Sunset yellow FCF	15985	Disodium salt of 1-p-sulfophenylazo-2-naphthol-6-sulfonic acid.
Ponceau 3R	16155	Disodium salts of a mixture of 1-aklyl-phenylazo-2-napthol 3, 6-disulfonic acids.
1	*2*	*3*
Amarnath	16185	Trisodium salt of 1-(4-sulfo-1-napthylazo) 2-naphthol 3, 6-disulfonic acid.
Erythrosine	45430	Disodium salt of 9-0-carboxyphenyl-6-hydroxy 2, 4, 5, 7-tetraiodo-3-isoxanthone.
Ponceau SX	14700	Disodium salt of 2-(5 sulfo-2, 4-xylyl-azo)-1-naphthol-4-sulfonic acid.
Brilliant Blue FCF	42090	Disodium salt of 4-(9-40(N-ethyl-p-sulfobenzylamino)-phenyl)-2-sulfonium phenyl)-methylene)-(1-(N-ethyl-N-p-sulfobenzyl)- 2, 5-cyclohexadienimine).

[1]Ins. by G.O.I. Notification No. GSR 753 (E) dt. 4.11.1999
[2]Ins. by G.O.I. Notification No. GSR 1183 dt. 17.8.1964.
[3]Renumbered as Part I by G.O.I. Notification No. GSR 11(E) dt. 7.1.1991.
[4]Subs. by G.O.I. Notification

Contd.

1	*2*	*3*
Indigocarmine	73015	Disodium salt of 5, 5′-indigotindisulfonic acid.
Wool Violet 5BN (Acid violet 6B)	42640	Monosodium salt of 4-(N-ethyl-p-sulfobenzylamino)-phenyl)-(4-N-ethyl-p-(sulfonium-benzylamine)-phenyl) methylene)-(N, N-dimethyl-2, 5 cyclohexadienimine)
Light Green SF Yellowish	42095	Calcium salt of 4-(4-(N-ethyl-p-sulfobenzylamino)-phenyl) (4-sulfonium-phenyl) methylene), (1-(N-ethyl-N-p-sulfobenzyl)-2, 5-cyclohexadienimine).
Alizarin Cyanine Green F	61570	Disodium salt of 1, 4-bis (O-sulfo-p-toluino) anthraquinone)
Quinazarine Green SS	61565	1, 4-bis-(p-Toluino)-anthraquinnone
Fast Green FCF	42053	Disodium salt of 4-(4-(ethyl-p-sulfobenzylamino)-phenyl) (4-hydroxy-2 sulphoniumphenyl) methylene)-(1-N-ethyl-N-p-sulfonium-benyl 2, 5, cyclohexadienimine).
Pyranine Concentrated	59040	Trisodium salt of 10-hydroxy-3, 5, 8-pyrene-trisulfonic acid.
Quinoline Yellow WS	47005	Disodium salt of disulfonic acid of 2-(2-Quinolyl)-1, 3-indandione.
Quinoline Yellow SS	47000	2-(2-quinolyl)-1, 3 indandiene.
Poneceau 2 R.	16150	Disodium salt of 1 xylylazo-2-naphthol-3, 6-disulfonic acid.
Lithol Rubin B.	15850	Monosodium salt of 4-(o-sulfo-p-tolylazo) 3 hydroxy-2-naphthoic acid.
Lake Red D.	15500	Monosodium salt of 1-0-carboxyphenylazo-2-naphthol.
Lake Red DBA	15500	Barium salt of 1-o-carboxyphenylazo-2-naphthol.
Lake Road DCA	15500	Calcium salt of 1-o-carboxyphenylazo-2-naphthol.
Toney Red	26100	[-p-phenylazophenylazo-2-naphthol.
Oil Red OS.	26125	[-Xylylazoxylylazo-2-naphthol
Tetrabromofluorescein	45380	2, 4, 5, 7-Tetrabromo-3, 6-flurandiol.
Eosin TS	45380	Disodium salt of 2, 4, 5, 7-tetrabromo-9-0 carboxyphenyl-6-hydroxy-3-isoxanthone
Bluish Orange T.R.	45457	1, 4, 5, 8, 15-Pentabromo-2, 7-dicarboxy-3, 6-fluorandiol.
Helindone Pink CN.	73360	5, 5-Dichloro-3, 3′ dimethyl-thio indigo
Briliant Lake Red R	15800	Calcium salt of 3-hydroxy-4-phenylazo-2-naphthoic acid.
Deep Maroon (Fanchon Maroon)	15880	Calcium salt of 4-(I-sulfo-2-naphthylzo (3-hydroxy-2-naphthoic acid).
Toluidine Red	12120	1-(o-Nitro-p-tolylazo)-2-naphthol.
Flaming Red	12085	[-(o-Nitro-p-tolylazo)-2-naphthol.
Deep Red (Maroon)	12350	3-Hydroxy-N-(m-nitrophenyl)-4-(o-nitro-p-tolylazo)-2-naphthamide.
Alba Red	13058	o-(p, B, B-Dihydroxy-diethylamino)-phenylazo)-benzoic acid.
Orange G	16230	Disodium salt of 1-phenylazo-2-naphthol-6-8-disulfonic acid.
Orange II	15510	Monosodium salt of 1-p-sulfophenylaxo-2-naphthol.
Dichlorofluorescein. NA	45365	Disodium salt of 9-o-carboxyphenyl-4, 5-dichloro-6-hydroxy-3-isoxanthone
Diiodofluorescein	45425	4, 5-Diiodo-3, 6-fluorandiol

Contd.

1	*2*	*3*
Erythrosine Yellowish NA	45424	Disodium salt of 9-o-carboxyphenyl-6-hydroxy-4, 5-diiode-3-isoxanthone.
Erythrosine Yellowish K	45425	Dipotassium salt of 9-o-carboxyphenyl-6-hydroxy-4, 5-diiodo-3-isoxanthone.
Erythrosine Yellowish NH	45425	Dipotassium salt of 9-o-carboxyphenyl-6-hydroxy-4, 5-diiodo-3-isoxanthone.
Orange TR	45456	4, 5, 15-Tribromo 2, 7-dicarboxy-3, 6-fluorandiol.
Alizarin	58000	1, 2-Anthraquinonediol.
Dibromodiiodofluorescein	45371	4, 5-Dibromo-2, 7-diiodo-3, 6-fluorandiol.
Resorein Brown	20170	Monosodium salt of 4-p-sulfophenylazo-2(2) 4, xylylazo)-1, 3-resorcinol.
Alphazurine FG.	42090	Diammonium salt of 4-(N-ethyl-p-sulfobenzyl) (amine-phenyl) 2-sulfoniumphenyl. Methylene (–1 (N ethyl-N-p-sulfobenzyl (2, 5-cyclohexandienimine).
Allarin Astro B.	1530	Monosodium salt of 1-methylamino-4-(o-sulfo-p-toluino)-anthroquinone.
Indigo	73000	Indigotin
Patent Blue NA	42052	Monosodium salt of 4-(4-(N-ethyl-benzylamino)-phenyl-5 hydroxy-4-sulfo-2-sulfoniumphenyl, methylene)-(N-ethyl-N-benzyl- 2, 5-cyclohexadienimine).
Curbrantherene Blue	69825	3-3-Dichlorolindanthrene.
Nibhihol Blue Black	20470	Disodium salt of 8-amino-7-p-nitrophenylazo 3-phenylazo-1-naphthol-3, 6-disulfonic acid.
Alizurol purple SS.	60725	I-hydroxy-4-p-tolhuno-anthraquinone.
Acid Red 89	23910	...
Acid Red 97	22890	...
Acid Blue 1	42045	...
Food Blue 1	42045	...
Natural Orange	75480	...
Solvent Blues 4	44045	...
Solvent Yellow 18	12740	...
Food Yellow 18	12740	...
Solvent Red 1	12150	...
Solvent Yellow 32	48045	...
Fanchon Yellow (Hansa Yellow G).	11680	(a) (o-Nitro-p-tolylazo) acetoacetanilide
[1][Part II-List of colours permitted to be used in Soaps].		
Phthalocyanine Blue	74160	(phthalocyninate (2..) copper
Iragalite Red CVPB Paste or Pigment Orange 5	12075	1-(2, 4-dinithro phenylazo-2-Naphthalenol.
Citrus Red No. 2	12156	1-2(2, 5-dimethoxy pehnylazo) 2-naphthol.

Contd.

1	*2*	*3*
Rhodamine B 500	45170	3-ethochlorde of 9-0 carboxy-ethenyl-6-diethylamino-3-ethylamine-3-isoxanthene.
Aqueous Green Paste	74260	Polychloro copper Phthalocyanine.
Pigment Yellow 3	11710	2-(4-Chloro-2-nitrophenyl)-azo-N-(-2-Chlorophenyl)-3-Oxobutamide.
Irgalite Carmine F-P Power or Pigments Red 5	12490	N-(5-Chloro-2, 4-dimethoxy-phenyl)-4-(CS-diathylamine) Sulfonyl-2-methoxyphenyl) azo-3-hyroxy-2-naphthalene carboxamide.
Monolite Red 4R HV Paste or Pigment Red 7	12490	N-(4-Chloro-2-methylphenyl-4-(-4-Chloro-2-methylphenyl) azo 3-hydroxy-2-naphthalenel Carboxamide.
[1]Ins. by G.O.I. Notification GSR 11(E) dt. 7.1.1991.		
Oil Red No. 1 or Solvent Red 24 or Oil Red 3R.	26105	4-0-Tolylazo-Toluidine azo 2-naphthalenol.

STANDARDS FOR COSMETICS

Standards for cosmetics in finished form (Schedule S): The following cosmetics have been placed under Schedule S to the Drug and Cosmetic Rules 1945 in finished form shall conform to the Indian Standards specifications laid down from time to time by the Bureau of Indian Standards (BIS).

1. Skin Powders
2. Skin Powder for infants
3. Tooth Powder
4. Toothpaste
5. Skin Creams
6. Hair Oils
7. Shampoo, Soap-based
8. Shampoo, Synthetic-Detergent based
9. Hair Creams
10. Oxidation hair dyes, Liquid
11. Cologne.
12. Nail Polish (Nail Enamel)
13. After Shave Lotion
14. Pomades and Brilliantine
15. Depilatories chemicals
16. Shaving Creams
17. Cosmetic Pencils
18. Lipstick
19. Toilet Soap
20. Liquid Toilet Soap
21. Baby Toilet Soap
22. Shaving Soap
23. Transparent Toilet Soap
24. Lip salve IS: 10284
25. Powder Hair Dye IS: 10350
26. Bindi (Liquid) IS: 10998
27. Kum Kum Powder IS: 10999
28. Henna Powder IS: 11142

COSMETIC prohibited under Sec.26.A of the Drugs and Cosmetics Act.1940:

Tooth pastes/tooth powders containing tobacco.-vide G.O.I. Notification No.G.S.R.444 (E) dt.30.04.2002.

Bureau of Indian Standards, based on information available within the country and overseas has classified raw materials of cosmetics in two categories, namely

- Generally recognized as safe (**GRAS**) refer to IS:4707 (Part I).
- Generally not recognized as safe (**GNRAS**) refer to IS:4707 (Part II).

SCHEDULE U (I)

[See rules 142 and 142-B of Drugs and Cosmetic Rules]

Particulars to be shown in the manufacturing records

1. Serial number
2. Name of the product.
3. Lot/Batch size.
4. Lot/Batch number.
5. Date of commencement of manufacture and date when manufacture was completed.
6. Names of all ingredients, quantities required for the lot/Batch size, quantities actually used.
7. Control reference numbers in respect of raw materials used in formulation.
8. Reference to analytical report number.
9. Actual production and packing particulars

indicating the size and quantity of finished packings.

10. Date of release of finished packing for distribution or sale.
11. Signature of the expert staff responsible for the manufacture.

Records of Raw Materials

Records in respect of each raw material shall be maintained indicating the quantity received, control reference number, the quantity issued from time to time, the names and batch numbers of the products for the manufacture of which the said quantity of raw material has been issued and the particulars relating to the proper disposal of the stocks.

Notes:

1. The licensing authority may permit the licensee to maintain records in such manner as is considered satisfactorily, provided the basic requirements laid down above be complied with.
2. The licensing authority may direct the licensee to maintain records for such additional particulars as it may consider necessary in the circumstances of a particular case.

OFFENCES AND PENALTIES

Different offences and penalties have been laid down in Drug and Cosmetic Act relating to import, manufacturing, sales and distribution of cosmetics. These are mentioned in Sections 13, 18 and 27A of the Act. Offences and penalties related to cosmetics in brief have been tabulated in the following table.

Table 26.2: Offences and Penalties Related to Cosmetics

S. No.	*Contravention in brief*	*Offence*	*Penalty*
1.	Import of spurious cosmetics or a cosmetic containing any ingredient which may render it unsafe/harmful under indicated directions.	13(1)(a)	Imprisonment for a term which may extend to three years and fine which may extend to five thousand rupees.
2.	Import of cosmetics, other than mentioned in S.No. 1, import of which is prohibited.	13(1)(b)	Imprisonment for a term which may extend to six months or fine which may extend to five hundred rupees, hundred rupees or with both.
3.	Import of cosmetics, import of which has been prohibited under Section 10-A.	13(1)(c)	Imprisonment for a term which may extend to one year or five thousand rupees or with both.
4.	Repeated offence under S.No. 1 or 3.	13(2)(a)	Imprisonment for a term which may extend to five years or fine which may extend to ten thousand rupees or with both.
5.	Repeated offence under S.No. 2.	13(2)(b)	Imprisonment for a term which may extend to one year or fine which may extend to one thousand rupees or with both.
6.	Manufacture for sale/distribution or stock/exhibit for sale/distribution or sale of a cosmetic not of standard quality or misbranded.	18(a)(ii)	Imprisonment for a term which may extend to one year or fine which may extend to one thousand rupees or with both.
7.	Manufacture for sale/distribution or stock/exhibit for sale/distribution or sale of a cosmetic containing any ingredient which may render it unsafe/harmful for use under indicated direction.	18(a)(v)	Imprisonment for a term which may extend to one year or fine which may extend to one thousand rupees or with both.

Contd.

S. No.	*Contravention in brief*	*Offence*	*Penalty*
8.	Manufacture for sale/distribution or stock/exhibit for sale/distribution or sale of a cosmetic in contravention with provisions of Chapter IV of Act or any rule made thereunder.	18(a)(vi)	Imprisonment for a term which may extend to one year or fine which may extend to one thousand rupees or with both.
9.	Sale or stock/exhibit for sale/ distribution of a cosmetic imported or manufactured in contravention with provisions of the Act or any rule made thereunder.	18(b)	Imprisonment for a term which may extend to one year or fine which may extend to one thousand rupees or with both.
10.	Manufacture for sale/distribution of a cosmetic except under, and in accordance with a license issued for the purpose.	18(c)	Imprisonment for a term which may extend to one year or fine which may extend to one thousand rupees or with both.
11.	Manufacture for sale/distribution or stock/exhibit for sale/distribution of a spurious cosmetic.	18(a)(ii)	Imprisonment for a term which may extend to three years and fine.
12.	Repeated offence under S.No. 6 to 11.	Respective	Imprisonment for a term which may extend to two years or fine which may extend to two thousand rupees or with both.

VELONA
ALOE VERA

Part

7

Herbal Cosmetics

27

Herbal Cosmetics

> *When you have only two pennies left in the world, buy a loaf of bread with one, and a lily with the other.*— Chinese proverb
>
> *By plucking her petals, you do not gather the beauty of the flower.*
> — Rabindra Nath Tagore

Traditional system of medicines date back to 5000 BC and remain critical source of everybody's health and livelihood for millions of people. Ayurvedic sages have said "*Nanaushadhi Bhootam Jagat Kinchit*", i.e. there is no plant in the world, which does not have medicinal properties.

Harsh chemicals and pollution fill our lives today. There is unrest all around and the simple and natural way of life has got lost everywhere. It's time that we looked back. Back in the time when Mother Nature took care, when herbs were the healing touch. Nature has endowed us with its everlasting treasure of versatile ingredients.

Words such as aloe, goose berry (amla), henna, apricot, *neem*, honey, turmeric and ginseng (to mention just a few) are found on labels of many of the packages. There truly seems to be **"back to nature"** attitude among consumers who appear to be attracted to products that contain botanical ingredients. Some are born beautiful and others are made beautiful. For both the groups of persons, cosmetics are vital. Use of herbal extracts has deep roots in history ranging from Hippocrates through the Renaissance, despite a slump in popular use during the age of the synthetics since World War II. Today's consumer is much more concerned about the environment, what foods are eaten by her family and what she spreads on the skin. Cosmetic ingredient labeling has educated her as to what is used in cosmetics. These factors have given rise to a revival of the naturals and use of herbal cosmetics in cosmetics.

In the earlier time, women used the kitchen as a beauty parlor, as many spices happen to be natural beauty aids, e.g. *malai* (milk cream), papaya peel, mashed cucumber, fermented curd, turmeric, lemon, *multani mitti* (Ayurvedic beauty treatment made of fuller's earth, cereal, milk and lime), etc. Even in the 21st century things have not changed but the style has changed. Cosmetics are primarily of two types:

1. Synthetic cosmetics
2. Herbal cosmetics.

Synthetic cosmetics are discussed in the previous chapters. Herbal cosmetics are in use and practice since thousands of years in India, without any other effects or side effects and are well proven and documented. The analysis of many herbal ingredients using modern scientific technologies has led to

the identification of phytochemical components in Indian herbs, which deliver functional benefits like anti-dandruff, deodorant, age-defying, etc. For almost any European herb containing certain phytochemicals, there are Indian alternatives containing identical and similar phyto-chemicals, e.g. witch-hazel-containing tannins are replicable with ascorbic acid. The ready presence of vitamin C in amla fruits in stable form enables us to use this amla fruit for better anti-wrinkle, anti-oxidant benefits and sesame oil for vitamin E content, etc. In view of the current trends of patenting many Indian herbs over the world, Indian companies dealing with herbal medicines strengthened their focus on research. This has led to identification of new applications of known herbs. All these inputs made the subject **"Herbal cosmetics"** richer.

According to the broad definitions in the preceding section, the various kinds of cosmetic products are best grouped in three principal classes:

1. For the skin,
2. For the hair, and
3. For the nails.

In form they may be dry powders, pastes, solid or liquid emulsions, or aqueous, alcoholic, or oily solutions. The methods of manufacturing are essentially those employed for the manufacture of other commercial products of similar form. In the following pages, therefore, specific information will be given only for those products that require special techniques, or notes on other points of special interest.

Advantages of herbal cosmetics on traditional cosmetics

1. They do not provoke allergic reactions and do not have any negative side effects.
2. They are easily incorporated with skin and hair.
3. These are very effective than other cosmetics with small quantity.
4. Extract form of the plants decreases the bulk properties of the cosmetics and gives appropriate pharmacological effects.
5. Easy to available and found in large of variety of plants.
6. They have more stability, purity, efficacy with their herbal constituents.
7. Easy to manufacture.
8. The storage and handling of herbal cosmetics is easier and for prolong period.
9. Chief in cost.

DEVELOPMENT OF HERBAL COSMETICS

Human beings from time being have been using plants (herbs) for different purposes like food, medicine and beautification. In considering the development of the cosmetics industry at each period, the principle points noted are

i. What items for beautification were available in the market,
ii. Who sold them—grocers, barber, perfumer, spicer, pharmacist, physician, beauty shop. Cosmetic were not invented by one single person. Cosmetic have been used by women from time immemorial. The reason for using cosmetics has always been to make women more beautiful and attractive. Different ideas of beauty have created different kinds of cosmetics by different people throughout the world.

The first ancient people whose concept of beauty was similar to that we have today were the Egyptian. They admired healthy, shining hairs. They thought a lady lips, cheeks, brows eyelids and lashes should be well developed and painted. They maintained that a woman should have good complexion and a slim figure. Egyptians were great users of perfumes.

The next people to make use of cosmetics were the ancient Greek. They also used to make their hair blond. When Romans conquered the Greeks they brought back with them the 'beauty doctors' thus acquiring the secrets of dying the hair special face washes and skin foods for complexion and various other beauty treatments. It is interesting to note that American-Roman ladies made masks out of beauty clays to get smooth and clear skin just as today's women do when they go to expensive beauty parlors.

DEVELOPMENTS IN WAR TIMES

Between the wars the cosmetics and toiletries industries flourished in America in spite of the belief that French cosmetics and perfumes were the best. In these favorable conditions some European companies expanded and set up in the USA. The availability of raw materials was a great advantage there. When new ingredients were discovered in other industries the potential for their possible use for cosmetics and toiletries in such large markets made their purification and human application economically feasible and cheap enough for importation to Europe. For instance soap less detergent used in textiles was marketed, such as 'Drene' by Procter and Gamble in 1934. They were manufactured in England from imported surfactants but Marshall Aid Ceased in 1950 so that Economic necessity was the stimulus for European production.

India has Medical code since 1000 BC in Ayurveda and used the native raw materials in Medicine, religious rites and for aesthetic use to alleviate the rigors of the hot climate. Excavation in the Indus valley has yielded cosmetic pots of clay, stone, ivory, from third millennium to the second millennium BC. They are believed to be Kohl pots and some to have been for perfumed oils and unguents.

Galena and Lamp black were used for eye make up and Kohl sticks of copper, bronze and wood have been found, also polished bronze mirrors have been found. Red Iron oxide used for rouge in small shells. The practice of coloring the soles of the feet, to nails and palms of the hand was prevalent. Heavy perfumes were used lavishly the most popular being sandal wood. Perfumed body oils to give a long lasting odour, women painted their faces with suns, moons flowers, stars and birds. By the time of Gupta Period, cosmetic hygiene was quite advanced.

CLASSIFICATION OF BOTANICALS

Extracts of over 45 medicinal plants are commonly used in Europe and they are also used extensively in Asia and in other countries. Many varieties of these plants are currently available in the United States. Different parts of plants like roots, rhizomes, stem, bark, leaves, inflorescence, flowers and seeds are used in cosmetics.

Classification of botanicals according to their functions

1. *Stimulating herbs:* Chamomile, thyme, peppermint, and arnica.
2. *Astringents, tannic acid containing plants:* Blood root sage, witch hazel.
3. *Saponins:* Horse chest nut, clover, nettle, yucca, agave.
4. *Salicylic acid containing plants:* Marigold, pansy, willow.
5. *Sulphur containing plants:* Coltsfoot, dandelion, garlic, onion.
6. *Mucin containing plants:* Plantain, mallow, coltsfoot.
7. *Plants containing mineral salts:* Ca, P, K, Fe, e.g. dandelion, valerian, rose hips.
8. *Essential oil plants:* Mints, lemon, wintergreen.

Selection of suitable plant for particular applications

Best possible choice of particular plant for particular products requires a great deal of experience and can be achieved only by specialists in phytology. The plant must be selected so that its known components and their effect provide, as closely as possible, the desired spectrum of action of the end product. Before using plant materials in cosmetics, the plant should be identified properly and then a standardized material like extract, tincture should be used. Following are suggestions to help selection of suitable plants for particular applications.

1. **Invigoration and toning of flabby skin:** Cucumber, hawthorn, horsetail, calendula, yarrow, cress.
2. **Regeneration of tired or reddened skin:** Chamomile, yarrow, comfrey, hops, horsetail, linden blossom, arnica, marshmallow, seaweed, burdock.
3. **Revitalization and strengthening the skin:** Cress, hawthorn, birch, nettle, eyebright, chamomile, horsetail.
4. **Improve greasy skin:** Hamamelis, chestnut, burdock, birch, ivy, calendula, seaweed.
5. **Oral care:** Neem, meswak (toothache tree), acacia (*babool*), pomegranate.
6. **Help dry skin:** Cucumber, marshmallow, linden blossom, hawthorn, cress, eyebright, chamomile, and arnica.
7. **Sun protection:** Aloe, hamamelis, horsetail, chamomile, marshmallow, cress, seaweed, comfrey.
8. **For dry, brittle hair:** Burdock, linden blossom, hops, henna, marshmallow, arnica nettle, yarrow.
9. **Improve greasy hair:** Ivy, chamomile, hamamelis, thyme, birch, arnica, yarrow, chestnut, and eyebright.
10. **Against dandruff:** Hawthorn, calendula, yarrow, burdock, comfrey, thyme, hops.
11. **For normal hair:** Chamomile, nettles, hayseed, cress, balm hawthorn.
12. **Sedative baths:** Balm, hops, hayweed, linden blossom, yarrow, burdock, chamomile.
13. **Stimulant baths:** Rosemary, sage, thyme, hawthorn, chestnut, cress.
14. **Baby care:** Olive oil, country mallow, zinc calx, sandal wood, khus-khus, almond, aloe vera, chickpea, hibiscus, winter cherry (*ashvagandha*) and licorice.

Contents and Action of Botanical Products

1. **Slenderizing products** are base on several extracts, but the main ones are ivy (because it contents hederagenin), seaweed (organic iodine), horse chestnut aescin and aesculin, butcher's broom (ruscogenin), coffee and mate (caffeine).
2. **Cleansing products** often contains soapwort and panama wood, both rich in saponosides, which have good cleansing properties.
3. **Bust firming products** may contain fenugreek and hops, both known for their positive effects on ski and bust. Hops according to a Chinese research group's report to the Barcelona congress, indeed helps the bust to become "pleasantly plump and erect." Horsetail also may be utilized for its concentration of organic silica, a basic component of collagen fibers.
4. **Slowing down** the aging process can be approached by incorporating soils rich in the scarce but essential fatty acids such as gamma

linolenic acid, extracted from evening prime rose oil and borage oil, the notion being that they have a positive effect on prostaglandin synthesis. Rope hips from Chile are rich in alpha linolenic acid, and contain about 80 percent linolenic acids, making it valuable in helping the skin recover from burns, wounds, scars. Also recommends is ginko biloba, extract of which are employed in cosmetics to prevent skin aging by inhibiting destruction of collagen and depolymerization of hyaluronic acid.

5. **Sun products** can be augmented with several approaches. Aloe ferox shows some UV absorbing properties because of its anthraquinonic components, but the extract has to be done so that these components aren't lost. Such extracts are dark green or nearly black. When water white aloe vera gel is treated so that mainly the mucilage and amino acids are extracted, these components show no UV absorption. All herbs are sun machines, which mean they probably contain chemicals with sun screening properties. Extracting them could thus conceivably yield new natural sunscreens in the UV B range. Among their potential sunscreens are paracoumaric acid, a biopolyphenol extracted from *Caffea arabica*.
6. **Hair care components** include ginseng titrated in ginsenosids, because research in Korea showed that ginseng had a positive effect on the hair, Stendhal launched a shampoo based on the natural saponoside of ginseng, other extracts used effectively in shampoos include chamomile, henna, common lime, and white dead nettle.
7. **Anti-inflammatory agents** are likely to include an extract of oak tree with soothing properties that proved to be effective as an after sun product ingredient. More generally, it also shows promise as an irritation counteractant for so called hypoallergenic products.
8. **Thickening and gelling agents** include a natural gelling agent that finds interesting applications in hair gels (semi-permanent for natural or synthetic hair).

All these examples suggest that herbal extracts have reached a high level of sophistication, and that they are used for both their proven efficacy and for their marketing appeal. They are rapidly emerging from the age of "black magic" into the scientific era.

Principle : During the past few decades, there has been a dramatic increase in the case of natural products in cosmetics. Only now are the antiseptics, anti inflammatory and general cosmetic effects of natural products on skin, hair, teeth and mucous membrane are being recognised. A wide range of active principles of various plants including vitamins, hormonies, phytoharmons, bioflavanoids, enzymes, tannins, fruit acids, amino acids, sugars, glycosides, essential oils and dye are being conceded useful in cosmetic formulations. The main principle of these cosmetics is to maintain skin resistant to infection and mechanical injury. Natural botanicals may be used to effects in their crude form or they may be extracted, purified or derived to render them more suitable for use in cosmetics.

Plant derived ingredients for use in cosmetics may be classified according to:

1. Their chemical constituents
2. Their morphology
3. Their cosmetic
4. Taxonomy of plants.

METHODS TO USE HERBAL PRODUCTS

Herbs have been used in dried and fresh forms from earlier times for medicinal and beautification purpose. Previously it was used by mashing and directly applying on body only. Nowadays, their decoction, infusions, extracts, tinctures, flower waters, steam distillates, etc. are used.

- **Decoction:** Decoction is prepared by boiling coarse powder of the herb with water in stainless steel pan. Gently boil and continue boiling for about 2–3 hours on gentle heat till the water is reduced to one quarter of its original volume.
- **Infusions:** Infusions are actually strong teas of herbs and can be prepared in stainless steel pan or vessels. Place the coarse powder of the herb in vessel and pour freshly boiled water in it and cover with lid and keep it for at least three to five hours aside. Strain and filter. The quantity of herb and solvent depend on whether herb is fresh or dried. Usually for 100 gm of

fresh herb or 50 gm of dried herbs required 575 ml of water. Aluminum vessel should not be used as it may spoil the infusions.

- **Tinctures and extracts:** Tinctures are prepared with either alcohol or hydro-alcoholic solvent mixtures with higher percentage of alcohol. Extracts are generally prepared with hydro-alcoholic solvents.

  Place coarse powder of herbs or flower in a macerator with alcohol or water-alcohol mixture. Close the macerator with lid and leave it overnight. Add more solvent and further macerate for two-three days, sometimes its takes one week to complete maceration. Strain the mixture with strainer and filter the tincture/extract. It should be stored in tightly closed container. Usually for 100 gm of fresh herb or flowers requires 575 ml of water, water –alcohol mixture or alcohol.
- **Flower waters:** It is made same way as infusions. The same proportions of herbs or flowers and water are used. The main difference in flower water and infusion is that solvent is allowed to remain in contact with flowers overnight in case of flower waters.
- **Oil soluble extracts:** It can be prepared by extracting herbs with petroleum ether. The coarse powder of herbs is placed in water overnight in stainless steel vessel. More water is added next day. This herb-water mixture is placed with oil and vessel is heated till all the water has been removed. The oil is allowed to cool and then filtered. Herbs soak water and its cells get swollen, with further soaking of water cells rupture exposing oil soluble principles. In this way oil soluble principles of herbs get into oil.

SELECTION AND EXTRACTION OF HERBAL PRODUCTS

Modern supplies of herbal extracts are able to produce materials that are easy to use and consistent in quality. First of course the herbs themselves must be carefully selected and tested. While a great many herbs are systematically cultivated, a large proportion of them, even today, still are found only growing in wild. In a typical production process, required parts of the plant are collected, allowed to air dry, then pulverized and added to the extracting solvent. Weight ratio of plant material to solvent, temperature, length of time, solvents and extraction method all are condition dependent upon kind of herb being extracted. The resulting mixture then is filtered and bulk of the solvent removed by continuous vacuum evaporation (or other similar type of processes) to yield a syrupy concentrate. Without further processing and refinement, however, this concentrate is not usable in most cosmetic preparations. Solubility is enhanced by blending concentrate with suitable solvents.

Common solvent systems can contain water, propylene glycol, ethoxydiglycol, butylenes glycol, and denatured alcohols. There are also special products developed by the industry for anhydrous cosmetic preparations. In these cases the solvents may include mineral oil, apricot kernel oil, and petrolatum, to mention a few.

Adequate preservation for the extract also is mandatory since these natural products often allows bacteria to grow and multiply. Most common preservatives for this purpose are parabens, although other cosmetic grade preservatives also can be used. Special attention must be given to preservation of final cosmetic formula since it does contain natural products. Whenever dealing with natural products of any type there must also be an affect by the supplies of the ingredient to standardize this product given the active variability problem encountered in nature. These standardization procedures are intended to offer a product which is consistent in colour, odour and major "active" chemical entities.

Although it is not practical to list the properties of all available herbal extracts following is a brief description of a few:

- **Arnica:** The extract is prepared from the dried flowers of *Arnica Montana*, a perennial herb found native in mountain of Europe and the American Rocky mountain region. Arnica is typically used as a topical stimulant.
- **Birch leaves and barks:** An extract of the leaves of the birch tree (*Betula alba* and other species) of which there are numerous species found on several continents. Major substances found in the birch leaf include saponine and

vitamin C. Popular use of the extract is in hair treatment products.

- **Chamomile (Camomile):** The flower heads of this annual herb (*Matricaria chamomilia* sp.) are used to prepare the extract. The fragrant plant is native to Europe, Asia and naturalized in North America. Whereas teas prepared from the flower heads are commonly used to treat many minor gastric problems, the extract also is used as a healing aid for damaged skin and hair.
- **Cinnamon bark extract:** Regulates oil (Sebum) producing glands.
- **Coltsfoot leaves:** An extract of the leaves of the Farfara plant, its use in cosmetics as a skin treatment and hair treatment is based on its protection action.
- **Gentian root:** A perennial herb native to Europe, and Western Asia, the root is extracted and used in cosmetics for its soothing properties.
- **Hay flower:** An extract of the flower of hay, common worldwide, is used in cosmetic skin treatment products of all kinds.
- **Ginseng:** An extract of the root is prepared. Ginseng is a perennial herb (*Panax quinque folium, P. schinseng*) with varieties found native to Manchuria and North America. Extensively cultivated and the extract is used in lotions, creams, soap and bath preparations.
- **Hamamelis (witch hazel):** The extract is prepared from the leaves, bark and twigs of this small tree or shrub (*Hamamalis virginiana*) native to North America. Witch hazel water is prepared by distilling water slurry of the macerated twigs, followed by addition of alcohol. It has astringent properties and becomes useful in cosmetics in shaving preparations and in skin and hair tonics.
- **Henna:** Extract of the leaves of this plant (*Lawsonia inermis* Linn., *L. alba* family: lythraceae), generally considered native to Africa and Asia but cultivated in many tropical regions of the world. From ancient times it has been used in many types of hair care products for its cooling effect, conditioning action and hair dying without any side effects. The henna contains lawsone, various phenolic glycosides, coumarines, xanthones, b-sitosterol and flavonoids. Lawsone (2-hydroxy-1, 4-napthoquinone) has shown analgesic, anti-inflammatory, anti-pyretic, anti-leprotic, antimicrobial, antituberculotic, cyto-toxic, fungicidal, and wound healing activity.
- **Hops:** The corn-like part of this perennial herb, cultivated worldwide, is extracted and in cosmetics it is used especially in skin care products for skin softening effects.
- **Juniper:** The extract of the berries of this evergreen shrub (*Juniperus communis*) is used mainly for skin treatment preparations.
- **Melissa (Lemon balm):** The leaves of this aromatic perennial herb are commonly used to prepare the extract, utilized in cosmetic preparations for its soothing effect.
- **Oak bark extract:** Combats free radicals
- **Pine needle:** Although the essential oil of the pine needle is extensively used in cosmetics as a fragrance material, the extract also is used as an additive in bath preparations.
- **Rosemary:** The leaves of this evergreen shrub (*Rosmarimes officinalis*) are extracted to yield a product found in hair care products.
- **Sage:** A small evergreen shrubby (*Salvia sclarea, S. lavandulaefolia, S. officianalis*) perennial, cultivated worldwide, products leaves extracted to yield product useful in shaving preparations and bath products.
- **Tea tree oil:** Has been proven to be an excellent natural anti-septic, anti-bacterial and helps in removing excessive oil build-up on the skin.
- **Valerian:** A common perennial herb native to Eurasia and naturalized in North America. The root is extracted and the extract used in bubble baths, skin creams and hand care products for soothing effects.
- **Yarrow (Milfoil):** A perennial herb (*Achillea millefolium*) native to Eurasia and naturalize in North America. The herb entirely above the ground is used for extraction and the extract added to bath preparation for a soothing action.

HERBAL COSMETICS FOR SKIN

We all are born with a gentle, soft and smooth skin, but many of us do not contain the same texture and tone even during our teens. This is because most of us deal with our skin carelessly when young, which results in various skin problems affecting beauty. Like any other living tissue, skin too really does respond to tender care and attention.

The herbal approach of proper skin care is principally based on three essential steps:

1. Cleanse
2. Nourish
3. Moistures.

Whatever may be the type of skin; these three steps are required for external care of the skin to protect from the constant effect of environment, stress and the skin's natural process of cell degeneration decay. Therefore in order to help the skin look young and radiant, our beauty products and treatments must provide:

- Exfoliation to remove dead skin cells.
- Epidermal stimulation for new cell growth.
- Antioxidant properties for cellular rejuvenation and repair.
- Improve capillary blood flow.
- Immune – stimulation.
- Penetrating moisture and nutrients to replenish all layers of skin.

The three step process of cleaning, nourishing and moistening of skin using only herbs and oils suitable to the individual skin type fulfils all three basic needs of healthy and rich skin.

PROPER CLEANING

Most of the herbal systems of medicine like Ayurveda prescribe herbal powders to clean and exfoliate the skin on a daily basis. The herb aids as a gentle scrub to clear away the dirt, toxins, pollutants and dead cells without washing away the necessary moisture ingrained in the skin. Face pack containing turmeric and lime juice removes dead skin cells and refreshes the face.

NOURISHING AND MOISTURIZING

Gentle massage of the skin with the essential oils helps improve blood circulation and straighten the connective tissue, thereby reducing wrinkles to keep the skin in good condition, avoid excessive exposure to sun, salt water, wind, cold weather and snow. Excessive use of oily cream clog pores and cause pulpiness. The use of soaps, detergent, chemical make up removers, heavy eye creams harsh scrubs, chemical powders, pumice stones, chemical astringents, products containing alcohol, very hot a very cold water, should be avoided. Drinking of oral liquids like water, coconut water, fresh fruit juices, chilled buttermilk are useful.

Herbal cosmetics for various types of skin

I. For Dry Skin

- **To Cleanse:** Mix 5 gm almond powder, 2.5 gm milk powders, 1 gm sucrose or glucose and make a paste with required quantity of warm water. Apply this freshly prepared paste all over the face and neck region and gently massage it for two minutes. Rinse well with lukewarm water.
- **To Nourish:** Mix 1 ml of sesame oil to 1 ml of coconut oil, 0.5 ml neem oil and 0.5 ml lemon oil. Mix 0.5 ml of this mixture (Nourishing oil with 0.5 ml of pure water. Gently massage this mixture well over the face and neck region for about two minutes.

Table 27.1: Herbal cleaning powders

Skin type	*Herbs*	*Fruit face mask*
I Dry skin	Rubia cardifolia (Manjista), sariva, triphla, kusta, tulsi, glyeyrihiza glabra (Yastimadhu Jeeraka), vacha, in sesame oil	Banana or avocado pulp
II Sensitive skin	Usheero, dhangaka, curcuma longa (Daru haridra), santalam alba, (Chanlana), triphala, azadiracta indica, nimba Mustaka, kamal in coconut oil.	Banana or pineapple pulp
III Oil skin	Tulasi, lodhra, jatiphata, nimba, terminalia arjuna (Arjuna), Triphala, curcuma longa (Haridara), mustaka in mustard oil, milk, sour milk, cream and buttermilk are used as gentle natured cleansers.	Strawberry or papaya pulp

Curd is also used as natural nourishment for dry skin. Apply fresh curd on face every morning and wash it off after a few minutes with cold water.

- **To moisturize:** Melt 15gm cocoa butter in a china dish in water bath and add 110 ml avocado or almond oil. Stir well and add 30 ml orange peel oil drop by drop with constant stirring. Add 0.5 ml each of coconut oil and rose oil after cooling.
- **Natural face cream:** Mix 5ml honey, 5 gm milk cream (malai), 2.5 ml olive oil 1 ml glycerin mix well and apply wash off after 30 minutes.

II. Sensitive Skin

- **To cleanse:** Mix 5 gm almond powder, 2.5 gm orange peel, 2.5 ml dry milk powder make a fine paste by adding a required quantity of rose water and apply the paste all over the face and neck region with gentle massage, rinse well after five minutes with cold water.
- **To nourish:** Mix 30 ml almond oil, 1 ml rose oil and 1ml sandal wood oil. It is a good nourishing oil.
- **To moisturize:** Melt 30 gm cocoa butter in a china dish in water bath and add 10 ml of sunflower oil, stir well and add 60 ml of rose water drop by drop and 1ml of sandalwood oil with constant stemming.

III. Oily Skin

- **To cleanse:** Mix 5 gm barley meal, 5 gm lemon peel, 2.5 gm dry milk powder and make a paste with required quantity of water.
- **To nourish:** Mix 30 ml sunflower oil, 1 ml lavender oil, 0.5 ml bergamot and 0.5 ml clay sage oil. It will make very good nourishing oil.
- **To moisturize:** Melt 30 gm cocoa butter in a china dish in water bath, add 90 ml almond or sunflower oil with stemming. Add 60ml rose water drop by drop. Add 0.1ml camphor oil and 0.5ml lavender oil after cooling.

Formula I

Herbal Sun Screen Cream

Lemon Juice	–	5ml
Xanthan powder	–	2.5 gm
Caster oil	–	5ml
Avocado oil/Hot shear butter	–	15 ml
Essential oil	–	3 ml
Titanium dioxide	–	0.2 gm
Hot distilled water	–	60 ml.

Procedure: Mix hot water, lemon juice and xanthan powder in a blender. Add castor oil and hot shear butter/avocado oil to the mixture and again blend for one minute cool to room temperature, add essential oil and again blend for thirty seconds. Add titanium dioxide then blend briefly. Keep overnight.

Formula II

Anti-Wrinkle Cream

(Orange flower)

Bees wax	–	10 gm
Emulsifying wax	–	10 gm
Almond oil	–	40 ml
Lanolin	–	20 gm
Coconut oil	–	20 gm
Orange flower water	–	30 ml
Tincture benzoin	–	0.5 ml
Orange oil	–	0.5 ml

Procedure: Melt the waxes and oils together and add orange flavor water drop by drop with constant stirring then add tincture of benison, and orange oil with constant stirring.

Formula III

Anti-Wrinkle lotion

Vitis vinifera (Grapes)	–	10 ml
Citrus lemon	–	5 ml
Solanum lycopersicum (tomato)	–	20 ml
Aloe-vera- fresh gel	–	50 ml

Procedure: Mix well and apply over the skin daily to prevent wrinkle formation.

Formula IV

Cream for Acne and Pimples

Salmalia malabarica (Silk cotton tree thorn powder)	–	10 gm
Melia azadirachta (Neem Patta Powder)	–	3.0 gm
Symplocus racermoso (Lodhra white powder)	–	1.5 gm

Curcuma Longa (Harda Powder) – 1.5 gm
Barberi's aristata (Daru haridra Powder) – 1.2 gm
Pteris Santalinus (Ratanjali powder) – 0.5 gm
Chandan (Sukhad Powder) – 0.5 gm
Andropogon muricatus (Sugandhi Khus powder) – 0.5 gm
Glycyrrhiza globra (Mulethi Powder) – 0.5 gm
Rubia Cordifolia (Manjistha Powder) – 0.5 gm
Randia dumetorum (Mindhol Powder) – 0.25 gm
Quercus infeeteria (Mayu Powder) – 0.25 gm
Acorus infectoria (Mayu Powder) – 0.25 gm
Acorus calamus (Vaj Powder) – 1.10 gm
Aloe vera gel (Kumari) – 2 gm
Alum Powder
Curd q.s. – 20 gm

Mix all the powder and make a paste with curd and apply on the face 2–3 times a day.

Formula V

Almond cream for dry skin

White wax – 30 gm
Almond oil – 60 ml
Rose water – 60 ml
Sodium benzoate – 3 gm

Melt white wax with almond oil and stir with rose water and add sodium benzoate.

Formula VI

Moisturising lotion

Aloe-vera juice – 100 ml
Wheat germ oil – 100 ml
Bees wax – 50 gm

Melt bees wax and add aloe vera juice and wheat germ oil with constant stirring. It has a powerful antioxidant that tones and softens the skin and prevent skin damage caused by ultraviolet rays and pollution. It is suitable for normal and oily skin.

Formula VII

Nourishing cream

Coconut oil – 45 ml
Olive oil – 30 ml
Almond oil – 15 ml
Cocoa butter – 30 gm
Emulsifying wax – 30 gm
Bees wax – 15 gm
Borax – 10 gm
Distilled water – 45 ml.

Mix coconut oil, olive oil, and almond oil in melted cocoa butter and emulsifying wax with constant stemming, dissolve borax separately in 45ml hot water. Then add this solution to the above mixture with constant stemming.

Formula VIII

All purpose face pack

Sandalwood fine powder – 5 gm
Usher fine powder – 5 gm
Nagarmotha fine powder – 5 gm
Ashwagandaha fine powder – 5 gm
Sariba fine powder – 5 gm
Maujistha fine powder – 5 gm
Ambahaldi fine powder – 5 gm
Turmeric fine powder – 5 gm

Mix this mixture in milk and multani and apply on the face.

Formula IX

Eye cream for Dark Circles under the Eyes

Lanolin – 15 gm
Almond oil – 22.5 ml
Soybean flour – 5 gm
Cold water – 10 ml

Add almond oil to melted lanolin, mix soybean powder and cold water to it.

Formula X

Moisturizing lotion for dry and Scaly Skin

Rubia Cordifolia (Manjstha) decoction – 50 ml
Aloe-vera juice – 50 ml

Mix well and apply on the skin of the body and face as required.

HERBAL CREAMS

Creams constitute the largest class of cosmetic preparations used on the skin. Although the variety may seem bewildering, every true cream consists basically of an emulsion of oily and watery substance in solid or liquid from. According to

Table 27.2: Herbal skin preparations		
Name	*Important Herbal Constituents*	*Uses/Claim*
Almond Lanolin under Eye Cream	Almond oil Lanolin date extract	Removes dark circles around the eyes, nourishes and clears skin
Vitaminsed Nourishing Cream	Wheat germ oil, cabbage, date extract	Luxurious night cream, provides skin a meaning full look without being unnecessarily greasy
Cactus Aloe vera Rehydant Cleansing Cream	Aloe vera known as the desert lily, lemon extract, carrot extract	Removes impurities from skin's surface, without disturbing its moisture balance
Vitaminised Whitening Cream	Almond oil, lactic acid, vitamin A and D, Corroti seed extract	Lightens the skin colour, act as powerful astringent porcelains the skin
Apricot Lustre Firming Cream	Pure apricot extract, rose oil	Gives lustre to dry and dull skin, Beautifying cream
Honey Intensive	Honey, sandal wood	Effective moisturizer, excellent softner, gives nourishment to the skin.
Precious Herbs	Tulsi, wild turmeric, date, apricot.	Moisturizer for fabled skin.

their function, creams fall into three principal groups: cleansers, emollients, and finishers. Most cleansing creams and all emollient creams are water-in-oil emulsions; most of the finishing creams are oil-in-water emulsions. The Toilet Goods Association has published standards for most of the raw materials used in the making of cosmetic creams.

Types of Herbal Creams

Cold creams: It is an emulsion in which the fat predominates, but the cooling effect produced when it is applied to the skin is due to the slow evaporation of the water contained. The base in general use is white beeswax, and traces of borax are occasionally added to aid emulsification. The perfume generally used is rose either as aqua rosea or by the addition of Otto. The method of manufacture is simple when borax is used and consists of melting the wax on a water bath adding the oil and warming the whole at about 80 °C. The aqueous portion containing the borax is heated to this temperature and stirred slowly. The perfume is added when gel and the cream is potted liquid if a brilliant white surface is desired.

Formula XI

Cold cream

Almond oil	550 parts
White wax	145 parts
Borax	10 parts
Water	290 parts
Rose water	5 parts

Vanishing creams: Vanishing creams are also called foundation creams because they disappears when rubbed into the skin. They consist of stearic acid partially saponified with alkali. The bulk of the fatty acid being emulsified by the soap thus formed. The main constituent is of course water and mucilage of tragacanth or agar agar to prevent the collapsing of the cream and the whole is preserved with a trace of an aldehyde.

Formula–XII

Vanishing cream

Stearic acid	130 parts
Borax crystals	28 parts
Sodium carbonate	12 parts
Water	740 parts
Rose granium oil	9 parts
Patchouli oil	1 parts

Cleansing creams: Are of three types: cold creams, **"quick-liquefying"** creams, and liquid cleaners. A satisfactory cleansing cream is a water-in-oil emulsion that melts at the temperature of the body and spreads readily over the skin. The oily ingredients should be light enough to penetrate a little to clear the skin of impurities, and flow away. It should not contain any appreciable amount of substances that might be retained by the skin. The product should not be sticky; after it is removed, the skin should feel smooth, clean but not greasy, and relaxed.

Formula XIII

Chamomile cleansing creams

Chamomile flower	50 parts
Distilled water	500 parts
Lemon Juice	5 parts
Sodium benzoate	2 parts.

Liquid creams: Practically any of the creams described here can be made in fluid form. Some manufacturers offer a entire line of liquid cleansing, emollient, and finishing creams (occasionally called **"milks"**); for advantages and disadvantages, these must be tested and judged on individual properties and personal preference. The compositions may contain up to 90% of water.

The most popular use of liquid creams is as so-called hand lotions, of which the long established **"honey-and-almond cream"** is typical. They are usually oil in-water emulsions, much improved in recent years by the use of the newer emulsifying agents, notably triethanolamine, polyhydric alcohol fatty acid esters, etc. As the stability of such emulsions is of paramount importance, the methods of manufacture frequently depend upon the nature of the emulsifying agents used, and are usually more complicated than those suitable for solid creams.

Formula XIV

Sun flower liquid cream

Lanolin	50 parts
Sun flower	50 parts
Wheat germ oil	5 parts
Witch hazel extract	25 parts
Sodium benzoate	5 parts

HERBAL LOTIONS

The oils and waxes in lotions are identical to those of an emollient cream but they are present in lower concentration. An o/w emollient lotion usually contains more water than the corresponding cream; a w/o type may have the same water content, with oily components replacing part of the wax like materials. These lotions are preferred for use during the day because they produce a lighter or less oily emollient film. However, they can be formulated to contain the same concentration of oil phase ordinarily used in creams. The sales appeal of emollient lotions derives partly from their convenience and partly from the greater variety of package design possible for liquid emulsions.

Types of Herbal Lotions

Face lotion: Are not so much in demand now as they were formerly, undoubtedly because of the vogue for tanning. At best the merits of such preparations are questionable. As the pigment-forming mechanism of the skin is situated at the base of the epidermis, any true bleaching agent strong enough to penetrate to the site of pigment and actually lighten the color of it could readily harm the epidermis itself.

These also called bleaching lotions actually mask, rather than lighten, the color of the skin. Like the whitening creams, these lotions usually contain alum, zinc oxide or titanium dioxide with various proportions of alcohol, glycerol, and water. As delicate flesh tints may also be added, such products often serve more properly as liquid powders.

Formula XV

Face lotion

Alum	10 parts
Zinc Sulfate	1 part
Glycerin	1 part
Tincture of benzoin	1 part
Essence of Rose	30 drops

Sun Burn Lotion: The purpose of sunburn lotion is to assist the skin in tanning without painful effects and the purpose of the anti-burn preparations is to minimize the harmful effects of sunburn. The materials which are used for the above purpose are known as sun tanning agents and sunburn preventive agents respectively. Combined these are known as sunscreens. Sun tanning agents are those sunscreens which absorb a minimum of 85% ultra-violet radiations of the wavelengths of 2900–3200 Å, but which transmit ultra-violet radiations of wavelengths of longer than 3200 Å and produce a light transient tan. Sunburn preventive agents are those sunscreens which absorb more than 95% of more of ultra-violet radiations of the wavelengths of 2900–3200 Å There is another type of sunburn preventive agents which scatter the sunlight. These include titanium dioxide, kaolin, talc, zinc oxide, calcium, carbonate and magnesium oxide.

Formula VI

Sun burn lotion

Zinc hydroxide	100 parts
Zinc carbonate	70 parts
Corn starch	30 parts
Glycerin	50 parts
Tincture of benzoin	50 parts
Benzyl cinnemate	2 parts
Heliotropin	5 parts
Tuberose absolute	1 part
Water to produce	1000 parts

Shaving lotions: Are offered for either before or after shaving. Before-shaving preparations may be of two types: (1) those intended primarily to soften the beard usually have as active ingredient a wetting agent in a base of water and glycerol, perfumed and tinted. Those to be applied before using an electric razor act like strong astringents, contracting the skin to make the hair stand up more straight. After-shaving lotions are usually similar in composition to astringent lotions, with a higher percentage also acts as a mild antiseptic to aid against infection of any abrasions. As many of the suggested ingredients may deteriorate, lotions of this type should contain preservatives.

DEODORANTS

Theoretically any compound which has antibacterial action can be used as deodorant.

Deodorants liquids: Many of the quaternary ammonium compounds have been found nontoxic and sufficiently non-irritating. Such compounds can be used in deodorant liquids.

Formula XVII

Deodorants Liquid

Chlorhexidine diaceate	0.5 parts
Propylene glycol	2.0 parts
Denatured spirit (50%) to make	100.0 parts
Perfume	q.s.

FACE POWDER

A face powder is basically a cosmetic product which has as its prime function the ability to complement skin color by imparting velvet like finish. It would enhance the appearance of the skin by masking the shine due to the secretion of the sebaceous and sweat gland.

A face powder must be a blend of specific raw materials, if it is to be a product which exhibits the particular characteristics desired. Therefore it would be well to list the basic ingredients normally employed as well as the properties each may impart to the finished powder formulation.

Table 27.3: Raw material used in face powder and their characteristics

Raw Materials	*Outstanding characteristics*
Talk	Slip
Kaolin	Absorbency, adhesion
Precipitated calcium carbonate	Absorbency, bloom
Magnesium carbonate	Absorbency, fluffiness
Zinc and magnesium stearates	Adhesion, water proofness
Rice starch	Absorbency, bloom
Silica and silicates	Absorbency
Zinc oxide	Opacity
Titanium dioxide	Opacity
Frosted-look materials (guanine bismutho-xychloride, mother-of-pearl, mica, aluminium, bronze)	Sparkle, pearly effect

Formula XVIII

Face powder

Talc	75 parts
Zinc oxide	10 parts
Rice starch	10 parts
Zinc stearate	5 parts
Perfume and color	q.s.

LIPSTICKS

Lipstick is used to make appearance of lips attractive by imparting colour. Narrow lips can be made to appear wider by applying lipstick above the upper lipline and broad sensual lips can be made to appear narrower by applying lipstick well within natural lipline.

Transparent Lipsticks

They do not contain any opaque pigment or lake, soluble or solubilized dyes are used in such preparation. Light can shine through these; staining action can be improved by using suitable compounds like manoalkyloamide or mixed fatty acid.

Formula XIX

Transparent Lipsticks

Vaseline	15 parts
Bees wax	10 parts
Spermaceti	400 parts
Carmine	6 parts
Perfume to suitable	q.s.

Lip jelly

A simple liquid lip gloss is prepared by perfuming and tinting castor oil with colour.

Formula XX

Lip jelly

Petroleum Jelly	45.0 parts
Anhydrous lanolin	55.0 parts
Perfume and colour	q.s.

HERBAL COSMETICS FOR HAIR CARE

Herbs are a wonderful gift of nature, compatible with both human skin and hair. They are used as powerful aids in treating almost any skin and hair problem. Unlike chemical based products, herbs are by nature more friendly to human body and therefore are completely safe, extremely effective and have almost no negative side effects.

The face of an individual is personal identity in which hair plays a significant role. Therefore, a perfect head with hair is an attribute of personality and beauty. Every strand of hair should be long, black, soft, shiny, and strong. In order to know how to keep one's hair healthy and wealthy, one should have knowledge about the factors responsible for hair growth, fall and other hair problems. Hair reflects the inside story of a person, as well as his/her general state of health. Because of growing importance of hair care, a new branch of cosmetology called Trichology has developed. This is derived from the Greek word **'trichos'** which means hair. Trichology deals in the science and study of hair or more specifically the science of physical, emotional and environmental causes of hair and scalp maladies.

Herbal hair products like herbal hair oils, herbal shampoo, and herbal conditioners are based on thousand of years old Indian Ayurveda system of medicine. They are the result of ancient knowledge on plants and modern scientific knowledge on herbs. Herbal hair care products, as they claim, are completely natural and 100% pure with absolutely no chemicals, preservatives and artificial fragrances. The composition of products is 100% pure dried powders of herbal plants mixed with essential oils.

Type of Herbal Hair Care Products

Hair oil: The hair oils used for dressing and nourishing the hairs and grace to appearance of hair. This preparation is generally used to increase the growth of hair and remains healthy.

Hair oil should have following properties:

1. They should give luster to the hair.
2. Retain them soft and flowing.
3. Invigorate their growth.
4. Prevent premature grayness.
5. Keep the brain cool.
6. Should not be sticky.
7. Should posses mild perfumes if required.

Various oils used for hair nourishment

1. Castor oil
2. Coconut oil
3. Sesame oil
4. Bella oil
5. Chameli oil (Jasmine)
6. Henna oil.

Formula XXI

Hair oil

Refined coconut oil	5 parts
Balsam Peru	2 parts
Sandal wood oil	15 parts

Alkanet root	15 parts
Extract of henna	5 parts
Oil rosemary	10 parts

Hair lotion: Hair lotion has a stimulating effect upon the hair follicles. They are generally perfumed with oil of rosemary and other ascenes as it posses a good stimulating property.

Formula XXII

Catharanthus lotion

Tincture cantharides	1 part
Aqua samburi	11 parts
Ess. Rosemary	5 parts

Shampoo: Shampoo is a preparation of a surfactant (i.e. surface-active material) in suitable form – liquid, solid, or power – which when used under the conditions specified will remove surface grease, dirt, and skin debris from the hair shaft and scalp without affecting adversely the hair, scalp or health of the user.

Formula XXIII

Coconut oil shampoo

Coconut oil	1000 parts
Potassium Hydroxide	300 parts
Distilled water	1000 parts
Potassium Carbonate	30 parts
Distilled water	2970 parts

Hair colorants: These are the preparation which are used for the coloring of the hairs. They enhance the attractiveness of the gray hair. They are applied externally on the hair with help of brush.

Formula XXIV

Hair colorant

Potash	7 parts
Ammonia	3½ parts
Glycerin	15 parts
Alcohol	12 parts
Rose water	550 parts

QUALITIES OF AN IDEAL HAIR PRODUCT

The ideal hair products must

- Protect the hair cuticle.
- Cleanse without stripping natural oils.
- Replace lost protein, moisture, and nutrients.
- Condition without weighing down the hair.
- Even out porosity and prevent moisture loss.
- Smooth abraded cuticle scales.
- Prevent intense drying from environment.
- Give an exceptional tactile quality or feel to the hair.
- Increase and fortify the strength and causticity of the hair.

Table 27.4: Some commonly used herbals for hair care

Latin Name	*Common Name*
Acacia concinna	Shikakai
Accaica arabica	Kikar
Arcticum lappa	Burduck
Arnica montana	Arnica
Betula pendula	Birch
Calendula officinalis	Marigold
Carthamus tinctoria	Safflower
Centella aciatica	Brahmi
Cocos nucifera	Coconut (Nariyal)
Cydinia oblonga	Bihi
Cuscuta reflexa	Amarvela
Eclipta alba	Bhringaraj (Ghangra)
Ginkgo biloba	Ginkgo
Glyccyrrhizy globra	liquorice (Yasthimadhu)
Haematoxylon camp	Pataing
Jiptuglans regia	Akhrot
Lawsonia inermis	Henna (Mehandi)
Morus Alba	Shahtoot
Nardostachys-jatamaus	Jatamansi
Phyllanthus emblica	Amla
Pilocarpus jaborandi	Jaborand
Pterrocarpus indica	Narra
Rubia tinctorum	Bacho
Sapindus mukorrossi	Ritha
Saussurea lappas	Kust
Sesamum indicum	Til
Terminalio belerica	Behara
Thymus serphyllum	Banajwain
Tinospora cordifolia	Giloe
Trigonella foenum Graecum	Methi
Utrica–dioica	Stinging nettle

Pure Shampoo

- 30 gm any four-five dried herbs powder from the table.
- 300 ml purified water.
- 60 ml liquid glycerin soap.
- 5 ml any base oil.
- 5 ml any essential oil.

Prepare decoction of the herbs in purified water mix liquid glycerin soap, base oil and essential oil. Shake well. Use the formula for every shampoo. Shake well before each use. Refrigerate the shampoo between uses for up to a week. (Egg yolk, aloe vera, honey, glycerin and apple juice can also be used as conditioner).

Herbal/Natural Shampoo

- **Health shampoo:** Take powder of reetha, shikakai, yellow clay (Multani mitti), nagarmotha, mehandi, and amla in equal quantities. Soak it in water for overnight and make a paste in the next morning. Use the paste to wash the hair. This is one of the best shampoos.
- **Regular shampoo:** Use of mixture of shikakai, reetha, and amla in equal ratio given hair better appearance than shampoo.
- **Protein shampoo:** Apply the white part of egg (egg yolk) once a week which provides protein to the hair root.
- **Hair fall preventive shampoo:** Paste of flower of *Til* (Sesame) and gokharu in cow's milk should be applied for seven continuous days which prevents hair loss and encourages growth.
- **Natural shampoos:** Nature has an abundance of goodness stored in product like *Shikakai and Reetha*. They act as cleaning agents, white henna to a good conditioner. Soaps don't clean hair effectively since they leave behind five granule deposits to calcium and magnesium, leading to an unhealthy scalp and makes hair dry and lifeless. The pH level of our hair is neutral, white soaps have pH about 10 which makes them highly alkaline. Therefore, hair requires much high quality care as skin does to keep it healthy.
- **Color improving paste:** Soak *mehandi* in tea water overnight and boil it until it becomes paste. Apply it on the hair and wait until it dries up before washing it with water. Do this process once in a week. This improves hair color and shine.

  Decoct calendula, chamomile are used for enhancing bold hair. Decoct rosemary, sage and black tea are used for enhancing dark hair, white decoct sandalwood is used for reddish brown tones and saffron is used for copper tones.

Shikakai Hair Wash

- *Shikakai* powders
- Dried *amla*
- Dried lime peel
- Green grams
- Dried curry leaves
- Fenugreek seeds (*Methi* leaves).

Mix powder of the above mentioned herbs in equal amount. Use it along with the requisite amount of water for cleaning and washing hair.

Anti-Dandruff Hair Wash

- Shikakai
- Mehandi
- Reetha
- Daru hariora
- Rind of lemon and orange
- Nimba tree bark (Margosa)
- Chandan
- Saindhava lavana.

Make a fine powder of the above herbs in equal amount soak 75–100 gm powder in water at night. In the morning, boil this mixture till it forms a paste. Wet hair and apply all over scalp. After 10–15 minutes, clean the scalp thoroughly.

Multipurpose Conditioner

Conditioner is essential for maintaining healthy, normal hair in its optimal condition. Aloe vera has tremendous moisturizing properties; white lemon juice is quite purifying and cleaning. Lavender and rosemary are used for relaxing and mint essential oils for stimulating effect.

Formula XXV

Aloe vera gel – 60 ml
Lemon juice – 10 ml
Essential oil – 0.5 ml

Mix *aloe vera* gel with lemon juice and add essential oil. Apply to freshly shampooed hair. Leave 3–5 minutes and then rinse thoroughly with water.

Kesh Raj Hair Oil

Formula XXVI

Bhringraj – 1 kg
Mehandi *(Lawsonia inermis)* – 1 kg
Amla *(Emblica officinalis)* – 1 kg
Vasaha *(Adhatoda vasica)* – 1 kg
Kamboji *(Adenanthera pavonina)* – 1 kg
Kachura – 100 gm
Usheera – 100 gm
Chandan – 100 gm
Musta – 100 gm
Sesame/Coconut oil – 4 liter

Make a paste of all these herbs (first five) and mixed with 4 liter of gingelly or Sesame oil/Coconut oil in the mixture evaporates. Thoroughly mix the remaining ingredients in the oil and heat over a gentle flame keep the container in an unheated over for two days. Strain the oil and store in a air tight bottle. This oil promotes hair growth, prevents hair fall and allergies.

Popular Hair Tonic

- Amla fruits
- *Anauthamula*
- Henna (*Mehandi*)
- *Bhringraj* leaves
- Gingelly oil/*Yastimadhu*
- Jasmine flower
- *Japa* flower
- Basil (*Tulsi*) leaves
- *Kamboji*
- *Medhika*
- *Kachura*
- *Jatamansi*
- *Vasaka*
- *Spirulina*

Coconut Oil

Heat all ingredients with oil on a low flame till all herbs are burnt. Strain the oil and store in airtight containers.

Spirulina (*Spirulina maxima, spirulina platensis*) the blue green alga (Cyanobacterium) has found a prominent place in the cosmetic industry. Japan is using *Spirulina* in lipsticks, eyeshades and eyeliners replacing potential carcinogenic coal tar dyes. *Spirulina* being rich in protein, beta-carotene and Vitamin E is being used in the preparation of anti-wrinkle cream, anti-pimple lotion and face masks.

Table 27.5: Herbal hair preparations

Name	*Important herbal constituents*	*Uses/Claim*
Herbal hair rinse	Mint, brahmi and other herbal extracts	Promotes hair growth used in combination with henna/amla shampoo.
Herbal hair conditioner	Amla, sandal wood brahmi, lichens and other precious herbs	A complete hair food treatment makes the hair strong healthy and lustrous prevents hair loss and acts as a scalp deep cleanser.
Herbal hair oil	Arnica, henna, shikakai and other herbal extract	Specially created to prevent hair loss and promote luxurious hair growth.
Herbal henna reinforced hair food treatment	Pure henna	Make hair lustrous manageable and silky acts as treatment for scalp disorder and gives a natural colour and sheen to hair.
Hair amla shampoo	Extract of amla dates armica and rare herbal extracts	For normal to dry hair, make the hair lustrous healthy and manageable. Cleansing the scalp while retaining and stimulating hair growth

DENTIFRICE

Dentifrices are preparations intended for use with a toothbrush for the purpose of cleaning the accessible surfaces of the teeth. They have been prepared in paste, powder, and to a lesser extent in liquid and block forms. In addition to enhancing personal appearance by maintaining cleaner teeth, brushing with a dentifrice reduces the incidence of tooth decay, helps maintain healthy gingivae, and reduces the intensity of mouth odors. Good dental health increases the possibility of good general health, a leading secondary result of cleaning teeth.

Formula XXVII

Dentifrice

Soap powdered	7.2 parts
Saccharin	0.24 parts
Amaranth (sol.)	1.0 parts
Oil of cinnamon	0.6 parts
Oil of peppermint	0.6 parts
Oil of clove	1.2 parts
Alcohol	90 parts
Distilled water q.s.	120 parts

HERBALS USED IN NAIL CARE

Fungi are everywhere in environment and can infect various parts of body. Fungal infections of nails are very common and notoriously difficult to cure. Nails are considered one of the body's hardest and strongest tissues. The conventional treatment of nail fungal infection is expensive and long term. Treatment may include topical preparations and / or oral anti-fungal medications. Nail fungus can be very stubborn to treat and people whose infection clears up often finds that it recurs soon after discontinuing the medication. However, all of the medications have the toxicity profile, including hepatotoxicity.

People prefer herbal and natural way to treat so that side effects are minimized. There are many herbs for curing nail infections such as *Melaleuca laterifolia, Lavendula officinalis, Cybopogon citrates, Syzygium aromaticum, C.zeylanicum, Thyme vulgaris, Allium sativum*, etc. These herbs have a deep penetration power which aids in the destruction of the nail fungus, since the fungus infects the root of the nail bed as well as the surface most of the conventional topical therapy for nail fungal infection are not yet sufficiently effective, and this failure may be due to poor penetration of the drug into the nail plate affecting the concentration of an applied coat from outer surface to inner surface. If these herbs utilize with penetration enhancers have little promise as accelerators of the nail permeability.

Nail Polish

A nail polish or enamel in order to be successful must contain a film former whose characteristics are ease of brushing application, fast drying and hardening and whose formulations should enhance adhesion to the nail without losing its resistance to chipping and abrasion. These characteristics can be accomplished by the proper formulation of the necessary constituents of nail enamel.

These are the following:
1. Celluloidal film former
2. Plasticizers
3. Solvents
4. Colorants
5. Specialty fillers.

Formula XXVIII

Nail polish

Celluloidal film, lut small	250 parts
Amylacetate	250 parts
Acetone	750 parts
Eosine A	q.s.

HERBAL ANTIOXIDANTS

Other organisms especially plants have evolved mechanisms for protection against oxidative stresses. These protective mechanisms provide interesting chemicals that can be extracted and added to topical antioxidant preparations for human use in the treatment of benign photodamage. The cosmetics and skin care industry is spending tremendous resources searching for botanical antioxidants that can provide human cutaneous protection from oxidative damage.

Antioxidant botanicals quench singlet oxygen and reactive oxygen species, such as superoxide anions, hydroxyl radicals, fatty peroxy radicals, and hydroperoxides. Many botanical antioxidants have been commercialized and are available from raw material suppliers. These botanical antioxidants are briefly discussed, as are the currently popular botanical antioxidants, such as soy, kinetin, curcumin, silymarin, and pycnogenol, which are commonly incorporated in cosmeceutical facial colored cosmetics.

Plant antioxidants are an important reservoir of active ingredients in the search for relevant human antioxidants. Currently, there is not enough scientific data to determine which botanical antioxidant shows the most promise; however, it is worthwhile for the dermatologist to become familiar with these substances in terms of their purported function and activity.

Categorizing Botanical Antioxidants

The botanical antioxidants are too numerous to mention in one article. However, most botanical antioxidants can be classified into one of three categories as flavonoids, carotenoids, and polyphenols. Flavonoids possess a polyphenolic structure that accounts for their antioxidant, UV protectant, and metal chelation abilities. Carotenoids are chemically related to vitamin A, which encompasses all of the naturally occurring retinol derivatives. Lastly, polyphenols compose the largest category of botanical antioxidants. The following is a list of commercialized botanical antioxidants and their chemical class based on the previously discussed categories:

- Astaxanthin (tomatoes)
- Carotenoids
- Chlorogenic acid (blueberry leaf)
- Diosmin (lemons, oranges)
- Ellagic acid (pomegranate fruit)
- Flavones
- Hesperidin (lemons, oranges)
- Hypericin (St. John's wort)
- Lutein (tomatoes)
- Lycopene (tomatoes)
- Mangiferin (mango plant)
- Mangostin (bilberry plant)
- Oleuropein (olive leaf)
- Polyphenols
- Quercetin (apples, blueberries)
- Rosmarinic acid (rosemary)
- Rutin (apples, blueberries)
- Soy
- Xanthones.

Soybeans

Soybeans are a rich source of flavonoids called isoflavones, such as genistein and daidzein. These isoflavones function as phytoestrogens when orally consumed and have been credited with the decreased rates of cardiovascular disease and breast cancer observed in Asian women. Some of the cutaneous effects of soy have been linked to its estrogenic effect in postmenopausal women. Topical estrogens have been shown to increase skin thickness and promote collagen synthesis. It is interesting to note that genistein increases collagen gene expression in cell culture; however, no reports of this collagen-stimulating effect in topical human trials have been published. Genestein has also been reported to function as a potent antioxidant, scavenging peroxyl radicals and protecting against lipid peroxidation *in vivo*.

Curcumin

Curcumin is a polyphenol antioxidant derived from the turmeric root. Tumeric is a popular natural yellow food coloring used in many products, from prepackaged snack foods to meats. It is sometimes used in skin care products as a natural yellow coloring in products that claim to be free of artificial ingredients.

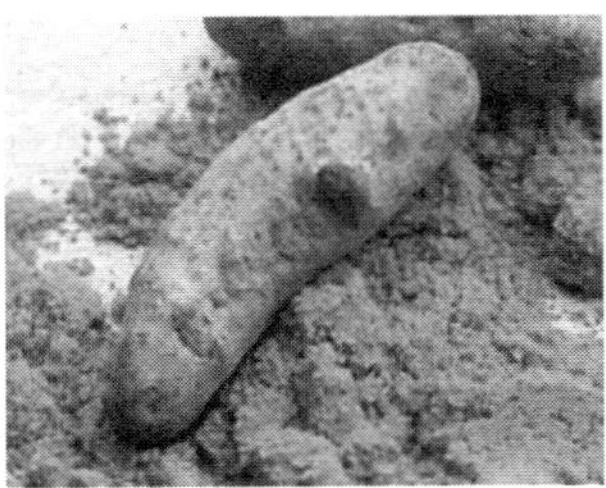

Curcumin is consumed orally as an Asian spice, frequently found in rice dishes to color the otherwise white rice yellow. However, this yellow color is undesirable in cosmetic preparations because yellowing of products is typically associated with oxidative spoilage.

Tetrahydrocurcumin, a hydrogenated form of curcumin, is off-white in color and can be added to skin care products not only to function as a skin antioxidant, but also to prevent the lipids in the moisturizer from becoming rancid. The antioxidant effect of tetrahydrocurcumin is reported by cosmetic chemists to be greater than vitamin E. Resveratrol, a chemical related to curcumin, is found in red wine, accounting for the antioxidant effect of this beverage.

Silymarin

Silymarin is an extract of the milk thistle plant (*Silybum marianum*), which belongs to the aster family of plants, including daisies, thistles, and artichokes. The extract consists of 3 flavonoids derived from the fruit, seeds, and leaves of the plant. These flavonoids are silybin, silydianin, and silychristin. Homeopathically, silymarin is used to treat liver disease, but it is a strong antioxidant, preventing lipid peroxidation by scavenging free radical species. Its antioxidant effects have been demonstrated topically in hairline mice by the 92% reduction of skin tumors following UVB exposure. The mechanism for this decrease in tumor production is unknown, but topical silymarin has been shown to decrease the formation of pyrimidine dimers in a mouse model. Silymarin is found in a number of high-end moisturizers for benign photoaging to prevent cutaneous oxidative damage.

Pycnogenol

Pycnogenol is an extract of French marine pine bark (*Pinus pinaster*), which is reported to function as a plant-derived antioxidant. It is a water-soluble liquid that contains several phenolic constituents, including taxifolin, catechin, and procyanidins. It also contains several phenolic acids, including p-hydroxybenzoic, protocatechuic, gallic, vanillic, p-coumaric, caffeic, and ferulic. It is a trademarked ingredient that is sold for oral consumption as a preventative for cardiovascular disease and as a topical skin antioxidant. It is a potent free radical scavenger that can reduce the vitamin C radical, returning the vitamin C to its active form. The active vitamin C in turn regenerates vitamin E to its active form, maintaining the natural oxygen-scavenging mechanisms of the skin.

Pycnogenol is the ideal antiaging additive because it demonstrates no long-term toxicity, no mutagenicity, no teratogenicity, and no allergenicity. It is consumed orally to enhance the production of nitric oxide, which inhibits platelet aggregation in coronary artery disease, thus it is also deemed safe for topical use. In short, pycnogenol is one of the new types of oral supplements sold for improving the appearance of benign photoaged skin from the inside, while topical application is said to augment this effect. As with many trademarked dietary supplements, validating the purported benefits is difficult.

Kinetin

Unlike the previously discussed compounds, kinetin is not a naturally occurring plant substance. It is a member of the N6-substituted adenine derivatives, known as cytokinins. In plants, this hormone has been shown to stimulate transcription and influence the cell cycle by stimulating growth. It is also a plant antioxidant. The specific cytokinin that is used in the commercial moisturizers currently marketed is N6-furfuryladenine. Kinetin is said to improve benign photoaging by decreasing fine wrinkles, improving pigmentation, and increasing skin smoothness. It is typically compared to the retinoids; however, human cells do not contain kinetin receptors while they do contain retinoid receptors. Whether moisturizers containing this active agent provide benefits above and beyond those attributed to moisturization alone is currently unknown.

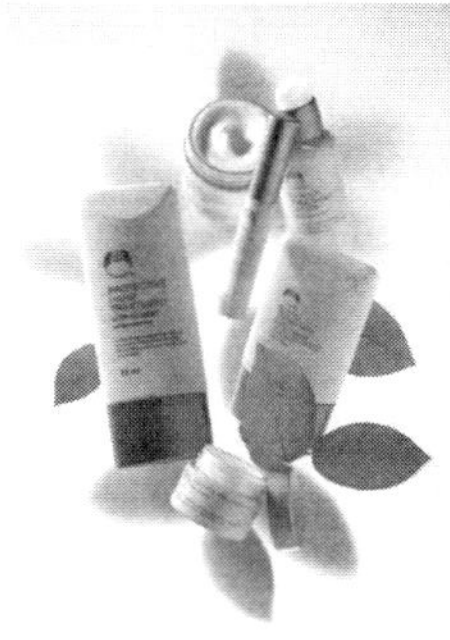

SPECIAL ACTION OF SOME OF THE NATURALLY OCCURRING BOTANICALS IN COSMETIC PREPARATIONS

Aloe Vera

Aloe is currently enjoying a resurgence of interest that began in the 1970's. Emphasis over the past few years on natural, healthful ingredients has benefited this quite ancient folk medicines *Aloe barbadensis* miller or *Aloe vera* Linn. is one of several hundred species of aloe. What may be confusing to some is the fact that this plant is completely different from Aloe vera Miller, which is not useful in cosmetics. In fact, this means that the common term **"aloe vera"** should be used with care.

Aloe is not a cactus but a member of lily (liliaceae) family. Aloe Vera is a kind of perennial succulent, multiferous medicinal herb gifted by nature. Aloe Vera is one of the oldest known medicinal plants. It is mentioned in Vedas, Quran, and Bible etc. It is unique plant with multipurpose and multi-nutritional qualities, possessing multi action and helps to maintain and restore health, promotes healing and maintain glow on the skin and its texture. Aloe Vera also known as **"The first aid plant"** as it has ability to heal, alleviate and eliminate skin disorders. Aloe Vera has cooling effect on skin and helps in healing cuts, burn and wounds. It penetrates the skin at cellular level and is a good moisturizer; it protects skin from harmful UV rays. It provides rich cocktail of nutritional elements like vitamins, enzymes, minerals, anthraquinones (aloin as laxative), lignins, saponins, fatty acids, salicylic acid, amino acids, proteins and sugar.

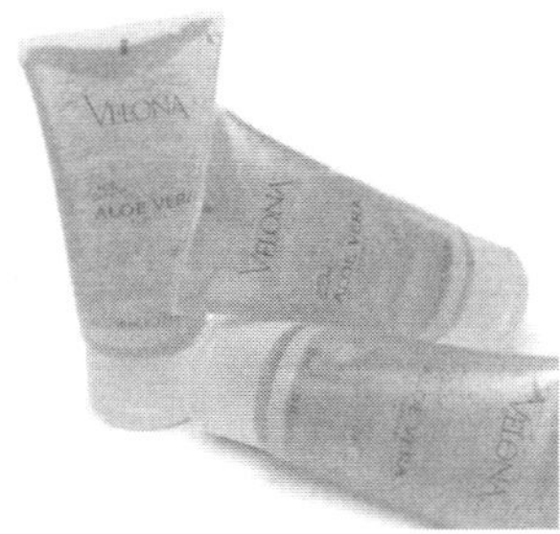

Now-a-days, a plethora of Aloe Vera products are available in the form of Aloe Vera health drinks, Aloe Vera skin gel, Aloe Vera after shave lotion, Aloe Vera cucumber cream, Aloe Vera herbal scrub, Aloe Vera hair gel, hair cleaner and hair richer, Aloe Vera herbal bath soap, hand wash and shampoo, Aloe Vera tooth cleaner, Aloe Vera herbal balm, Aloe moisturizing cream and lotion, Aloe Vera sun screen lotion, and multi-purpose natural cleaning solutions.

Applications: The drug USP is the "dried juice obtained from the pericyclic cells beneath the leathery skin" while aloe gel is obtained from the "inner, thin walled mucilaginous cells of the plant and does not contain anthraquinone glycosides as aloe USP except may be in trace amounts." Upon purification the drug aloe yields aloin and aloe extracts, both of which are used in cosmetics. Aloe vera gel has been added to a wide spectrum of cosmetic products. Skin care products containing aloe vera gel including products (suntan lotion, sunscreen gel, aftershave moisturizer, sunburn relief lotion and gel); moisturizers (emollient lotion, cream, night cream, baby lotion, hand cream) and cleansers (cleaning cream, peel off mask, skin freshener, facial scrub, baby wipes, wet wipes). Hair care products including general purpose and antidandruff shampoo, baby shampoo gel, and hair conditioner gel. Nail care products include nail crème and nail polish remover. Miscellaneous cosmetics containing aloe include aftershave lotion, sports rub gel, lip balm.

For the formulator, aloe comes in a variety of forms. The gel can be purchased as pure gel, in a natural or decolorized state. It also comes as a liquid concentrate (in 10, 20 or 40 fold concentrations) again either natural or decolorized). The gel can be concentrated and dried under vacuum to produce spray and freeze dried powders, and there are also oil extractions and separated pulp. Form chosen for cosmetic depends upon such factors as miscibility with product, appearance or color of finished product, and shipping, storage and handling costs of the aloe material. Fully modernized production facilities like those of Aloe crop are capable of supplying all these forms to accommodate a wide range of formulator needs.

Formula XXIX: **Powder Mask Activator**

Material	% w/w
Deionized water	95.3
Croquet M (Cocodimonium hydrolyzed animal protein)	0.5
Incronan AL-30	0.5
Honey	1.2
Aloe vera gel liquid 1:10	1.0
Lipofruit cucumber	0.5
Germaben II	1.0

Procedure: Dissolve Germaten II in water, when uniform, add Incronan AL-30. Next dissolve aloe, honey and cucumber extract. Mix well, add Croquat M, adjust pH to 4.2±0.1 with 10% TEA. Immediately prior to application, mix the two components to form thick paste.

Formula XXX: **After Sun Cooling Gel**

Material	**Parts**
Lubrajel MS	50
Deionized water	44
Aloe vera gel (4x)	5
Sorbitol	1
Propylene glycol	1
Methyl paraben	0.1
Propyl paraben	0.02

Procedure: Premix aloe, sorbitol and water. Add Lubrijel MS with low shear mixing until homogeneous. Premix methyl paraben, propyl paraben and propylene glycol. Mix into formulation and package.

Rose In Perfumery and Cosmetics

Probably the most universal flower, appreciated in eastern as well as western civilizations, is the rose.

In ancient Rome, rose pomade was used for festive occasion and in rituals, and dried rose petals were used in home made cosmetics. Although a primitive distillation of roses was already known in ancient India, invention of distillation is attributed to the Arabs. The famous rose water was used lavishly in Arab world, and became known in other countries.

In the Middle Ages rose water was known as **"Saracen novelty"** in France and Germany. In the 16th century rose water was used in conjunction with other spices and aromatic materials as medicine against plague and pestilence.

Persia became the center of the rose industry between the 10th and 17th centuries. From there it spread to North Africa and eventually to Spain. During the crusades the rose found its way to Southern France, and Turkish invasions spread the rose to the Balkans in the early 17th century.

By the 19th century, Bulgarian rose plantations were well developed, and because of the high quality of their rose oil acquired almost a monopolistic position. Turkey also began producing rose oil, though the quality was inferior. In the early 19th century *Rose centifolia* began to be cultivated in Southern France.

Morocco now is the newer source of rose oil. India, Pakistan, Japan, China and Soviet Union are among other countries which produce rose oil. Of the more than 10,000 varieties of rose known to the horticulturist, and many different colour rose flowers enjoyed by the layman, it may come as a surprise that two are mainly used in perfumery.

1. *Rose damascene* Mill, a pink rose grown in the Bulgaria and Turkey.
2. *Rose centifolia* L., also known as Rose de Mal is cultivated mainly in Southern France and Morocco.

Of these two varieties *Rose damascene* Mill is processed by distillation of the flowers, and yields the essential oil known as otto of rose (or attar of rose). *Rose centifolia* L. extracted with volatile solvents, because distillation gives a poor yield, rose concrete and absolute are obtained. With modern selective solvent techniques purer absolutes are now obtained. The distilled rose otto has more top note, the absolute pure oil more fixative power, in colour rose otto is light yellow, sometimes with a greenish tint, and absolute rose is yellow-orange to orange-green.

According to some procedures, 40,000 rose flowers are needed to obtain one ounce of rose oil. The yield of *Rose centifolia* L., using petroleum ether as solvent is 0.24 to 0.265 percent of the concrete, the concrete gives from 55 to 65 percent of the absolute oil. Average yield of *Rose damascena* is 1 kg of oil per 4,000 kg of flowers, when distilled. Modernized processing methods (i.e. subjective roses to acid hydrolysis or enzymatic hydrolysis) have increased the yield of rose oil.

Until 1950 only major rose oil components and few minor ingredients were known. Guenther lists the following constituents identified in Bulgarian rose oil:

l-citronellol, nerol, germiniol, l-Binelcol, phenyl ethyl alcohol, farnesol and esters nonyl aldehyde and possible other higher aldehydes citral, eugenol, methyl eugenol, carvone, a small amount of azulenogenic sesquiterpenes.

First breakthrough came in 1957 with the identification of rose oxide. By 1958 a total of 39 Bulgarian rose oil components were known. Improved instrumentation made possible identification of other components present at concentration below 1 percent. Today 275 rose oil constituents are known. Among the more important minor ingredients discovered within the last 15 year are rose furane, nerol oxide, β-damascone and β-damascenone. But there is still a long way to go to reconstitute natural rose oil.

Synthetic Compounds

Rose lends to infinite variation since it is the most universal perfume component. It is compatible with many other floral notes, hyacinth, jasmine, lilac lily of the valley, mimosa, narcissus, orange flower, tuberose and violet, for instance, rose blends well with woody and spicy notes, geranium, lavender, and green-herbal notes. Rose compounds share common ingredients, i.e. citronellol and esters, geraniol and esters, hydroxycitronellal, linalool, phenylethyl alcohol and esters, nerol, rhodinol and esters, phenyl acetaldehyde and its dimethyl acetal, ionones and derivatives alcohol and aldehydes C-8 to C-11, aldehydes C-14 and C-16, among others. The classical synthetic rose compounds are: Rose d'Orient (otto), rose centifolia, red rose, white rose, rose geranium. The base of these rose compounds usually is a mixture of the main rose alcohols, i.e. rhodinol, geraniol, citronellol, nerol and phenylethyl alcohol which form the bulk of the natural rose oil.

Application of Rose Compounds

There is hardly a fragrance which would not have a rose compound, or at least a rose component, as part of its formula. The much copied Rose Marechal Niel of the past, used in extract form, was a rose compound which contained vativer acetate, methyl ionone, santalol cinnamyl esters, musk, iris and natural rose oil. Another favorite of the past was Tea rose which was characterized by the use of various cinnamic esters, quaiacwood and its acetate, as well as reseda compounds. Tea rose has better popularity, its formula being modernized by the use of damascenones and other new rose oil components.

Rose is also used in variety of cosmetic, toiletry and soap fragrances. It was included in various milks, brown spot removal preparations, hair colorants and used in a combination with orange flower, to which a small amount of sodium borate was added, it was even found beneficial for red rose. Today, rose is a popular fragrance in creams as well as shaving creams, and it still used in baby talc and soaps.

Rose fragrance is known for its ability to mask base odors and depilatories and permanent lotions; in the latter, more modern rose compounds may include a small amount of 2-cyclohexyl-cyclohexanoate, of a herbal-mint odor tonality.

Many recent cosmetic and toiletries preparations contain proteins, and some rose fragrance used in such products may have to be adjusted, since it is known that phenylethyl alcohol associates with the protein substrate.

Formula XXXI Sulphur lotion

Material	Quantity
Precipitated sulphur	10gm
Spirits of camphor	10 cc
Alcohol	80 cc
Methyl cellulose (2Y solution)	30 cc
Rose water to make	240 cc

Acne usually occurs in individuals whose skin is oily, so it is advisable to use a non-greasy base for creams containing sulphur.

Formula XXXII Sulphur skin cream

Material	Quantity
Precipitated sulphur	2gm
Salicylic acid	2 gm
Cream base (see below)	26 gm
Oil of Rose	0.25 cc

Formula XXXIII Cream base

Material	Quantity
Sodium lauryl sulphateCetyl	0.80 gm
alcohol	15 gm
Glycerin	5 cc
White petrolatum	14 gm
Water	35 cc

MARKETING BRANDS OF HERBAL COSMETICS

- **Himalaya herbal healthcare:**
  - i. *Herbal hair care products:* Anti-dandruff hair oil, anti-dandruff shampoo, dandruff hair cream, hair loss cream, protein conditioner, protein hair cream, protein shampoo, protein shampoo and conditioner and revitalizing hair oil.
  - ii. *Herbal products for skin care:* Acne-n-pimple cream, anti-wrinkle cream, astringent lotion, deep cleansing milk, face moisturizing lotion, fairness cream, gentle foliating apricot scrub, gentle foliating walnut scrub, gentle face wash cream, gentle face wash gel, gentle refreshing toner, intensive face moisturizing lotion, lip balm, moisturizing almond soap, moisturizing liquid soap, neem face pack, nourishing skin cream, peel-off face mask, protective sunscreen lotion, purifying mud pack, purifying neem face pack, refreshing cucumber soap, refreshing fruit pack, revitalizing night cream and soothing body lotion.
  - iii. *Health care herbal product:* Antiseptic cream, anti-stress massage cream, ayur slim capsules, blood purifier capsules and syrup, cold balm, daily health capsules, foot care cream muscle and joint rub, pain balm, pain massage oil.
  - iv. *Baby care herbal products:* Baby cream, powder and lotion, diaper rash cream, gentle baby shampoo, moisturizing baby soap, nourishing baby oil .

HERBS AND OTHER NATURAL SOURCES USED IN HERBAL COSMETIC

A. Herbs

Plants are indispensable to human being for survival has been proved since antiquity. They are the source of herbal cosmetic. The major part of herbs is being consumed in the cosmeceuticals industry and is realized by going through their estimated demand in the US market. At present United States annual market for cosmeceuticals has been estimated to be $2.5 US billion, where as $5.0 million for European countries.

Cosmeceuticals are the cosmetic products, which contain biologically active principles or ingredients of plants origin having effects on users, or they are combination products of cosmetics and pharmaceuticals, intended to enhance the health and beauty of the skin. They differ from cosmetics, since cosmetics are the inert substance which cleanse or used to treat or prevent the diseases or are intended to affect the physiological structure or function of the body (Table 27.6).

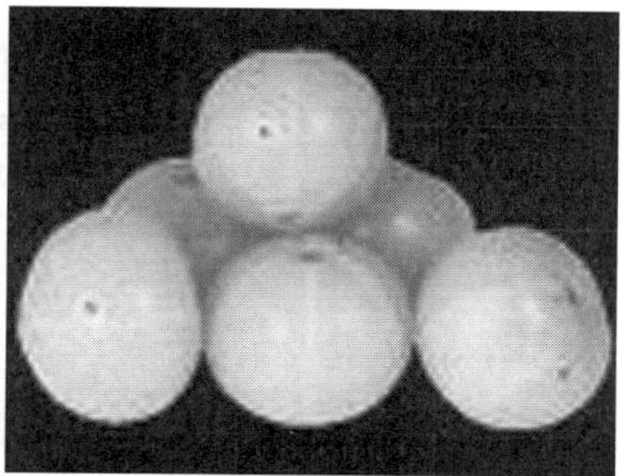

Amla
(*Emblica officinails*)

Aloe vea (*Aloe barbadensis*)

Arjuna
(*Terminalia arjuna*)

Bitter almond
(*Prunus amygdalus*)

Coconut oil
(*Cocos mucifera*)

Citronella
(*Cymobpogon gardus*)

Eucalyptus
(*Eucalyptus globulus*)

Evening primrose
(*Oenothera Biennis*)

Henna powder
(*Lawsonia inermis*)

Hibiscus
(*Hibiscus rosa sinensis*)

Jojoba oil (*Simmondisa chinesis*)

Kokum butter
(*Garcinia indica*)

Lemon
(*Citrus limonis*)

Neem
(*Azadirachta indica*)

Olive oil
(*Olea europoea*)

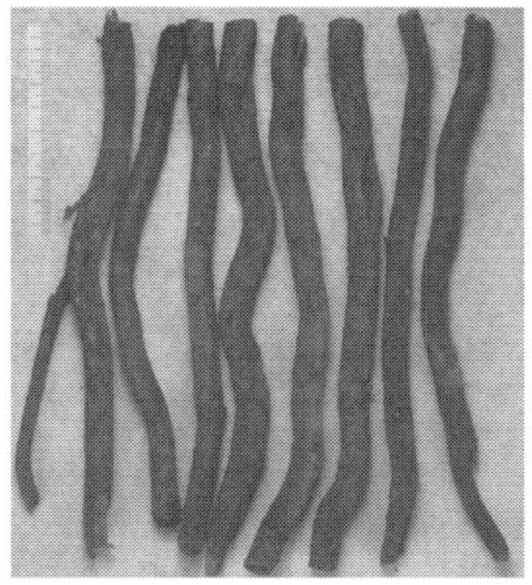

Quillaia bark
(Quillaja saponaria)

Reetha
(Sapindus trifoliatus)

Shikakai
(Acacia concinna)

Satawari
(Asparagus racemosus)

Sesam oil
(Seasamum indicum)

Sandalwood oil
(Sentalum album)

Turmeric
(Curcuma longa)

Tulsi
(Ocimum sanctum)

Rosemary
(Rosmarinus officinalis)

Vanilla
(Vanilla griffithii)

Table 27.6: Herbs used in cosmetics

Herbs	*Botanical name*	*Part used*	*Chemical constituent*	*Cosmetic use*
Amla	*Emblica officinalis (Euphar-biaceae)*	Dried fruits (Powdered form and volatile oils)	1. Natural source of Vit. C 2. Fruit contain 0.5% fat, phyllembin and 6% tannin 3. Rich materials of phosphorus, iron and calcium 4. Fixed oils	Conditioner, refrigerant hair oil and shampoo promoting hair growth
Aloe vera	*Aloe barbaden-sis (Lilaceae)*	Juice of leaves and mucilaginous tissue of leaves	Mono and polysaccharides tannins, steroles, cycloxygenase, saponin vitamins and minerals Lipid contains cholesterol, gamolenic acid and arachidinic acid	In ointment creams to assist healing of wounds, burns, eczema and in psoriasis in hair gel, lotions, scrubs etc.
Arjuna	*Terminalia arjuna (Com bretaceae)*	Dried stem and bark (powder)	Ellagic acid, β-sitosterol	In skin disease, face cream and lotion
Bitter almond	*Prunus amygdalus (Rosaceae)*	Dried ripe seeds (Volatile oils)	1. Fixed oil (50%) 2. Protein (20%) 3. Volatile oils (0.5%) 4. Amygdalin glycoside 5. Benzaldehyde (80%)	Demulcent skin, in lotion, hair oil, hair conditioners
Castor oil	*Risinus communis*	Dried seeds	Ricinoleic and isoricinaleic acid	Lubricants and emollients. In creams and lotions
Coconut oil	*Coccos mucifera (palmaceae)*	Dried fruit	Triglycerides	Has a abundant lather and used in the manufacturing of hair oil, toilet and shaving soaps
Citronella	*Cymobpogon gardus (Germinae)*	Fresh leaves (Volatile oil)	Geraniol (40%) Citronellal (5–20%) Camphene, Limonene elimicin	Flavoring agents, in liniments and lotion, perfumes for soaps
Eucalyp-tus	*Eucalyptus globulus (Myrtaceae)*	Dried fresh leaves (Volatile oil)	Cineole, pinene, camphene, phellan-drene, citronella, geranyl acetate	Counter and emollient, effect used in cold creams and liniments
Evening primrose oil	*Oenothera Biennis (Onagraceae)*	Dried ripe seeds (Volatile oil)	Gamma linoleic acid and linolenic acid	Nutritive and emollient effect used in cold creams and liniments
Henna	*Lawsonia inermis (Lythraceae)*	Dried leaves (Powdered form)	Lawsone (0.5% to 1%), Gallic acid, white resins, sugars and tannins, xanthines	Used in hair dye, several hair products like rinses conditioner, antibacterial and antifungal action
Hibiscus	*Hibiscus rosa sinensis*	Fresh flowers (Volatile oil)	Volatile oil	Natural foaming agent, removes dirt, conditioning agents for hair, promote hair growth and used in hair preparation
Jatamansi	*Nordostachys Jatamansi (Velerianaceae)*	Dried Rhizomes (Volatile oil)	Yellow volatile oils, isovaleric aster, Jatamansi acid, ketones Jatamansone, Nardosstachone	Flavoring agents also promotes the growth of hair, in parts the bleaches of hair
Jojoba oil	*Simmondisa chinesis (Buxaceae)*	Dried seeds (Volatile oil)	Mixture of long chain esters of fatty acids	Lubricants and emollient used in hair oil, creams and ointments
Kokum butter	*Garcinia indica (Guttiferae)*	Dried Roasted seeds (fixed oils)	Glycerides or stearic, palmitic acids, linolenic acid	Demulcent, emollient effect, used in creams, lotion

Contd.

Herbs	*Botanical name*	*Part used*	*Chemical constituent*	*Cosmetic use*
Lemon	*Citrus limonis*	Dried peels (outer parts of pericarp) of ripe fruits	Hesperidin, pectin, limanen, geranyl acetate, sitral	Perfuming and flavoring agents
Neem	*Azadirachta indica (Meliaceae)*	Dried leaves and stems (Powder and volatile oil)	Leaves—Azadirachtin meliantral slanin, Seeds—Nimbin, nimbidin, azadirachtin Barl—Margolone, margalonone Fruit—Deacetyl azadirach tinal	In eczema and skin condition, widely used in beauty enhancer, antibacterial action used in face masks. Body wraps, face and body scrubs also
Olive oil	*Olea europoea (Oleaceae)*	Dried ripe fruit (Volatile oil)	Olein, palmitin and linolein	Emollient effect used in creams and vaseline
Quillalia	*Quilaja saponaria (Rosaceae)*	Dried inner part of bark (powdered)	Triterpenid saponia glycosides, sucrose and tannins	Emulsifying agent, detergent, used in preparation of shampoo and soap
Reetha	*Sapindus trifoliatus*	Dried fruit (powdered form)	Foaming and soaping agents like, saponins and other fixed oils	Used as an excellent hair tonic, washing and cleansing so widely used in facial scrubs creams
Satavari	*Asparagus racemosus (Liliaceae)*	Dried roots and leaves (Powdered form)	Roots-Steroids saponins, Shatavari I-IV, it contains glucose, rhamnose moieties attached with sars apopgenin	Used in ointment to stimulate hair growth
Shikakai	*Acacia concinna*	Dried pods (Powdered form)	Cleansing agents antioxidants	Cleanser for hair, astringent, reducing dandruff and creating shiny, soft and clean hair, cooling effect, used for hair growth
Sandal wood	*Sentalum album (Santalaceae)*	Dried stems (powdered and volatile oil)	Oils of santalum contains 1. α-santalol, β-santalol 2. Santene, santenone, teresantol, sentalone and santelene	Antimicrobial, soothing, catanious, inflammation, alleviating itching, skin softener, used in creams face scrubs, perfumes
Sesame	*Seasamum indicum (Pedaliaceae)*	Dried seeds (fixed oils)	Fatty acids like linoleic, palmitic, stearic and arachidic acids, Pheno-seasamol, sesamin and seasamolin	Demulcent, emollient used in liniments, ointments, soaps skin and hair tonic
Turmeric	*Curcuma Longa (Zingi-beraceae)*	Dried Rhizome (Powdered form)	Volatile oil, resin,curcumin, colouring substances curcuminoids α and β pinene, camphor, camphene	Used in eczema, skin infection, in face packs, used in ointment creams, as colouring agents
Tulsi	*Ocimum sanctum*	Dried leaves (Volatile oil)	Volatile oils, eugenol, carvacrol, engenolmethyl-ether, caryophyllin, seeds fixed oils	Antibacterial, insecticides, removes dandruff
Rosemary	*Rosmarinus officinalis (Labiatae)*	Fresh flowering tops (Volatile oil)	Volatile oils, Borneol, Bornyl acetate, Camphor, Cineol	Rubefacient, flavoring agents, used in hair lotions and soaps
Vanilla	*Vanilla griffithii (Orchidaceae)*	Dried leaves (Volatile oils)	Gluco vanillin	Hair to produce thick and healthy growth
Pumpkin	*Cicirnota Pepo (Cucurbitaceae)*	Ripe fruits		Juvenile acne and acne vulgaris, pimples, black heads also anti seborrhea agents

B. Other Natural Sources

1. **Waxes:** Waxes are unctuous, fusible, variably viscous solid substances, with characteristic waxy lustre. These are esters of fatty acids with high weight monohydric alcohol, such as cholesterol, cetyl alcohol, melissyl alcohol, etc. They are insoluble in water, but soluble in most organic solvents. The esters in waxes are generally more resistant to saponification than the glycerides of oils and fats .They are obtained from vegetable and animal sources. (**Table 27.7**)

Beeswax

Natural source: secreted from the underside of the honey bee (*Apis mellifica*) and used to form the wall of honey comb cells.

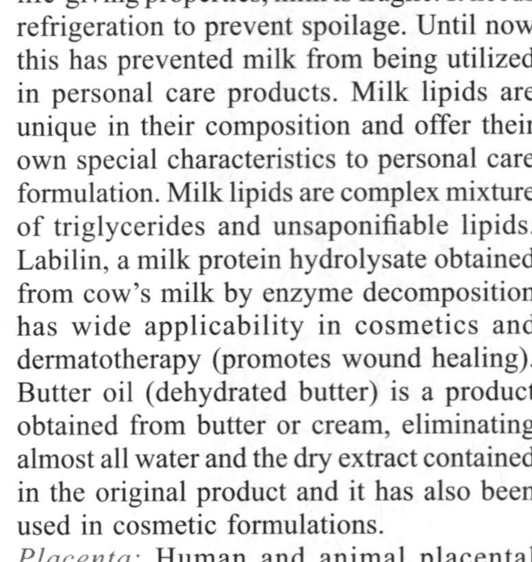

Uses: In floral fine fragrance water in oil emulsions and cosmetics creams.

2. **Mineral origins:** Majority of natural drugs are derived from plant and animal origins, a few of them obtained from mineral sources are of paramount significance in herbal cosmetic. The summary of mineral is given below. (**Table 25.8**)

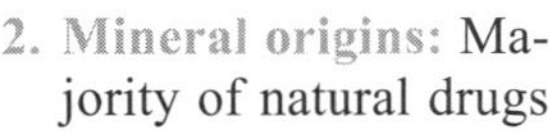

3. **Others**

- *Milk:* Interest in milk has gone beyond its role in diet. The image of Cleopatra in a bath of milk has been used time and again to communicate beauty and elegance. Today many consumer products (Dove, Camay) attempt to capitalize on these images but without the ability to actually incorporate milk. Despite all its life-giving properties, milk is fragile. It needs refrigeration to prevent spoilage. Until now this has prevented milk from being utilized in personal care products. Milk lipids are unique in their composition and offer their own special characteristics to personal care formulation. Milk lipids are complex mixture of triglycerides and unsaponifiable lipids. Labilin, a milk protein hydrolysate obtained from cow's milk by enzyme decomposition has wide applicability in cosmetics and dermatotherapy (promotes wound healing). Butter oil (dehydrated butter) is a product obtained from butter or cream, eliminating almost all water and the dry extract contained in the original product and it has also been used in cosmetic formulations.

- *Placenta:* Human and animal placental extracts in cosmetics are claimed to have a softening and hyperaemizing effects on the skin. Water-soluble placental extract can be examined for Glutamate-Oleate-Transaminase (GOT), Glutamate-Pyruvate-Transaminast (GPT), Lactate Dehydrogenase (LDH) and Alkaline Phosphatase (AP) activities.
- *Egg:* Egg oil is regarded as an excellent vehicle to flavour absorption of the polylipoamino acid following local application in the form of face creams, hair lotions and hair dressings. Egg oil extracts dried egg yolk. It has a pale yellow colour due to the natural content of carotenoids.

- *Hormones and Estrogens:* These when incorporated into cosmetic (250–230 IU./gm) are suggested to retard aging of the skin and to make the wrinkles less apparent. Studies have also been conducted on various proteolytic enzymes in cosmetic beauty masks with results claiming a deep physiological cleansing property including loosening of follicular plugs.
- *Canada balsam*

  Natural source: The balsam fir, *Abies balsamea* Miller and the hemlock spruce.

Uses: Excellent fixative for soap, perfumes containing a high percentage of citrous oils may be used upto as much as 10% in cologne and verbena compounds.

- *Carragenan*
  Natural source: *Chondrous cripus*, stack house, N.O. Gigertinaceae
  Uses: A small percentage of the mucilage is used in vanishing creams, thus increasing the adhesive qualities of the product.

- *Castor oil*
  Natural source: *Ricinus comunis, Linn* N.O. Euphorbiaceae.
  Uses: In lipsticks, imparting a spreading quality and acting as a partial solvent for eosin, also as important constituent of transparent soaps.

- *Coconut oil, Copra oil*
  Natural source: The coconut palm, *Cocus mucifera* and *C. butyraceae Palmaceae.*
  Uses: It has an abundant lather and is used in the manufacture of toilet and shaving soaps.

- *Cucumber juice*
  Natural source: *Cucumis sativus*, a trailing annual of N.O. Cucumbitaceae.
  Uses: In the preparation of glycerin, cucumber, creams and face packs.

- *Curcuma oil*
  Natural source: *Curcuma longa*
  Uses: It is used by the Egyptians to perform the red slippers sold in the bazaars. Sometimes used for colouring face powders.

STANDARDISATION AND QUALITY CONTROL OF HERBS IN COSMETIC

In order to produce quality products, it is essential that standardization and validation of botanical materials and process is done and documented. It is an exceedingly complicated job to standardize thousand of plant extracts with respect to their medicinal value and constituents. The quality in yester years was based on physical aspects of the plant material such as identification, colour, odour, size, type, age, etc. Today, there is an additional requirement, of knowing the exact chemical composition of the botanical raw material, along with the different physical tests. Standardization of natural products is a complex task due to their heterogeneous composition, which is in the form of whole plant, parts or extracts obtained there of. To ensure reproducible quality of herbal products, proper control of staring material is utmost essential.

India's share in the global trade of herbals is very poor due to lack of quality control and standardization measures. Quality control of herbal

Table 27.7: Waxes used in cosmetics

Name	*Biological sources*	*Active constituents*	*Cosmetic uses*
Bees wax	Apis mellifica (Apidae), Obtained from: Honey comb of bee	Myricyl palmitate, cerotic acid	Hardening agent, ointment bases
Spermaceti	Pyster macrocephalus (Physeteridae), Obtained from: Head sperm of whale	Cetyl palmitate	Cosmetic creams and vaseline
Lanolin (Hydrous wool fat)	Ovis aries (Bovidae), Obtained from wool of sheep	Esters of cholesterol and carnaubic and oleic acids	Absorbent and hardening in ointment

Table 27.8: Minerals used in herbal cosmetics

Name	*Main constituents*	*Uses*
Bentonite	Zinc oxide and 5% of ferric oxides	Emulsifiers, bases of lipsticks, creams
Calamine	Zinc oxide and 0.5% of ferric oxide	Skin protectant, astringent locally
Talc (Talcum)	Hydrated magnesium silicate	Lubricant, dusting powder
Chalk	98% calcium carbonate and magnesium carbonate	Used in tooth powder, dusting powder
Fullers earth (floridin)	Aluminium, magnesium silicate	Dusting powder, decoloring of oils

preparation is determined by the quality of the starting material, development, in-process quality control, GMP controls and specifications applied to them throughout development and manufacturing process. Various factors for quality assurance of herbal products: proper identification, good collection practice, trace the trade channel, right processing, right packaging, right storage, proper labeling, test against Ayurvedic Pharmacopoeia, re-testing and honest approach.

A. FACTORS INFLUENCING QUALITY OF HERBS

Authentication of Name and Source of the Herb: Majority of the herbs in India come from wild sources and are collected by poor, illiterate tribal and local people, without any attention to botanical identification and authentication. Thus, the material supplied is mostly adulterated, either intentionally or unintentionally. Each crude drug has different names in vernacular languages and local trade occurs in these vernacular names only. Botanical/Latin names are rarely used for these purposes.

Adulteration: The likelihood of adulteration of raw botanical material is very high. It could be unintentional as there are many crude drugs which when fully dried look morphologically very similar to each other. On the other hand, it could be deliberate, for example, unwanted foreign organic and inorganic material may be added to increase weight. In such circumstances, it becomes mandatory to resort to complete manual cleaning of raw material prior to processing.

Phytochemical Variation Due to Environmental Factors: There are many intrinsic factors, which govern the growth and medicinal quality of herbs. This is largely due to the change in their chemical constitution, which often leads to change in their bioactivity. Due to these inherent variations, standardization will help control the quality. The major environmental factors are:

- Seasonal changes
- Geographical variations
- Age of the plant at time of the harvest
- Genetic factors (ploidy and variety)
- Edaphic factors (soil pH, soil composition, macro and micronutrients).

Inadequate knowledge of active principles: Each herb is a complex mixture of hundreds of organic compounds containing different types of secondary metabolites and mostly the active principal is not identified.

Lack of reference standard: Reference standard are required at two stages:

1. Crude drug reference standard
2. Active principle/marker compounds.

Authentic reference standards, both for crude drug as well as for active principles are not easy to obtain for all the drugs, but still many Herbal Research Companies and Laboratories are making efforts to develop a library of standards of the phytochemicals.

Inadequate number of official monographs: There are hardly any pharmacopoeias available to the nature product scientist for guidance. Indian Herbal Pharmacopoeia and The Indian Pharmacopoeia, 4th edition, 1996, Vol. I and II, addendum 2003, has analytical monographs on very few medicinal plants. Though Ayurvedic Pharmacopoeia of India (Vol. 1–3) covers about 240 medicinal plants, but the parameters are not adequate for export purpose.

B. DIFFERENT METHODS OF STANDARDIZATION OF HERBS IN COSMETICS

1. **Development of in-house crude drug references standards:** For the purpose of identification of the crude drugs, it is best to possess the authentic reference standard of those particular crude drugs. This can be done by generating in-house reference standard for crude drugs. Fresh plants can be collected, from right geographical locations, in the right season, at the right age. Its taxonomic identity may be verified, followed by preparation of herbarium sheets and documenting the relevant collection details. Then they should be care fully collected, dried under controlled conditions and stored appropriately so as to serve as reference standards.
2. **TLC finger printing:** Many of the crude drugs have look-alikes and it becomes difficult to differentiate the genuine ones from the fake drugs after they are powdered. This then necessitates the application of thin layer chromatography (TLC). TLC is one of the easiest and cheapest tools available for phytochemical analysis and is the method of choice for crude drug authentication. The TLC fingerprint of the sample under test is compared with the TLC finger print of the crude drug reference standard. Thus, availability of the crude drug reference standard becomes very crucial for Quality Control.
3. **Quantification of markers/actives:** Markers are compounds unique to the plant in question and are preferably present in detectable amounts and can be easily isolated .A marker compound can be quantified and the quantification can be carried out by High Performance Thin Layer Chromatography (HPTLC) or with High Pressure Liquid Chromatography (HPLC). HPTLC is well suited to obtain a detailed fingerprint of herbal extract or product. Such a fingerprint comprises of scanning in UV light, fluorescence and photographic images in ultraviolet light (254 nm and 366 nm) and occasionally in visible light after derivatization. For plants where actives/ markers are not known phytochemical profiling of the category compounds can be carried out. Plants contain different categories of molecules, viz., alkaloids, terpenoids, glycosides, saponins, flavonoids, tannins, etc. For the quantification of these categories of compounds the gravimetric method can be adopted for crude drugs.
4. **Residual analysis:** One of the very important aspects of standardization is to ensure the absence of objectionable amounts of pesticide residues, heavy metal residues, organic solvent residues, mycotoxin residues and bioburden.
5. **Biological standardization methods:** A bioassay called Brine Shrimp Lethality Bioassay is being used as a routine biological quality control tool along with finger printing.
6. **Physicochemical and psychrometric evaluations:** The physicochemical parameters for herbal creams and lotions are like pH, acid value, ester value, saponification value, spreading test, skin irritation and sensitivity

test, thermal stability, microbial sensitivity and sun protection factor (SPF).

LIMITATION OF HERBS IN COSMETICS

The major problems in the use of herbs and natural products in cosmetics remain that of product uniformity or constituent variability, quality control, coupled with formulation stability and preservation. The other difficulty, which is encountered in using phytoconstituents is seasonal variation of plant constituents.

Natural extract are not as consistent as products based on synthetic chemicals. Inherently, they tend to be less stable, often requiring the addition of a suitable preservative, varying in consistency from batch to batch, and be subject to seasonal availability which is often reflected in fluctuating price and doubtful continuity of supply.

These problems are not necessarily as serious as to preclude the use of botanicals but are aspects that must be taken into consideration when including these substances in a new product formulation.

The makers of cosmetics and perfumes have long experimented with different plants for their scents beautifying properties. The beautician can produce any number of cosmetics containing some kind of herbal ingredient. In cosmetology, it has now become fashionable to use "biological vegetable complexes" or to use "herbal components" to produce preparations which apart from the cosmetic effect, claim to have additional advantages (say, of nurturing and curative or preventive effects).

Sufficient evidence exists supporting the use of plants in folk medicine as well in cosmetology, both in India, and rest of the world.

However, it is strongly felt that the acclaimed use of any herb (or herbal ingredient) in cosmetics be based on rational, controlled experiments, and tests be also conducted so as to examine the presence of medicinal effects.

CONCLUSION

Herbs are wonderful gift of nature, compatible with both human skin and hair. They are used as powerful aids in herbal cosmetic in treating almost any skin and hair problem. Unlike chemical based products (synthetic), herbs are by nature more friendly to human body, mild, biodegradable and therefore are completely safe, extremely effective and have almost no negative side effects (low toxicity profile). They are used in cosmetic for enhancing attractiveness and cleansing purposes.

These herbs may be used to effects in their crude form or they may be extracted, purified or derivatized to render them more suitable for use on cosmetic.

The standardization of the herbs and the quality control of the herbs are very necessary in the cosmetic. The standardization generally based on physical aspects of plant material such as color, odor, size, type, age, etc. The standardization of herbs with the extract form of the crude drug. Different methods applied in the standardization like HPTLC, TLC finger printing, etc.

The major problem in use of herbs in cosmetic that of product uniformity, constituent, variability, quality control, stability and preservation.

The cosmetic use is dependent on a number of factors economic condition population trends, fashions, scientific progress, and government involvement. Most of the factors that have been responsible for recent growth of the industry will continue to have a favorable effect; increasing population, particularly in the 15 to 34 years age group. Increasing affluence resulting in greater spending for personal care items, more women in the work force, the continuing promotion and advertising. Also these are indications that exports of cosmetics and toilet goods will grow at a faster rate.

The field of herbal cosmetic users and uses, should continue increasing. With increased population with more money to spend and more time to spend it, there is every to believe that the cosmetic and toilet industry will flourish.

Avaliable Market Products

- Himalaya Herbal Product range
- Lotus Herbal Product range
- Dabur Herbal Product range
- Nature Care
- Natural Organics.

Appendix, Glossary, Question Bank and Bibliography

Appendix

The pain passes, but the beauty remains.
— Pierre Auguste Renoir (1841-1919), French impressionist painter

VARIOUS FAMOUS COSMETIC COMPANIES WORLD WIDE

Amway
Anita Roddick
Aquaunite
Aveda
Aviance Beauty Solutions
Avon glimmersticks
Babyliss
BeneFit (cosmetics)
Bobbi Brown
Bodycare
Boots
Bourjois
Braun
Brihans Natural Products
BVLGARI
C.F.HKO professional
Cartier
Chambor-true colour
Chanel
Chanel No. 5
Chanel-coco
Charles of the Ritz
Christopher Sheldrake
Chypre
Clarins
Clinique
Colorbar
Cosmetics
Coty, Inc.
Creed (perfume)
Dabur
Diana of London
Diana smart
Dior
DKNY
Dolce And Gabbana
Dr Hauschka
Eau perfumée
Edmond Roudnitska
Elizabeth Arden
Erno Laszlo
Escada
Estée Lauder Companies
Eugene Rimmel
Eugène Schueller
Eve Lom
Exceon
fcuk
Forest essentials
Forever Living Products (Ind.) Pvt. Ltd.
François Coty
Garnier

GHD
Gillette
Guerlain
Hard Candy (cosmetics)
Heaven's garden
Helena Rubinstein
Hermès
Himalaya cosmetics
Horst Rechelbacher
Iman Cosmetics
Jean Despres
Jean Paul Gautier
Kenneth Cole Signature
Kerastage
Kiehl's
Kohl (cosmetics)
Kräuter healthcare ltd.
L'Oréal India
La galena
Lakmé
Lancome
Lancôme
La Roche-Posay
Lotus herbals
Lux
LVMH
M cont.
MAC Cosmetics
Make-up Studio
Mary Kay
Mascara
MaxCare products
Max Factor
Max Factor, Sr.
Maybelline
Miessence
ModelCo
Naturence Herbals
Neals Yard Remedies
Nectar Naturals
Nicky Clarke
Nina Ricci
Nivea
No7 by Boots
Olay
Oriflamme
P&G (Max factor)
Parfums Caron
Parfums Christian Dior
Paul Smith
Philosophy
Ponds
Pure Roots Herbals
Remington
Revlon
Rimmel
Roc
Sedu
Serge Lutens
Skin whitening
Skincare (H_2O oasis)
Space.NK
Stila
Sugan
Tanorexia
Temptation
Tenderils herbal baby care
The Body Shop
Tigi
Toni And Guy
Urban Decay
Versace
Vichy laboratories
Virgin Vie
VLCC Personal care
Wella
YSL (*Y*ves *S*aint *L*aurent)

PROTECT YOURSELF AGAINST THE DANGERS OF COSMETICS

- Never share make-up. Always use a new sponge when trying products at a store. Insist that salespersons clean container openings with alcohol before applying to your skin.

- Keep make-up out of the sun and heat. Light and heat can kill the preservatives that help to fight bacteria. Do not keep cosmetics in a hot car for a long time.
- Don't use cosmetics if you have an eye infection, such as pinkeye. Throw away any make-up you were using when you first found the problem.
- Never drive and put on make-up. Not only does this make driving a danger, hitting a bump in the road and scratching your eyeball can cause serious eye injury.
- Keep make-up containers closed tight when not in use.
- Never add liquid to a product unless the label tells you to do so.
- Avoid color additives that are not approved for use in the eye area, such as "permanent" eyelash tints and kohl (color additive that contains lead salts and is still used in eye cosmetics in other countries). Be sure to keep kohl away from children. It may cause lead poisoning.
- Don't deeply inhale hairsprays or powders. This can cause lung damage.
- Throw away any make-up if the color changes or it starts to smell.
- Never use aerosol sprays near heat or while smoking, because they can catch on fire.
- Some skin and hair care products can cause acne. To help prevent and control acne flare-ups, take good care of your skin. For example, use a mild soap or cleanser to gently wash your face twice a day. Choose "non-comedogenic" make-up and hair care products. This means that they do not close up the pores.

Trade Names of Cosmetics Raw Materials

Trade name	*Chemical nature*	*Remark/manufacturer*
Acetol	Liquid blend of acetylated lanolin alcohol	Malmstrom
Acetulane	Acetylated lanolin alcohols, liquid fraction	Amerchol
AEPD	Aminoethylpropanediol	
Aerosol OT	Dioctyl sodium sullonsuccinate	American Cyanamid
AHR 483		Robins
Albagel 4444	Inorganic hydrophilic colloid Gel forming	Whittaker, Clark and Deniels
Alcholan	Lanolin sterol Absorption base	Robinson-Wagner
Alcolec and Alcolec DS	Vegetable lecithin products with a carrier of fatty oil DS is a highly bleached fluid grade	American Lecithin
Alkanol OA	Surfactant	du Pont, Dyes and Chemical Division
Alkanol OJ (Merpol O)	Ethylene oxide condensation product	du Pont, Dyes and Chemicals Division
Alrosol C	Condensation product of diethanolamine and capric acid	Geigy
Amerchol CAB	Solid multi-sterol extract	Amerchol
Amerchol L-101	Liquid multi-sterol extract	Amerchol
Amerlate LFA	Lanolin fatty acids	Amerchol
Amerlute P	Isopropyl lanolate 100%	Amerchol
AMP	Aminomethyl propanol	

Contd.

Trade name	*Chemical nature*	*Remark/manufacturer*
Aqualose L 30	Alkylated liquid lanolin	Westbrook-Marriner
Aqualose L30	Ethoxylated USP lanolin	Westbrook-Marriner
Aqualose W20	Ethoiylated lanolin alcohol	Westbrook-Marriner
Aqualose L75	Ethoiylated USP lanolin	Westbrook-Alarriner
Arctic Syntex M	Salt of sulfated fatty esters	Colgate
Argobase L1	Lanolin-derived extracts of sterols and steryl esters	Westbrook-Marriner
Argonol ACE-5	Fractionated and acetylated lanolin alcohols	Westbrook-Marriner
Argonol ACE-20	Fractionated and acetylated lanolin alcohols	Westbrook-Marriner
Argonol ACFE-50 super	Liquid lanolin	Westbrook-Marriner
Argonol-60	Liquid lanolin	Westbrook-Marriner
Argowax	Wool wax alcohols B.P.	Westbrook-Marriner
Arfacel C	Sorbitan sesquioleate	ICI-Atlas
Arfacel 20	Sorbitan monolaurate	ICI-Atlas
Arfacel 60	Sorbitan monostearate	ICI-Atlas
Arfacel 80	Sorbitan monooleate	ICI-Atlas
Arfacel 83	Sorbitan sesquioleate	ICI-Atlas
Arfacel 85	Sorbitan trioleate	ICI-Atlas
Arfacel 165	Glyceryl monostearate and polyoxyethylene stearate Acid stable, self-emulsifying	ICI-Atlas
Arfacel 186		ICI-Atlas
Arlatone T		ICI-Atlas
Aromox DM-18 DW-L-25	Aliphatic amine oxide	Armak
Arquad 2-HT-75 and Arquad DM 18-B	Quaternary ammonium compounds	Armak
Atlas G-263	Cetyl ethyl morphilinium ethosulphate 35% aqueous solution	ICI-Atlas
Atlas G-1425	Polyoxyethylene 20 sorbitol lanolin der.	ICI-Atlas
Atlas G-1441	Polyoxyethylene 40 sorbitol lanolin der.	ICI-Atlas
Atlas G-1471	Polyoxyethylene 75 sorbitol lanolin der.	ICI-Atlas
Atlas G-1702	Polyoxyethylene 6 sorbitol beeswax der.	ICI-Atlas
Atlas G-1726	Polyoxyethylene 20 sorbitol beeswax der.	ICI-Atlas
Atlas G-1790	Polyoxyethylene 20 lanolin der.	ICI-Atlas
Atlas G-2135 (Brij-35)	Polyoxyethylene 23 lauryl etherder.	ICI-Atlas
Atlas G-2152 (Myrj-52)	Polyoxyethylene 50 stearate	ICI-Atlas
Atlas G-2160 and 2162	Polyoxyethylene 24 propylene glycol stearate	ICI-Atlas

Contd.

Trade name	*Chemical nature*	*Remark/manufacturer*
Bi-ron	Bismuth oxychloride, Pearl pigment	Rona
Bital	Bismuth oxychloride coated talc, Pearl pigment	Rona
Brij 30	Polyoxyethylene 23 lauryl ether	ICI-Atlas
Brij 72	Polyoxyethylene 2 stearyl ether with antioxidants	ICI-Atlas
Brij 93	Polyoxyethylene 2 oleyl ether with antioxidants	ICI-Atlas
BTC	Alkyl dimethyl benzyl ammonium chloride	Onyx
Butyl cellusolve	Ethylene glycol monobutyl ether	Union carbide
Butyl cellusolve acetate	Ethylene glycol monobutyl ether acetate	Union carbide
Butyl parasept	Butyl-p-hydroxyl benzoate	Heyden
Cab-o-sil	Silica Fine-particle material with high surface area	Cabot
Carbitol	Diethylene glycol monoethyl ether	Union Carbide
Carboset 414, 514	Acrylic film forming resins	Good rich
Carbowax 200, 1500	Polyethylene glycols	Good rich
Carolate (Cetina)	A spermaceti der. Self-emulsifying synthetic	Robeco
Ceepryn	Cetyl pyridinium chloride	Merrell
Cellosize QP-3, QP-4400	Grades of hydroxyethyl cellulose	Union carbide
Cellosolve	Ethylene glycol monoethyl ether	Union carbide
Cera emulsificans	10% sulphated cetyl and stearyl alcohols and 90% of the free alcohols	Official product of BP
Ceralan	Lanolin alcohols	Robinson-Wagner
Ceraphyl 28, 31, 50, 140, 140-A, 230, 424	Cetyl lactate, butyl lactate, myristyl lactate, decyl oleate, diisopropyl adipate, myristyl myristate respectively	Van Dyk
Cerasynt D	Hydroxyethyl stearyl amide	Van Dyk
Cerasynt MN	Ethylene glycol monostearate Self-emulsifying	Van Dyk
Cerasynt WM	Glyceryl mono stearate Acid stabilized	Van Dyk
Cerasynt 840	Polyethylene glycol 1000 monostearate	Van Dyk
Cerasynt 945	Glyceryl monostearate Acid stabilized	Van Dyk
Cetab (Cetrimide)	Cetyl trimethyl ammonium bromide	
Cheelox BF-13	Anionic chelating agent	GAF
Citroflex A-2	Plasticizer	Pfizer

Contd.

Trade name	*Chemical nature*	*Remark/manufacturer*
Clerate Extra	Bleached lecithine	Cleary
CMC-7 LP	Grade of sodium carboxy methyl cellulose	Hercules
Cosmetic liquid 585	Water-white saturated hydrocarbons	Humble
Crodafos N. 10 Neutral	Oleylether phosphate	Croda
Crodamol	Isopropylmyristate	Croda
Duna # 1	Dispersing agent	Vanderbilt
Deriphat 170 C	Sodium salts of N-lauryl-p-iminodipropionic acid	General Mills
Dicrylan 325-50	Acrylic copolymer dispersion	Glba-Giegy
Drewmulse 1128	Glyceryl monostearate Acid stabilized	Drew
Dry-Flo	Modified corn starch	National Starch
Duponol C	Sodium lauryl sulphate USP	du Pont Dyes and Chem div.
Duponol WA paste	Sodium lauryl sulphate USP Aqueous form	du Pont Dyes and Chem div.
Duponol WAQ	Sodium lauryl sulphate USP Aqueous form	du Pont Dyes and Chem div.
Duponol WAT	Alky alkylolamine sulfate	du Pont Dyes and Chem div.
Emcol E-607	Lapyrium chloride	Witco
Emcol E-607S	Stearoyl analog of Emcol E-607	Witco
Emcol MAS	A fatty amide	Witco
Emcol 14	Polyglycerol fatty acid easter	Witco
Emulphor EL-719, VN-430	Dispersing and emulsifying agent	GAF
Emulsynt 610-A	Polyethylene glycol 400 monolaurate	Van Dyke
Emulsynt 1055	Polyoxyalkylene oleate laurate	Van Dyke
Escalol 106, 506	Glyceryl p-aminobenzoate, amyl-p-dimethyl aminobenzoate Sunscreens	Van Dyke
Ethomen TD/25	Ethoxylated aliphatic amine	Armak
Ethoxylan 50,100	Water and oil soluble anion in 50%, and 100% active form Aqueous solution	Malmsrom
Ethoxylols 5,16 (Nimcolan and Mimcolan S)	Decoction product of lanolin alcohols with 5 and 16 moles respectively of ethylene oxides Nonionic emulsifiers	Malmsrom
Ethoxylols 16 R (Super Nimcolan S)	Decoction product of lanolin alcohols Nonionic emulsifiers	Malmsrom

Contd.

Trade name	*Chemical nature*	*Remark/manufacturer*
Forlan	Solid absorption	RTA
Freon	Fluorinated hydrocarbon propellants	du Pont Feron products division
G-4 & G-11	Dichlorophene and hezacholophene	Givaudan
Gafanol E-5550B	Xanthan gum product	GAF
Gantrez AN-119 esrer, ES-225, and ES-425	Film formers	GAF
Keltrol	Xanthan gum product	Kelco
Kesso PEG-600 distearate	Fatty acid ester	Armak
Lanacet	Acetylated lanolin	Malmstron
Laneto 50 and 100	Completely water-soluble lanolin	RITA
Methocel 60 HG	hydroxypropylmethylcellulose	Dow
Methyl cellusolve	Ethylene glycol monomethyl ether	Union Carbide
Myri 52	Polyoxyethylene 40 stearate	ICI-Atlas
Plexiglas	Acrylic resin Plastic sheet and molding powder	Rohm and Haas
Pluronic F68 and P123	Polyoxyalkylene glycols	Wyandotte
Resyn 28-3307, 2261	Polyvinyl acetate/crotonic acid teropolymer, containing a copolymerized benzophenone monomer	National Starch
Solulan	Ethoxylated lanolin alcohols	Amerchol
Span 60, 80	Sorbitan monostearate and monooleate	ICI-Atlas
Triton X-100 and X-200 (720)	Wetting and emulsifying agent	Rohm and Hass
Tween series (20,60,61,80,85)	Polysorbates	ICI-Atlas
Veegum	Magnesium aluminium silicate	Vanderbilt
Wecobee R	Hydrogenated coconut oil	Drew
WSP-X-250	Collagen-derived protein hydrolysate	Wilson
Zephiran	Alkyl dimethl benzyl ammonium chloride	Winthrop

List of Cosmetic Companies				
Company	*City*	*State*	*Country*	*Description*
Adore Beauty	Carlton North	_	Australia	Sells products for skin care, makeup, fragrance, bath and body, her and men
Adoro Cosmetics	Los Angeles	California	USA	Nails, nail accessory, implements, beauty accessory, eyelashes, adhesives and wigs
Adriel International	Rocky River	Ohio	USA	Products for eyebrow design, waxing and other professional needs
Adrien Arpel	New York	New York	USA	Offers a wide range of beauty products
Aesop	Melbourne	_	Australia	Sells skin, hair, body and fragrance products
AFE Cosmetics and Skincare	Powell	Wyoming	USA	Collection of quality cosmetic, skincare, body care and hair care products
All Naturals Cosmetics Inc	Toronto	Ontario	Canada	Providing customers with high performing plant-based products
Aloette	Atlanta	Georgia	USA	Sells skin care, makeup, bath and body products
Aloe Vera Products	Vista	California	USA	Manufactures a wide variety of aloe vera based skin care and beauty products
Anastasia	Beverly Hills	California	USA	Specializing in eye brows, eyes, face, lips, brushes and tools
Anna Sui Beauty	New York	New York	USA	Fashion designer with line of cosmetics
Arbonne International	Irvine	California	USA	Skin care products developed in Switzerland over 25 years ago
Avalon Natural Products	_	_	_	Creates organic bath and body products
Avea	London	_	UK	Organic skincare, cosmetics and perfumery
Aveda	Blaine	Minnesota	USA	Manufactures plant-based hair care, skincare, makeup, Pure-Fume™ and lifestyle products
Avon	New York	New York	USA	Lipsticks, fragrances and anti-aging skincare sales
Bare Escentuals	San Francisco	California	USA	Bare minerals makeup line of cosmetics and treatments for skin care
Barry M	London	_	UK	Offers over 400 different colors in a wide range of cosmetic products
BeautiControl.com	Dallas	Texas	USA	Spa-quality beauty products
Beauty of New York	Brooklyn	New York	USA	Over 100,000 beauty supplies and hair care products
Beauty Scene	_	_	USA	Sells brand name cosmetics and fragrances
Bejar	Barcelona	_	Spain	Lines of perfumes, eaux de toilette, and related products
BeneFit Cosmetics	San Francisco	California	USA	Offers makeup, skin and body care products

Contd.

Company	*City*	*State*	*Country*	*Description*
Best Organics For Health.com	_	_	USA	Organic cosmetics and natural health and beauty products
Biotherm	Savage	Maryland	USA	Selection of innovative natural beauty products for both women and men
Bobbi Brown Cosmetics	New York	New York	USA	Offers makeup, skincare products, tools and accessories and fragrance
Body Shop, The	_	_	UK	global manufacturer and retailer of naturally inspired beauty and cosmetics products
Bonne Bell	Lakewood	Ohio	USA	cosmetics for eyes, lips and face
Borghese Cosmetics	New York	New York	USA	innovative products combine time honored botanicals and cutting edge technology
Branded J Collections	Sherman Oaks	California	USA	makeup brushes which are not manufactured from animal products
Cargo Cosmetics	Toronto	Ontario	Canada	sells beauty and cosmetic supplies
Chanel	Paris	_	France	makeup and fragrance line from famous fashion designer
Cheap Cosmetics Australia	_	_	Australia	designer label perfume, cosmetics and skincare
Christian Dior	Paris	_	France	offers a fragrance, makeup and skincare line
Clarins	Paris	_	France	offers a full line of cosmetic products
Clinique Laboratories	New York	New York	USA	skin care and makeup products that are all allergy tested and 100% fragrance free
Colecciones de Raquel	Beverly Hills	California	USA	cosmetic collections for golden complexions
Color Lab Cosmetics	Rockford	Illinois	USA	_
Color Me Beautiful	Manassas	Virginia	USA	cosmetics brand based on seasonal color analysis
Colose	Stockholm	_	Sweden	Scandanavian cosmetic company
Cosmetech Laboratories	Fairfield	New Jersey	USA	laboratory for the cosmetic, personal care and topical pharmaceutical industries
Cosmetic Company, The	Mississauga	Ontario	Canada	produces makeup, skincare, waxing, and esthetic supplies for schools, salons and spas
Cosmetics Kitchen	Providence	Rhode Island	USA	sells natural and organic beauty products
Cover Girl	Cincinnati	Ohio	USA	Proctor and Gamble company offering a full line of cosmetic products
Covermark Cosmetics	Northvale	New Jersey	USA	products for concealing and correcting various skin imperfections
Crabtree and Evelyn Products	Woodstock	Connecticut	USA	beauty supplies for hands and feet, bath and body and facial skin care

Contd.

Company	*City*	*State*	*Country*	*Description*
Dermablend Coverage Cosmetics	–	Illinois	USA	cosmetic problems for various skin conditions
Dermalogica	Carson	California	USA	skin care system researched and developed by The International Dermal Institute
Duri Cosmetics	Brooklyn	New York	USA	manufacturer and distributor of full nail polish and nail treatment lines, hair removal wax
EI Solutions	Santa Ana	California	USA	Provides make-up, skin care, body and hair products
Elizabeth Arden	New York	New York	USA	Prevage, skincare, fragrance, makeup and men's products
Erno Laszlo	Seattle	Washington	USA	Sells skincare products
Estee Lauder	New York	New York	USA	Collection of skincare, makeup and fragrance products since 1946
Esthederm	Paris	–	France	Skincare cosmetic products made by this French company
Face Stockholm	Hudson	New York	USA	200+ lip shades, 150+ eye shadows, blushes, foundations, powders, makeup tools
Fashion Fair Cosmetics	Chicago	Illinois	USA	Makeup and skincare for women of color
Fresh Cosmetics	New York	New York	USA	Spanning skincare, makeup. fragrance, hair care, body care and more
Garden Botanika	Redmond	Washington	USA	Products made from botanically based formulas
Giovanni Cosmetics	–	–	–	Organic hair care and body care
Guerlain	Paris	–	France	Skincare, makeup and fragrance cosmetic products
Guinot	Paris	–	France	Provides a range of beauty care, covering 3 specific categories- face, body and sun
Gustavo Cosmetics	Arlington	Virginia	USA	Organic skin care, mineral cosmetics and makeup school
Hard Candy Cosmetics	Newport Beach	California	USA	Products for eyes, lips, face, face, body and nails
Hazel Keller Cosmetics	Charlotte	North Carolina	USA	Cosmetic products which contain vitamins and herbs for skin health
Helena Rubinstein	–	–	Germany	Offers skin care, make up and sun care products
Help Me Rhonda	Atlanta	Georgia	USA	Physical appearance service and product provider for media and private use
House Kwang Sang Hong	Wan Chai	–	Hong Kong	Sun block, skincare, makeup, bath gel and more

Contd.

Company	*City*	*State*	*Country*	*Description*
Iman	New York	New York	USA	Cosmetics, skincare and fragrance beauty company founded by famous model
Inouvi Cosmetics	_	_	_	Specializing in makeup products, skin care and beauty tools
JAFRA Cosmetics	Westlake Village	California	USA	Offers skin care, in-home spa, hair care and color collections
Janet Sartin	New York	New York	USA	Provides a range of skin care, body and color products
Jason	Boulder	Colorado	USA	Offers a wide range of natural personal care products
Jenulence	Staten Island	New York	USA	Source for beautiful mineral makeup, natural mineral cosmetics and organic skin care
Jordane Cosmetics	Toronto	Ontario	Canada	Specializes in providing Beauty Professionals with all products and tools
Josie Maran Cosmetics	Hollywood	California	USA	Sells a wide range of beauty products for all body parts
Judith August Cosmetics	Las Vegas	Nevada	USA	offers corrective cosmetic products
Jurlique	Irvington	New York	USA	Environmentally friendly skincare products
Kose Products	_	_	Japan	Japanese cosmetic company offering Awake and Sekkisei brands
L.A. Girl Cosmetics	Ran. Cucamonga	California	USA	Innovative products, brought to you in trendsetting colors and styles
Lancome	Paris	_	France	Products for skincare, makeup, fragrances, body and sun care for men
La Prairie	Montreaux	_	Switzerland	Scientifically advanced Swiss skincare products
Laura Mercier	Stafford	Texas	USA	Sells a full line of cosmetic and beauty products
Lavera Skin Care	_	_	_	All natural organic makeup, skin care, anti-aging, sunscreens
Lee Fran Beauty Imports	Queensland	_	Australia	Distributing many overseas and local branded cosmetic products
Les Naturelles	_	_	Switzerland	Manufacturer of Swiss cosmetic products
Lip Ink International	El Segundo	California	USA	Sells cosmetic products for lips
L'Oreal	Paris	_	France	French cosmetic firm created in 1909
LoveLula.com	_	_	UK	Over 800 natural and organic skincare and beauty products
MAC Cosmetics	New York	New York	USA	Offers a full range of cosmetic products
Magique Kiss	Longueuil	Quebec	Canada	Offers lip products, located outside of Montreal

Contd.

Company	*City*	*State*	*Country*	*Description*
Manic Panic	Long Island City	New York	USA	Cosmetic seller started business during late '70's punk era
Marlene Klein Cosmetics	Lyndhurst	Ohio	USA	Cosmetics for lips, eyes, cheeks and hues
Mary Kay Cosmetics	Dallas	Texas	USA	Global cosmetic firm handling a wide variety of products
Maybelline International	New York	New York	USA	Full range of cosmetic products and tools
McCuaig Solutions	Winnipeg	Manitoba	Canada	Manufacturer of liquids, creams, ointments, lotions, gels, etc.
Merle Norman Cosmetics	Los Angeles	California	USA	Skin care and color products for a wide range of complexions and skin types
Molton Brown	London	–	UK	Line of skincare, hair care, lotions and bath and body products
NARS Cosmetics	New York	New York	USA	Cosmetic company selling a full line of products
Naturismo	Datchet		UK	Organic cosmetics with a respectful approach to skin and nature
Neutrogena Cosmetics	Los Angeles	California	USA	part of the Johnson and Johnson family of companies
Nickel	Paris	–	France	face and body products, perfume and spa treatments
Noevir	Irvine	California	USA	skincare, body, hair care and other cosmetics offered
No Miss Nail Care Products	Boca Raton	Florida	USA	manufacturer of healthy alternative hair care products
Nu Skin	Provo	Utah	USA	offers a full range of cosmetic products since 1984
Olive Organic	–	–	–	huge range of Natural and Organic skincare products for you and your family
Oriflame	–	–	Holland	Dutch cosmetic company
Origins Natural Resources	Groveport	Ohio	USA	provides organics, skin care, bath and body, color, hair care and more
Paula's Choice	Seattle	Washington	USA	offering for skin care, make up, body and her care, brushes
Philosophy	Phoenix	Arizona	USA	full range of cosmetic and beauty products offered
Physicians Formula	–	California	USA	sells innovative products made especially for sensitive skin
Phyt's Aromatic	–	–	France	organic cosmetics manufactured in France
Pola Cosmetics	Carson	California	USA	offers a full range of cosmetic and beauty products

Contd.

Company	*City*	*State*	*Country*	*Description*
Principal Secret	Arden	North Carolina	USA	anti-aging cosmetics and skin cream
Profaces	Cherry Hill	New Jersey	USA	manufacturer of eye and lip cosmetic products
Rad Cosmetics	Mississauga	Ontario	Canada	fresh cosmetic products in distinctive and varying shades
Ramy Beauty Therapy	New York	New York	USA	products and brushes for the eyes, face, lips and man in your life
Real Purity	Crossville	Tennessee	USA	products are derived from certified organic and wild crafted herbs
Redheads Fancy	Southington	Connecticut	USA	lip pencil, lipstick and nail polish for redheads
Revlon	New York	New York	USA	full range of cosmetic products from this global firm
Rubiglo Cosmetics	_	_	USA	specifically formulated to work harmoniously with one another
Saffron Rouge	Guelph	Ontario	Canada	source for organic skin care and aromatherapy
Sally Hansen	Uniondale	New York	USA	provider of head-to-toe problem-solving beauty treatment products
Scarlett Cosmetics	New Hope	Pennsylvania	USA	line of colorful cosmetic products
Shiseido Japan	_	_	Japan	skincare, makeup, fragrance, body care and sun care products
Shu Uemura	Tokyo	_	Japan	offers a line of skincare, face, eye and lip products
Silk Skin	Tarzana	California	USA	beauty, skin care, makeup and anti-aging products
Smashbox Cosmetics	Los Angeles	California	USA	sells a full range of cosmetic and beauty products
Somerset Cosmetic Company	Issaquah	Washington	USA	supplier of cosmetic ingredients and containers for individuals and small manufacturers
So Organic	London	_	UK	offers a large selection of natural and organic cosmetics
Sothys	Paris	_	France	innovative cosmetic products
Suncoat Products	Guelph	Ontario	Canada	complete line of water-based nail polish products and natural organic colour cosmetics
Sweet Cheeks	_	_	_	cosmetic powders and creams for the face
The Balm.com	San Francisco	California	USA	cosmetics for the lips, eyes, cheeks and face
The Organic Salon.com	_	_	UK	seller of natural and organic beauty products
Three Custom Color Specialists	New York	New York	USA	30 years combined experience in the world of beauty

Contd.

Company	*City*	*State*	*Country*	*Description*
Tonnie Cosmetics Co Ltd	Tainan City	–	Taiwan	cosmetics manufacturer specializing in full ranges of cosmetic products
Too Faced	Irvine	California	USA	cosmetic products for the lips, face and eyes
Townley Girl Cosmetics	New York	New York	USA	cosmetics for a younger clientele
Urban Decay Cosmetics	Costa Mesa	California	USA	cosmetic products for eyes, lips, face and body
Victoria Jackson Cosmetics	–	California	USA	sells cosmetics and offers beauty tips
Youngblood	Simi Valley	California	USA	mineral cosmetics
Yves Rocher	–	–	France	a leader in botanical beauty care
Za New York	New York	New York	USA	offers skincare and makeup products
Zhen Beauty	Anoka	Minnesota	USA	sells cosmetics for the face, lips eyes and skin

Cosmetics are big business and the cosmetic industry worldwide witnesses rapid growth annually. However, in India, cosmetics like lipsticks, eyeliners and kajal and rouge are the only cosmetic products that have been popular with the masses so far. The Indian consumer has only recently started to become aware of other cosmetic products and hence the potential for the market for these in India is huge.

Despite this lack of awareness within India however a look at the Indian directory of cosmetic companies will show you that there are many companies within India which produce cosmetic products both for the Indian market as well as to cater to the demands for other countries. Indian herbal and natural organic cosmetics are also in great demand these days and many Indian cosmetic manufacturers like the well-reputed Balsara group produce cosmetics for recognized brands like The Body Shop.

It is, hence, not difficult to find cosmetic manufacturing companies within India but each company specializes in a different range of products so you need to consider your needs before looking at which companies to do business with.

A list of cosmetic companies in India would include names like Emami and Lakme which are well-recognized in India and also a variety of other suppliers and manufacturers of cosmetic products, cosmetics companies in India, some of which are named below.

AVM Enterprises, based in New Delhi, manufactures and exports Neem products like antiseptic cleansing milk, hair-growth enhancers and hair tonics, facial massage gels, anti-acne masks and creams and skin moisturizers and toners.

Gayatri Herbals PVT LTD based in the Thane Township near Mumbai, are suppliers of herbal cosmetics. The range of their products include herbal face packs and cleansing lotions, facial creams, hair gels and conditioners, aloe Vera cosmetic products, orange peel scrubs, papaya skin gels and herbal Shikakai powders. They can be contacted at their website, which is gayatriherbalsindia.com.

Radico is a cosmetic manufacturing and exporting company which deals with the production and export of various Ayurvedic shampoos, Ayurvedic bath gels, Ayurvedic hair dyes, Ayurvedic body oils, hair conditioning henna, henna tattoos and henna tattoo stencils and a vast variety of other personal care products. The company is based in Faridabad and can be contacted at their website, 'www.radico.com'.

MK Industries, based in Jamnagar manufactures and exports various cosmetic toiletries like soaps, hand gels, etc.

Weldon industries are manufactures of cosmetics and cosmetic items. Their range of products include various cosmetic brushes like powder or blush-on brushes, eye-grooming brushes, lip liners and various facial make-up application brushes. The company is based in Moradabad.

Classic creations based in Noida, are involved in the manufacture and export of a vast range of herbal cosmetics like hair care products. They manufacture and export henna products, hair care oils, shampoos, cleansing milks, face packs, face wash lotions, after-shave balms, and a variety of other cosmetic gels and lotions. They can be contacted at their website, classiccreationsindia.com.

WEIGHTS AND MEASURES

1 Fluid ounce (Fl oz)	29.6 ml (aprox. 30 ml)
1.6 Fluid ounce (Fl oz)	50.0 ml
4.1 Fluid ounce (Fl oz)	125.0 ml
6 Fluid ounce (Fl oz)	180.0 ml
6.8 Fluid ounce (Fl oz)	200 ml
8.3 Fluid ounce (Fl oz)	250 ml
1 Pound (lb)	454 g
1 ml	16.2 minims
1 pint	473 ml./0.568 Lt.
1.76 pint	1 Lt.
1 grain (gr.)	0.065 g
15.43 grain	1 g
2.205 lb	1 kg.
1 Ounce (oz)	28.35 gr.
1 Gallon	4.546 Lt.
1 cc (1 cm^3)	1 ml
1 Drop	0.6 minims
1 Teaspoonful	= 5 ml

Glossary

Abscess : A localized collection of pus in a cavity formed by the disintegration of tissue.

Acid Balance: The acid balance refers to natural pH level of the skin's moisture.

Acid Mantle: A protective oily layer on the skin's surface which functions as a protectorent against environmental impurities and helps regulate moisture loss and water retention.

Acid Perm: An acid perm produces permanent hair waves with curls that are actually softer than an alkaline perm. It also has a pH from 6.5 to 8.0.

Acidophilus: Helps prevent fungus, acne, and bad breathe.

Acne : Acne is an inflammatory skin disease that's made evident by pimples that can appear on almost any part of the body, but are usually on the face.

Acne vulgaris: A chronic inflammatory disease of the pilosebaceous follicles, the lesions occurring most frequently on the face, chest and back.

Acrid: Pungent; producing an irritation.

Acrylic: Something that has the property of being able to repel water or moisture.

Active ingredient: The ingredient that is responsible for producing the desired effect of a mixture of ingredients and for giving the product its main characteristic. The active ingredient is not necessarily the most common ingredient in a product.

Allantoin : Allantoin, extracted from a plant, is used in creams and skin preparations to heal and soothe.

Allergen: An allergen is something that produces an allergic reaction.

Almond oil: Almond oil is vegetable oil made with almonds. It's used in cosmetic products that soften the skin.

Aloe Vera: Aloe Vera is a plant that has wonderful healing and softening properties. Many people have an Aloe Vera houseplant, because you can break off a piece and use the extract from inside to soothe and heal cuts, burns and abrasions. It's a very common ingredient in cosmetics because of its softening properties.

Alpha hydroxy acid: Mild acids presents in plants that prevent wrinkles by exfoliation.

Amino acid: The basic building block of protein, crystalline amphoteric compounds of significance in protein biosynthesis. All amino acids contain an amino (NH_2) end, a carboxyl end (COOH) and a side group (R). In proteins, amino acids are joined together when the NH_2 group of one forms a bond with the COOH group of the adjacent amino acid. The side group is what distinguishes each of the amino acids from the others.

Antioxidants: Agents that prevent oxidation.

Antipruritic: Substance which prevents itching.

Antiseptic : Substance which prevents putrefaction and inhibits growth of microorganisms.

Astringent: An astringent, as part of the facial cleansing system, is commonly known as toner, and it controls oily skin and lowers the pH of the face after cleansing. Basically, it draws tissues together.

Azo Dyes: Used in non-permanent hair rinses. The dyes contain mild acids but can cause allergic reactions. People who are sensitized to permant hair dyes containing paraphenylene diamine also develop a cross sensitivity to azo dyes. Azo dyes are absorbed through the skin.

Beeswax: Beeswax has been used in cosmetics for centuries, mostly for its emulsifying properties.

Benzoyl Peroxide: Benzoyl peroxide is a common ingredient found in acne treatments, used for its antibacterial properties.

Bergamot: The Bergamot tree is grown predominantly in southern Italy. The Italians make oil from the rind of the bergamot tree's citrus fruit, which is good as an antiseptic, a cleanser, and a deodorizer.

Blackhead: A blackhead is a type of pimple that forms when the pores of the skin get clogged with oils and impurities.

Botanical: Botanical products are those that are made from plants.

Botox: Botox is simply the trade name for botulinum toxin A. It is a product that's becoming increasingly popular for getting rid of wrinkles in the face, which is performed by injections; known as Botulinum toxin. When used in tiny amounts it can temporarily paralyze a muscle and reduce or eliminate wrinkles or frown lines without harm.

Bronzers: Bronzers are products that darken the skin to make it look naturally tanned.

Buffer: A buffer is a pH-balanced cleanser that makes the skin shinier and softer.

Butcher's Broom: Butcher's Broom is a plant extract that's used to "sweep away" redness in the face.

Callus: A buildup of tough layers of skin is referred to as calluses. They can be removed with a pumice stone, or you can get them professionally removed by an esthetician.

Candela wax: Candela wax is combined with oil to give more body to lipsticks.

Carcinogen: A substance that is known to cause cancer.

Caustic: Destroys or disorganizes living tissues.

Cell: The smallest organized structure of a living organism which can survive on its own.

Cellulite: Cellulite is the word used to refer to the fatty deposits that cause a dimpled or uneven appearance of the skin, usually around the thighs and buttocks.

Ceramides: Ceramides is a substance that protects the skin against moisture loss. It's also synthetically reproduced in skin care products.

Chamomile: It's a great anti-inflammatory agent when used in lotions. Also used as chamomile tea.

Chitin: Chemically made up of N-acetyl-2-glucoseamine units; nitrogen containing poly saccharides present in covering organs/ tissue of insects and lower organisms.

Citric acid: Citric acid has many uses. It's used in skincare products as an astringent and an antioxidant.

Clarifying: Clarifying lotion is the toner part of a cleansing system that balances the pH of the skin after cleansing.

Collagen: Collagen is a natural part of the skin, but it can be increased in volume with injections to plump up a certain area. Those women who want that full pouty look can have collagen injected into their lips.

Concealer: An opaque makeup used to cover darkness under eyes, redness of the skin, or anything irregular in the skin's color or texture. Comes in a waxy stick, cream, or opaque liquid formula.

Cordial: Sweetened non-alcoholic drink typically made from fruit juice which is used to destroy odour.

Cream rinse: A mixture of wax, thickeners, and a group of chemicals used to coat the hair shaft and detangle the hair. Generally applied after shampooing.

Crow's-feet: Wrinkles in the skin around the outer corner of the eye.

Cucumber juice : Cucumber juice is used on the face to tighten the pores and stop the skin from drying out.

Cuticle : A fatty layer consisting of cutin formed on outer walls of epidermal cells.

Cytotoxic: Destructive to cells.

Deionized water: Water purified by removing highly active ions especially positively charged cations like calcium (Ca^{++}) magnesium (Mg^{++}) and iron (Fe^{++}) and (Fe^{+++}).

Dematitis: Inflammation of the skin.

Demineralized water: Water run through active resin beds to remove metallic ions and filtered through a sub micron filter to remove suspended impurities. Denaturant

Demulcent: Agent to soften and sooths or relieves irritation. e.g. of mucous membrane.

Dental carries: Decay of the teeth.

Depilation : Depilation is the process of removing hair from the skin.

Depilator, epilator, depilatory: A cosmetic for temporary removal of undesired hair

Dermabrasion: A procedure in which the skin is sanded to improve its texture. Microdermabrasion uses a type of sand to abrade the skin, usually done by an esthetician or a cosmetic doctor.

Dermatophyte: A fungus pathogenic for the skin.

Desquamation: The sloughing of dead corneocytes (cells located in the epidermis that are packed with fibrous protein called Keratin) from the horny cell layer of the epidermis (stratum corneum). The final stage in the ongoing cycle of skin cell birth, maturation, and death.

Detangler : Agent that detangles hair.

Dihydroxyacetone (DHA) : DHA is a cosmetic product that bronzes the skin to resemble a tan.

Eau de perfume: Less concentrated than Parfum, but more concentrated than any other form of fragrance, with a stronger, longer-lasting scent. Because there are more perfume oils (more oil and less alcohol) in the formula, the cost to purchase Eau de Perfume is generally more expensive than other forms of fragrances.

Eau de toilette: A less-concentrated fragrance containing less oil and more alcohol.

Echinacea: Echinacea is a natural product that's most commonly used to reduce flu or cold symptoms. But it is also used in some skin care products to stop itching and soothe the skin.

Eczema: Eczema is an inflammatory condition that causes the skin to become red, scaly and itchy.

Elasticity: The ability of hair or skin to stretch without breaking/morphing and then return to their original shape.

Elastin: Elastin is used in cosmetics to protect the skin from getting dry.

Electrolysis: Electrolysis is a process where an electric current is applied to hair roots to kill them.

Emollient: A substance that makes something soft or supple; also, soothing especially to the skin or the mucous membrane.

Emulsifier: An emulsifier is a substance added to a product to thicken it.

Epilation: Epilation is the process of removing hair from beneath the surface of the skin.

Esthetician: An esthetician is a professional trained to give beauty treatments like facials, manicures, and pedicures.

Evening primrose: Evening primrose is a natural substance used as a toner or moisturizer.

Exfoliant: An ingredient or cosmetic tool used to help slough away the dead skin cells.

Exfoliate: When you exfoliate your skin you remove layers or scales.

Eyelash curler: A device used to curl ones eyelashes.

Eyeliner: Makeup used to emphasize the contour of the eyes.

Finishing spray: A hairspray with medium hold used to maintain the hair's shape and hold hygroscopic. Capable of absorbing and retaining moisture and used in cosmetics as part of moisturizing ingredients in creams and lotions.

Foundation: Cosmetic used to cover slight imperfections of ones complexion. Foundation is typically liquid and tinted and applied all over the face to provide an "even" appearance of skin tone. Foundation is usually worn over base/primer and under powder.

Glycerin: Glycerin can be found in moisturizers; it holds water particles together.

Glycolic peel: A glycolic peel is a method used to exfoliate or remove dead layers of skin.

Grease: Rendered animal fat; a thick lubricant; oily matter

Gynaecomastia: Excessive development of male mammary glands.

Highlighter: A cosmetic used to highlight the eyes or cheekbones

Humectant: A substance, such as glycerin, that absorbs or helps another substance retain moisture; a humectant is a substance that helps retain moisture; a substance that can absorb water from moist surroundings.

Hydrate: When you hydrate skin, you add moisture to it.

Hydrophilic: Describes a substance that absorbs, dissolves in or is attracted to water.

Hydroquinone: Hydroquinone is a white crystalline compound used in skincare products as a bleaching agent.

Hyperpigmentation: Hyperpigmentation is a skin condition caused by ultraviolet light from the sun, which darkens the skin.

Inflammation: The reaction of living tissue to injury, infection or irritation characterized by pain, swelling, redness and heat.

Integument system: Outer covering, e.g. the skin.

Inunction: The administration of a drug in ointment form applied with rubbing with the purpose of causing absorption of the active ingredients.

Irritant: Causes an inflammatory response when applied to skin.

Itching : An uncomfortable sensation of irritation of the skin or mucous membranes that causes scratching or rubbing of the effected parts.

Jojoba: Jojoba is a multi-purpose substance, used as a moisturizer, as well as to reduce wrinkles and stretch marks.

Karite: Karite is a natural ingredient that keeps the skin healthy.

Keratin: Keratin is the natural substance that gives nails and hair their resiliency.

Keratin: The tough protein that is the major component of stratum corneum cells, hair, hoof, horn, and nails. A surface protective agent with film-forming and moisturizing action.

Keratinization: The process by which the epidermis forms its outer protective layer, the stratum corneum; conversion into keratin or keratinous tissue.

Kojic acid: Kojic acid is used to lighten the color of the skin.

Lactic acid: Lactic acid is a natural ingredient in the body that moisturizes the skin.

Lakes: Colour lakes are pseudo pigments produced by precipitation of water soluble colourants with a water insoluble salt such as aluminum hydroxide.

Lanolin: Lanolin is a fatty substance made from wool, which moisturizes and emulsifies, as well as absorbs water.

Lash primer: A clear or white coat worn under mascara to make lashes appear thicker and longer.

Lead carbonate: A naturally occurring white amorphous powder with a chemical formula of $PbCO_3$. Used in exterior paints, ceramics, cements, processing of parchment and as a laboratory reagent.

Licorice: Licorice, in its natural form (not the kind we eat), is good for treating acne and soothing skin irritated by allergies.

Lip liner: Makeup used to emphasize or enhance the contours of the lips.

Lipophilic: Describes a substance that dissolves in or is attracted to fats, oils or other lipids. Lipophilic functional groups or molecules prefer to be in an environment where there is no water.

Lipstick: A waxy, solid, and usually colored cosmetic in stick form that is used to color the lips.

Lubricant: A substance, such as grease, that is capable of reducing friction, heat, and wear when introduced as a film between solid surfaces; something that lessens or prevents friction of difficulty.

Makeup base: Or primer is worn under foundation. Typically a base/primer is used to extend the staying power of the foundation or to blot excess oil for those with oily skin.

Make-up, war paint: Cosmetics applied to the face to improve or change your appearance

Malic acid: Malic acid comes from raw fruit, like apples, cherries and tomatoes, and is used as a glycolic agent.

Mallow: Mallow is a softening agent made from plants, which helps reduce inflammation, age lines, and eye swelling.

Mascara: A cosmetic for coloring and coating the eyelashes available in various colors and formulations. Some formulations include : waterproof, curling, thickening, lengthening, etc.

Melanin: Melanin is the dark pigment in hair and skin.

Mien: Person's appearance or bearing, especially as an indication of mood, etc.

Moisturizer: A substance that imparts or restores moisture to (something); to supply moisture, in cosmetics, moisturizer is a cream that hydrates the skin.

Mousse: Thick creamy liquid put on hair to shape it or improve its condition, e.g. styling/conditioning mousse.

Nail enamel, nail polish, nail varnish: A cosmetic lacquer that dries quickly and that is applied to the nails to color them or make them shiny.

Non-comedogenic: Cosmetic products that are non-comedogenic (or they may be referred to as non-occlusive) don't plug the pores, so do not cause skin irritation or pimples.

Notes: When a fragrance is applied, the element smelled is known as notes. It is top (volatile material), middle (floral scents) and bottom or base notes (least volatile, e.g. vanillin, woods, etc.).

Occlusion: The act of closing, shutting or stopping up. In cosmetics this usually refers to a shield or film that is spread onto the skin to slow or prevent moisture evaporation. This shield or film is usually made up of materials, such as oils and waxes that cannot be penetrated by water.

PABA : PABA stands for para-aminobenzoic acid. It's a part of the vitamin B complex and is used in some sunscreen lotions.

Panthenol: Panthenol, or vitamin B5, is used as a moisturizer.

Patch test: Test used to identify allergies due to chemicals coming in contact with skin. Individual chemicals are applied separately to the skin and then the skin is observed for reactions over a few days.

Pencil: A cosmetic in a long thin stick; designed to be applied to a particular part of the face; "an eyebrow pencil".

Petrolatum: Petrolatum is another word for petroleum jelly. It's used in creams as a lubricant, to soften and soothe the skin. It also seals in moisture.

pH: pH is an abbreviation for percentage of hydrogen. The pH scale is used to measure the strength of acids and bases (or alkalis). The acid strength in the human stomach is about pH 2. Alkalis such as caustic soda and basic household cleaners have a pH of about 12 to 14. Neutral is pH 7, (i.e., neither acidic nor alkaline). The scale is logarithmic, so pH 4 is ten times as acidic as pH 5 and pH 2 is ten times as acidic as pH 3, and so on. In cosmetics, it measures the level of acidity.

pH balance of hair: Balance of acidity and alkalinity of hair, keeping it healthy.

Photoaging: Photoaging is damage to the skin caused by too much exposure to the sun.

Photosensitivity: Photosensitivity is a skin condition, manifested in rashes or swelling, that results from applying or eating certain chemicals or foods, then exposing the skin to sunlight.

Phytocosmetics: Phytocosmetics are cosmetics that are made with natural ingredients from plants.

Pilosebaceous: Pertaining to the hair follicle and the sebaceous gland opening into it.

Plasticizer: Any of the group of agents added to other organic synthetic substance to make them soft and flexible. Poisonous or unpleasant substance added to alcoholic cosmetics to make them undrinkable.

Polymer: Polymers are large molecules that are made up of many units (monomers) linked together in a chain. There are naturally occurring polymers (e.g. starch and DNA) and synthetic polymers (e.g. nylon and silicone).

Polysaccharide: A carbohydrate made up of a long chain of simple sugar molecules joined together. Starch and cellulose are examples of polysaccharides.

Pore: A pore is a tiny opening in the skin that serves as an outlet for sweat.

Pruritis: Itching

Pseudomembrane: False membrane.

Psoriasis: A chronic skin disease in which erythematous areas covered with adherent scales; psoriasis is a skin disease that produces dry, itchy red patches.

Retinol: Deficiency in retinol, another name for vitamin A, results in a hardening and roughening of the skin. Retinol is the main ingredient in products like Retin-A and Renova, which help to reduce wrinkles and heal acne.

Rosacea: Rosacea is a skin condition that produces red oily skin and acne.

Royal Jelly: Royal jelly is a natural product that's taken from bees. It has many medicinal purposes and creates generally good health, including healthy skin.

Rubefacient: An agent which causes a reddening of the skin by increasing blood circulation where applied to provide heat.

Sage: Sage, a member of the "parsley, sage, rosemary and thyme" group, is not just an herb for cooking. It also helps to disinfect and heal wounds.

Salicylic acid: Salicylic acid dissolves layers of the skin and is used in the treatment of eczema.

Salve: Oily substance used on wounds, sores or burns.

Scab: A crust formed by coagulation of blood, pus, serum, or a combination of these on the surface of an ulcer, erosion, or other type of wound.

Scabies: An eruption due to *Sarcoptes scabiei* var. *hominis*.

Sebum: Oily secretions from the sebaceous glands just below the skin.

Shea butter: Shea butter is a fat made from the seeds of the shea tree. It's used in creams and lotions to soothe soften and moisturize the skin.

Skin allergy: An allergy is a hypersensitivity to substances in the environment which do not bother most people. Allergy to cosmetics usually manifests as a rash on the skin where the product has been applied. This condition is known as allergic contact dermatitis, and is often due to fragrances and preservatives in the cosmetic product.

Spasm: A sudden, involuntary muscular contraction as a result of some irritant.

SPF: SPF is an acronym for sun protection factor. Sunscreen products have an SPF; the higher the SPF, the more protection you get from sunburn.

Spider veins: These enlarged blood vessels, also known as telangiectasia, appear near the surface of the skin on the face and legs as blue or red veins that often are short. They occur more frequently in women than men due to hormone fluctuation.

Styptic: Agent that stops hemorrhage.

Sunblock: Sunblock, as opposed to sunscreen, is a lotion that actually blocks the ultraviolet rays of the sun. It's more effective than sunscreen.

Sunscreen: Sunscreen is a product, usually a cream, that's applied to the skin to protect it from getting sunburned. The amount of protection is governed by its SPF.

Surface tension: A property of liquid surfaces that causes the surface layer to behave like a thin elastic 'skin'. Molecules in a liquid have attractive forces that hold them together. Molecules on the surface are attracted to molecules from all sides and below, but not from above. This results in a downward and sideways pull on molecules on the surface layer.

Surfactant: A surfactant is an ingredient that promotes an efficient mixture of oil and water.

Tartaric acid: Tartaric acid comes from apples and is used to promote the texture and tone of the skin.

Tea tree oil: Tea tree oil is a natural preservative that's used in soap, shampoo and skin care products to clean and disinfect.

There are 20 common amino acids: alanine, arginine, asparagine, aspartic acid, cysteine, glutamic acid, glutamine, glycine, histidine, isoleucine, leucine, lysine, methionine, phenylalanine, proline, serine, threonine, tryptophan, tyrosine, and valine.

Thyme: Thyme, another herb primarily used for cooking, is also useful as an antiseptic and antibacterial agent. You'll find it in some soap.

Tocopherol: Tocopherol is another name for the fat-soluble vitamin E, which aids in the skin reproduction process.

Toilet article, toiletries, toiletry: An artifact used in making your toilet.

Toner: The toner is a part of the facial cleansing system. It's used after the cleanser, and cleans any last traces of dirt that may be left, as well as returning the skin to its natural pH.

Triglycerides: Combination of glycerol with three different fatty acids.

Ulcer: An open sore of skin or mucous membrane with inflamed tissue.

Unctuous: Oily appearance.

Undertone: Undertone is a term used to refer to the tone of the skin, like warm or cool.

UV (Ultraviolet) rays: Rays from sources like the sun that causes pigmentation of the skin.

Vaginitis: Inflammation of the vagina.

Vesicant (epispastics): Substance which produces skin blisters.

Vitamin A: Vitamin A is a fat-soluble vitamin. A deficiency in vitamin A can cause toughening and hardening of the skin.

Vitamin C: Vitamin C is a water-soluble vitamin. It's used in anti-aging creams because of its preservative and antioxidant properties.

Vitamin D: Vitamin D is a fat-soluble vitamin. It promotes strong bones and teeth, but can be detrimental if collected in the skin as a result of too much exposure to the sun.

Vitamin E: Vitamin E is a fat-soluble vitamin. Its antioxidant properties make it useful for promoting healthy skin and hair.

Wheat germ oil: Wheat germ oil, an ingredient found in natural cosmetics, aids in renewing skin cells. It also has a large quantity of vitamin E in it.

Zinc oxide: Zinc oxide is a chemical compound that was originally used as a whitening face powder around the beginning of the 20th century. It's quite often found in sun protection products because of its ability to protect. It also soothes and heals the skin.

Question Bank

1. Define and classify cosmetics. Write a detail notes on face powder.
2. What are compacts? Discuss in detail.
3. What are talcum powders : write a detail note on it.
4. What are creams? Compare the function and formulation aspects of cold and vanishing creams.
5. What are foundation creams? Discuss their function along with one formula.
6. Write a note on:
   i. Bleaching creams
   ii. After shave lotion
   iii. Acid creams
   iv. Barrier creams
   v. Shaving cream
7. What are deodorants? Discuss their formulation and functions.
8. Discuss the function and formulations of the following preparations.
   i. Lipsticks
   ii. Mascara
   iii. Eye shadow
   iv. Rouges
   v. Nail polish
9. Write a detail notes on colour cosmetics.
10. How lipsticks are formulated and evaluated?
11. What are shampoos? Discuss their functions and formulation.
12. Write a detail note on depilatories and discuss their formulation factors.
13. Write a note on hair straighteners and hair waving preparations.
14. Write a note on shaving media.
15. Write about nail polish and nail polish remover.
16. Define dentifrices and discuss their formulation aspects.
17. Write a note on cosmetic packaging.
18. Write a detail notes on herbal cosmetics.
19. How herbal cosmetics are standardized.
20. Write a note on:
   i. Bathing preparations
   ii. Brilliantines
   iii. Eyebrow pencils
   iv. Hand lotions and creams
   v. Deodorants
21. Write ideal properties of the following:
   i. Lipsticks
   ii. Dentifrices
   iii. Skin creams and lotions
   iv. shampoos
   v. Shaving cream
   vi. Face powder

22. Define cosmetics under the Drugs and Cosmetics Act 1940.

23. Classify cosmetics according to the part or organ of the body on which they are used.

24. Classify cosmetics according to their physical nature

25. Write a detail notes on history of cosmetics.

26. Write a note on dental cosmetics.

27. Write about mouth wash and gargles.

28. What are misbranded, adulterated and spurious cosmetics?

29. Write note on colouring agents, preservatives, flavouring agents used in cosmetics.

30. Write evaluations of the following:

i. Dentifrices
ii. Face powder
iii. Lipsticks
iv. shampoos
v. Skin creams and lotions

31. Write newer aspects of cosmetic preparations.

32. Discuss anatomical and cosmetically aspects of skin, hair, nail, eye and tooth.

Bibliography

1. "Creams and Lotions Formulary." Cosmetics and Toiletries. November, 1986;139–70.
2. "Deodorants, Antiperspirants and Shaving Products Formulary." Cosmetics and Toiletries. April, 1990; 75–87.
3. "Flexibility is the hallmark of fluid packaging." Drug & Cosmetic Industry (June 1996) : 98.
4. "Gillette deodorants are fit to fill upside-down." Packaging Digest (September 1993) : 54.
5. "Hair Dye Study." FDA Consumer, May 1994, p. 4.
6. "Makeup Formulary." Cosmetics & Toiletries, April, 1986;103–22.
7. Agrawal A. "Critical issue in quality control of natural products", *Pharma times*, vol-37, no-6, June 2005, 9–11.
8. Andrews Edmund L. "Patents : A Nail Polish That Dries Fast." New York Times, March 7, 1992;40.
9. Andrews Edmund L. "Patents: Quick-Dry Coating for Nail Polish." New York Times, June 13, 1992;36.
10. Angeloglou Maggie. A History of Make-up. The Macmillan Company, 1970.
11. Asha R. "Herbal Indian Perfumes and Cosmetics", Sri Satguru publication, 122–157.
12. Aucoin Kevyn. The Art of Make Up. Harper Collins, 1994.
13. Balsam MS. ed. Cosmetics: Science and Technology. Krieger Publishing, 1991.
14. Balsam MS, Edward Sagarin. Cosmetics Science and Technology. John Wiley & Sons, 1972.
15. Birmingham DJ: Clinical aspects of cutaneous irritation and sensitization, Toxicol, Appl., Pharm, 7:54(1965).
16. Breuer Hans. "Depilatories" Cosmetics and Toiletries 105 (April 1990): 61–66.
17. Brooks Geoffrey J. and Fred Burmeister. "Preshave and Aftershave Products." Cosmetics and Toiletries. April, 1990;67–69.
18. Brumber, Elaine. Save Your Money, Save Your Face. Facts on File Publications, 1986.
19. Charegaonkar, D., "Standardization of Botanicals by HPTLC", *Pharma times*, vol.37, no.6, June 2005;12–24.
20. Chattopadhyay PK, "Herbal cosmetic and Ayurvedic medicines", ISBN-81-86.
21. Chemistry of Soap, Detergents and Cosmetics. Flinn Scientific, 1989.
22. Chopra, R et al. "Indigenous drugs of India", 2nd Ed., Academic publishers 1982;82–85.
23. Dallal, Joseph, and Colleen Rocafort. Hair Styling/ Fixative Products, in Hair and Hair Care. Marcel Dekker, 1997.
24. DeNavarre, M.G. The Chemistry and Manufacture of Cosmetics. Van Nostrand 1962.
25. Donsky, Howard. Beauty Is Skin Deep. Rodale Press, 1985.
26. Evans WC, Trease and Evans, "Pharmacognosy", 13th Ed., Balliere Tindall, 1357–1359.
27. Flick Ernest W. Cosmetic & Toiletry Formulations. 2nd ed., Noyes Press, 1992.
28. Foltz-Gray, Dorothy. "Declare Your Right to Dye." Health, May-June 1996;54–57.
29. Harry RG. "The principles and practices of modern cosmetic", Leonard Hill, London, 2nd vol., 1962, 122–125.
30. Ikeda T, T. Kobayashi, C. Tanaka. "Development of Highly Safe Nail Enamel." Cosmetics and Toiletries, April, 1988;59–60+.
31. Iverson, Annemarie. "Pigment of the Imagination." Harper's Bazaar, May 1995;160–164.
32. Jellinek JS. Formation and Functions of Cosmetics, Wiley-Inter Science, New York, USA.
33. Kaushik P. "Indigenous medicinal plants including microbes and fungi", 127–133.
34. Keiltheler, W.R., "The formulation of Cosmetic and Cosmetic specialties" Drug and cosmetic industry New York, 1956;180–82.

35. Kintish, Lisa. "A Clear Advantage. The emergence of clear technology has given a much-needed boost to the antiperspirant and deodorant market." Soap-Cosmetics-Chemical Specialties (July 1997): 29.
36. Knowlton, John and Steven Pearce. The Handbook of Cosmetic Science and Technology. Oxford : Elsevier Science Publishers, 1993.
37. Kokate CK, Purohit AP, Gokhle SB: "Textbook of Pharmacognosy", 29th Ed., Nirali Prakashan, 109–165.
38. Laden Karl. Antiperspirants and Deodorants New York: Marcel Dekker, 1988.
39. Lubowe Irwin I. Cosmetics and the Skin. Reinhold Publishing Corp., 1964.
40. Men's Shaving Products Market. Frost & Sullivan, 1990.
41. Meyer Carolyn. Being Beautiful: The Story of Cosmetics from Ancient Art to Modern Science. William Morrow and Company, 1977.
42. Mithal BM, Sahu RN, "A Hand book of cosmetics", Vallabh Prakashan, Delhi, 110–117.
43. Miwa T, "Herbal aspects of jojoba oil -a unique liquid wax", cosm. perf. (88), 1973;39–41.
44. Nanda Sanju, Arun and Khar Roop K. "Cosmetics in Ayurveda", The Indian pharmacist, July 2003;7–12.
45. Panda H. "Herbal Cosmetic Handbook", ISBN-81-80.
46. Poucher, W.A., "Perfumes cosmetics and soaps", vol.-III, Chapman and Hill, Ltd, London, 1947; 170–174.
47. Schemann Andrew. Cosmetics Buying Guide. Consumer Reports Books, 1993.
48. Schlossman Mitchell L. "Nail Cosmetics." Cosmetics & Toiletries, April, 1986;23–4+
49. Schoen Linda Allen, ed. The AMA Book of Skin and Hair Care. J.B. Lippincott Company, 1976.
50. Schueller, Randy, and Perry Romanowski. Beginning Cosmetic Chemistry. Allured Publishing, 1999.
51. Sharma PP. "Cosmetic, formulation, manufacturing and quality control", Second ed., Vandana publication, 319–331.
52. Springer Neil, Helga Tilton. "Staying Power." Chemical Marketing Reporter (August 9, 1993) : SR12.
53. Sroboda R, "Ayurveda, life, health and longevity", penguin books 1992;80–82.
54. Strandberg Keith. "Antiperspirant & Deodorant Update." Soap-Cosmetics-Chemical Specialties (April 1993): 30.
55. Thomsson EG, "Modern Cosmetics Drug and Cosmetic Industry", New York, 1947;15 18.
56. Tyler VE, Brady LR. Robbers JE, "Pharmacognosy", 8th Ed., Sea and Febiger, Phladelphia, 1981;122–128.
57. Umbach, Wilfried. Cosmetics and Toiletries Development, Production, and Use. New York : Ellis Horwood, 1991.
58. Wells FV, Irwin I, Lubowe MD, eds. Cosmetics and the Skin. Reinhold Publishing Corp., 1964.
59. Wetterhahn, Julius. "Eye Makeup," in Cosmetics : Science and Technology. M. S. Balsam and Edward Sagarin, ed. John Wiley & Sons, 1972.
60. Wikinson JB, Harry's Cosmetology, Leonard Hill Book, London, UK.
61. Winter Ruth. A Consumer's Dictionary of Cosmetic Ingredients, Crown, 1989.
62. Young Anne, Practical Cosmetic Science, Mills and Boon Ltd., London, UK.

Index

Face is the index of mind

— Plato